AF355974

Remote Sensing and GIS for Monitoring Urbanization and Urban Health

Remote Sensing and GIS for Monitoring Urbanization and Urban Health

Guest Editors

Xinhu Li

Shihong Du

Peng Jia

Franz W. Gatzweiler

Jinchao Song

Basel • Beijing • Wuhan • Barcelona • Belgrade • Novi Sad • Cluj • Manchester

Guest Editors

Xinhu Li	Shihong Du	Peng Jia
Hangzhou City University	Peking University	Wuhan University
Hangzhou	Beijing	Wuhan
China	China	China
Franz W. Gatzweiler	Jinchao Song	
Chinese Academy of Sciences	University of Michigan	
Xiamen	Ann Arbor	
China	United States	

Editorial Office
MDPI AG
Grosspeteranlage 5
4052 Basel, Switzerland

This is a reprint of the Special Issue, published open access by the journal *Remote Sensing* (ISSN 2072-4292), freely accessible at: https://www.mdpi.com/journal/remotesensing/special_issues/urbanization_and_urban_health.

For citation purposes, cite each article independently as indicated on the article page online and as indicated below:

Lastname, A.A.; Lastname, B.B. Article Title. *Journal Name* **Year**, *Volume Number*, Page Range.

ISBN 978-3-7258-2851-7 (Hbk)
ISBN 978-3-7258-2852-4 (PDF)
https://doi.org/10.3390/books978-3-7258-2852-4

Contents

Article

MSL-Net: An Efficient Network for Building Extraction from Aerial Imagery

Yue Qiu, Fang Wu *, Jichong Yin, Chengyi Liu, Xianyong Gong and Andong Wang

Institute of Geospatial Information, Information Engineering University, Zhengzhou 450001, China
* Correspondence: wufang_630@126.com

Abstract: There remains several challenges that are encountered in the task of extracting buildings from aerial imagery using convolutional neural networks (CNNs). First, the tremendous complexity of existing building extraction networks impedes their practical application. In addition, it is arduous for networks to sufficiently utilize the various building features in different images. To address these challenges, we propose an efficient network called MSL-Net that focuses on both multiscale building features and multilevel image features. First, we use depthwise separable convolution (DSC) to significantly reduce the network complexity, and then we embed a group normalization (GN) layer in the inverted residual structure to alleviate network performance degradation. Furthermore, we extract multiscale building features through an atrous spatial pyramid pooling (ASPP) module and apply long skip connections to establish long-distance dependence to fuse features at different levels of the given image. Finally, we add a deformable convolution network layer before the pixel classification step to enhance the feature extraction capability of MSL-Net for buildings with irregular shapes. The experimental results obtained on three publicly available datasets demonstrate that our proposed method achieves state-of-the-art accuracy with a faster inference speed than that of competing approaches. Specifically, the proposed MSL-Net achieves 90.4%, 81.1% and 70.9% intersection over union (IoU) values on the WHU Building Aerial Imagery dataset, Inria Aerial Image Labeling dataset and Massachusetts Buildings dataset, respectively, with an inference speed of 101.4 frames per second (FPS) for an input image of size $3 \times 512 \times 512$ on an NVIDIA RTX 3090 GPU. With an excellent tradeoff between accuracy and speed, our proposed MSL-Net may hold great promise for use in building extraction tasks.

Keywords: building extraction; semantic segmentation; group normalization; deformable convolution; remote sensing images

Citation: Qiu, Y.; Wu, F.; Yin, J.; Liu, C.; Gong, X.; Wang, A. MSL-Net: An Efficient Network for Building Extraction from Aerial Imagery. *Remote Sens.* **2022**, *14*, 3914. https://doi.org/10.3390/rs14163914

Academic Editor: John Trinder

Received: 8 July 2022
Accepted: 10 August 2022
Published: 12 August 2022

Publisher's Note: MDPI stays neutral with regard to jurisdictional claims in published maps and institutional affiliations.

1. Introduction

As the main gathering places for human production and living, buildings are crucial indicators for monitoring urbanization, and the extraction of buildings is playing an increasingly notable role in the study of urbanization [1]. Currently, various data sources, such as satellite images, aerial imagery, and point cloud data, are available for building extraction tasks. Among them, aerial images have high spatial resolutions and are easy to obtain. Building extraction from aerial imagery is critical for urban planning [2], population estimation [3] and digital cartography [4].

In building extraction tasks, considerable human labor will be consumed if all buildings are manually annotated. Therefore, how to extract buildings with algorithms rather than human experts is an immediate challenge to be addressed. Traditional extraction methods can generally be divided into feature detection-based methods [5–7], area segmentation-based methods [8–11] and auxiliary information-combined methods [12–18]. However, based on handcrafted features such as spectral, shadow, and texture features, these traditional methods can only process the low- or mid-level information contained in images, and their building extraction results usually have poor accuracy and integrity [19]. Traditional methods are not sufficiently intelligent and often require tedious parameter tuning steps. With aerial image acquisition becoming easier, if buildings can be automatically extracted by algorithms in real time, the efficiency of the extraction task will be significantly improved, and human labor costs will be dramatically reduced.

Considerable semantic segmentation methods paired with deep learning have been developed in recent years. Compared with traditional methods, semantic building segmentation methods based on deep learning approaches are capable of obtaining and utilizing the high-level features contained in images. Applicable for fully automatic semantic segmentation, trained deep learning models hold great promise in building extraction tasks. In fully convolutional networks (FCNs) [20], the fully connected layers in the convolutional neural network (CNN) structure are replaced with convolutional layers, and some researchers [21,22] have used variants of FCNs to automatically extract buildings, as they eliminate the jagged edges encountered when segmenting blocky regions and achieve obviously improved segmentation accuracy. However, as early semantic segmentation models, FCNs are limited to utilizing the contextual information contained in an image, leading to discontinuities and holes in the segmentation results.

To fully obtain and utilize the features in images, two main approaches are currently available for improving semantic segmentation models.

(1)   For feature maps, feature pyramids can be applied to enlarge their receptive fields and obtain multiscale target features. The pyramid scene parsing network (PSPNet) [23] fuses four different scales of feature maps in parallel via a pyramid pooling module, improving the network's ability to obtain multiscale information. In DeepLabv3 [24], an atrous spatial pyramid pooling (ASPP) structure is adopted to enlarge the receptive fields and, thus, has a significant advantage in large object segmentation. To restore more building contour information, Xu [25] enhanced the combination of an encoder and a decoder based on a DeepLabv3+ [26] network embedded with an ASPP module.

(2)   For input images, skip connections are applied to fuse different levels of feature maps to obtain multilevel image features. The level of image features increases as the network layers deepen. Low-level features provide the basis for object category detection, and high-level features facilitate accurate segmentation and positioning. U-Net [27] uses long skip connections to integrate low-level features with high-level features and has high performance in medical image segmentation. Improved from U-Net, networks such as IEU-Net [28], HA U-Net [29] and EMU-CNN [30] have performed well. The MPRSU-Net [31] was constructed by combining long and short skip connections, alleviating the holes and fragmentary edges in the segmentation results obtained when extracting large buildings.

Researchers [32–34] have also considered both types of approaches and constructed new building segmentation models by combining feature pyramid modules and skip connections, notably enhancing the efficiency of the building extraction task and the generalization capacities of the developed models. Nevertheless, the majority of existing building extraction methods fail to address model applicability, resulting in considerable computational complexity and tedious parameter tuning steps. These problems limit their deployment in practical applications such as disaster/emergency response [35], damage assessment [36,37], and military reconnaissance [38] that require high algorithmic efficiency. Therefore, to facilitate the practical application of our method, we reduce the model complexity and propose a network with an "encoder-decoder" structure called MSL-Net, which is capable of obtaining and utilizing both multiscale and multilevel features, where "M" represents the prefix "multi", "S" represents "scale", "L" represents "level", and "Net" represents "Network ". The key contributions are as follows.

1.   In the encoding stage of MSL-Net, we introduce the MobileNetV2 [39] architecture to extract multilevel features. The inverted residual blocks in MobileNetV2 are constructed as bottlenecks using depthwise separable convolution (DSC) [40] and

group normalization (GN) operations [41], which noticeably reduce the model complexity while improving its training and inference speeds. The multiscale features are extracted by an ASPP module to enhance the ability of the model to recognize multiscale buildings.

2. In the decoding stage of MSL-Net, long skip connections [42] are applied to establish a long-distance dependence between the feature encoding and feature decoding layers. This long-distance dependence is beneficial for obtaining the rich hierarchical features of an image and effectively preventing holes in the segmentation results [31]. Before performing pixel classification, a deformable convolution network (DCN) layer [43] is added to ensure strong model robustness even when extracting buildings with irregular shapes.

2. Materials and Methods

2.1. MSL-Net Architecture

The encoder and decoder in MSL-Net are shown in Figure 1. The function of the encoder is to extract image features in a layer-by-layer manner. As the network layers gradually deepen, the feature map becomes more abstract; nonetheless, the extracted semantic information becomes richer, which is beneficial for classifying each pixel in the input image. In MSL-Net, the lightweight MobileNetV2 network with inverted residual blocks is introduced as the backbone to extract the original image features. We embed a GN layer in the inverted residual block, and three levels of feature maps with channels × height × width values of 24 × 128 × 128, 32 × 64 × 64, and 320 × 64 × 64 are output. Since the image downsampling operation during feature extraction lowers the feature map resolution and causes a partial loss of spatial information, the downsampling rate is limited to 8, and only three downsampling operations are performed. An ASPP module is used to extract multiscale features from the output high-level feature maps.

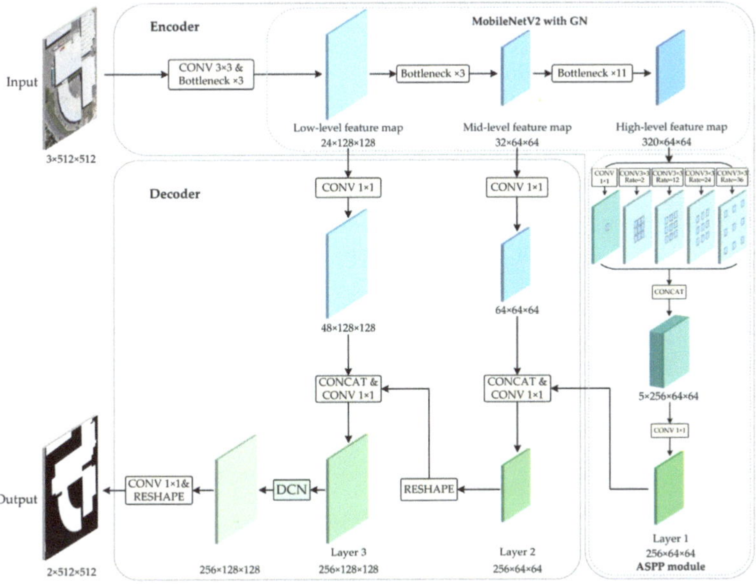

Figure 1. Overall architecture of MSL-Net. The encoder produces multilevel semantic feature maps with different spatial resolutions. The decoder parses them to the segmentation mask.

The decoder outputs prediction results with the same size as that of the original input image; these results are expressed as the final binary building segmentation map in the building semantic segmentation task. First, layer 1 is output by the ASPP module and concatenated with the mid-level feature map output by the backbone through a long skip connection. After the channels of the concatenated feature map are adjusted to 256 through a 1×1 convolution, we obtain layer 2. Second, layer 2 is resized by bilinear reshaping (denoted RESHAPE) to the same size as that of the low-level feature map extracted by the backbone, and then a concatenation operation (denoted CONCAT) and a 1×1 convolution (denoted CONV 1×1) are executed to obtain layer 3 with 256 channels and a size of 128×128. Thus far, multilevel image features and multiscale feature map features have been extracted. Third, the features of irregular building shapes are extracted through a DCN layer, whose output feature map has the same number of channels and size as its input feature map. Eventually, after a 1×1 convolution and a bilinear reshaping, a semantic segmentation image of the buildings with 2 channels and a size of 512×512 is output.

2.2. Feature Extraction Backbone in the Encoder

Typically, the deeper a network is, the richer the extracted features and the better the model performs. However, a deep learning model does not always perform better after simply stacking the layers of the network. Instead, the weight matrix may degrade, causing the network performance to deteriorate. The residual structure in ResNet [44] allows certain layers to be connected to each other through short skip connections, which weakens the strong correlation between two adjacent layers and mitigates network degradation. The inverted residual structure in the MobileNetV2 feature extraction backbone is based on the residual structure of ResNet, while the feature extraction sequence of "downscaling-convolution-upscaling" is changed to "upscaling-convolution-downscaling", and the middle normal convolution is replaced by a DSC.

2.2.1. DSC in the Backbone

As shown in Figure 2, a DSC consists of two steps, a depthwise convolution and a pointwise convolution, which are performed separately in the spatial and channel dimensions to capture spatial information and fuse cross-channel depth information. Suppose that the input feature map size (length $\times$ width $\times$ channels) is $D_F \times D_F \times M$, the output feature map size is $D_F \times D_F \times N$, and the normal convolution kernel size is $D_K \times D_K \times M$. Then, the ratio of the number of DSC parameters to the number of normal convolution parameters is:

$$\frac{(D_K \times D_K \times 1) \times M + (1 \times 1 \times M) \times N}{(D_K \times D_K \times M) \times N} = \frac{1}{N} + \frac{1}{D_K^2} \tag{1}$$

which indicates that the DSC can exponentially reduce the required number of parameters, and this advantage becomes increasingly apparent as the number of layers increases.

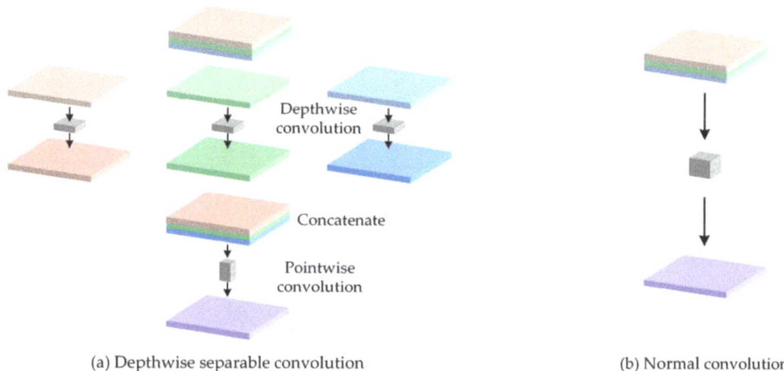

Figure 2. Schematic diagrams of the DSC and the normal convolution. The left figure (**a**) shows the DSC, while the right figure (**b**) shows the normal convolution.

2.2.2. GN in the Backbone

Batch normalization (BN) [45] has been widely used in existing deep learning algorithms. MobileNetV2 contains three BN layers in each inverted residual structure. Each BN layer takes the overall statistics for inference, imposing constraints on the search spaces of the system parameters, accelerating network convergence, and alleviating the overfitting problem. However, due to the stacking effect of BN in the network, the input distribution deviation between the training and test sets causes BN estimation bias to accumulate, which adversely affects the test performance of the model [46]. Note that GN can prevent the accumulation of such estimation bias. For this reason, we replace the second BN layer in the original inverted residual structure (shown in Figure 3c) with a GN layer to prevent network performance degradation due to distribution bias and ensure the robustness of the network.

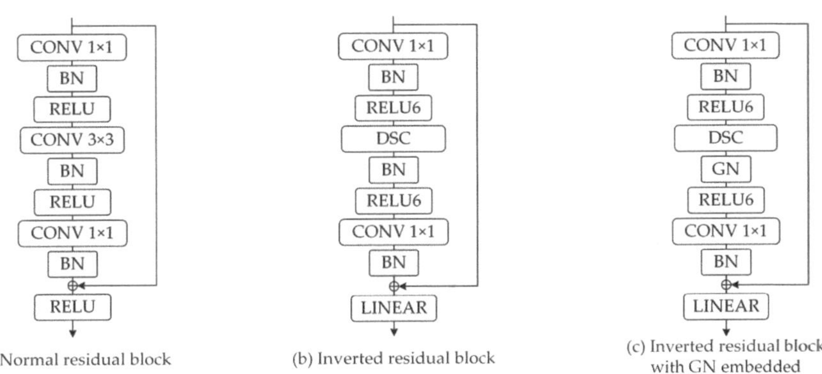

Figure 3. Schematic diagram of (**a**) the normal residual block, (**b**) the inverted residual block, and (**c**) the inverted residual block with GN embedded.

The general feature normalization formula is,

$$\hat{x}_i = \frac{1}{\sigma_i}(x_i - \mu_i) \tag{2}$$

For two-dimensional images, x is a computed feature derived from the feature map, and $i = (i_N, i_C, i_H, i_W)$ is a four-dimensional vector of features indexed in the following order: "batch axis, channel axis, spatial height axis, spatial width axis". μ and σ are the mean and standard deviation, respectively, computed using the following equations:

$$\mu_i = \frac{1}{m}\sum_{k \in S_i} x_k, \quad \sigma_i = \sqrt{\frac{1}{m}\sum_{k \in S_i}(x_k - \mu_i)^2 + \varepsilon}, \tag{3}$$

where ε is a constant with a small value, S_i is the set of pixels used to compute the mean and standard deviation, and m is the size of S_i. Then, formally, the set of groups normalized computation sets is defined as,

$$S_i = \left\{ k \,\middle|\, k_N = i_N, \left\lfloor \frac{k_C}{C/G} \right\rfloor = \left\lfloor \frac{i_C}{C/G} \right\rfloor \right\}. \tag{4}$$

here, G and C are the numbers of groups and channels, respectively, and C/G is the number of channels in each group. $\lfloor \cdot \rfloor$ is the floor operation, and $\left\lfloor \frac{k_C}{C/G} \right\rfloor = \left\lfloor \frac{i_C}{C/G} \right\rfloor$ indicates that both indices i and k are in the same channel group, assuming that each channel group is stored sequentially along the channel axis. GN computes μ and σ along the spatial height axis, the spatial width axis and a group of C/G channels. Specifically, we use the same μ and σ to normalize the pixels in the same group.

2.3. ASPP in the Encoder

Atrous convolutions [47] introduce the concept of "dilation rates" based on the normal convolution, as shown in the ASPP module in Figure 4. Atrous convolutions with different dilation rates insert corresponding zero values into the normal convolution kernel to achieve convolution dilation, thereby increasing the receptive fields; this is similar to pooling operations. The scale features extracted by atrous convolutions with different dilation rates are also different. Additionally, since the inserted zero values are not calculated, the calculation counts do not increase, and the spatial resolution of the output feature maps does not decrease.

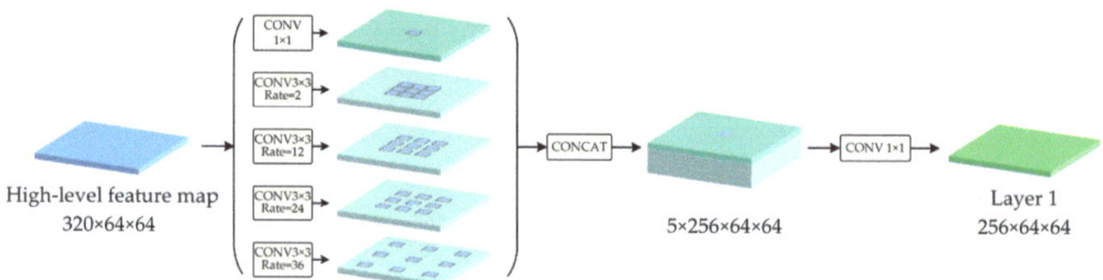

Figure 4. Schematic diagram of the ASPP module. The ASPP module concatenates and fuses five feature maps to obtain multiscale feature map features.

We replace the original ASPP branch with an atrous convolution possessing a dilation rate of 2 in our experiments. The ASPP module in MSL-Net concatenates and fuses the five feature maps output by a 1×1 convolution and four atrous convolutions with dilation rates of 2, 12, 24 and 36, obtaining both large-scale global information and small-scale local detail information. After a 1×1 convolution, the number of channels in the concatenated feature map is compressed to 256.

2.4. Deformable Convolution in the Decoder

Buildings in images are susceptible to different degrees of deformation due to external conditions such as the attitude of the equipment and weather conditions. In addition, due to the diversity of the shapes of buildings, a normal convolution has difficulty extracting the shape features of buildings, while a deformable convolution is able to adaptively adjust according to the deformation of the object in an image and efficiently extract robust features from objects with different shapes and directions.

Figure 5 depicts schematic diagrams of a normal convolution and a deformable convolution in the two-dimensional plane. Figure 5a represents the normal convolution, where the convolution kernel size is 3×3, and the sample points are organized in a regular pattern. Figure 5b represents the deformable convolution, where each sample point has position offsets, and the arrangement becomes irregular.

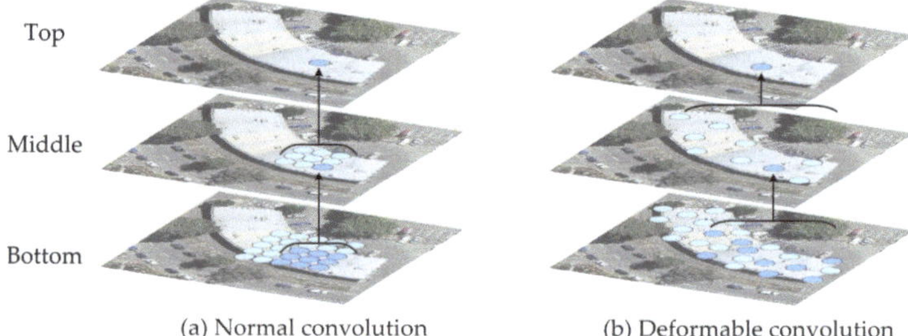

Figure 5. Schematic diagrams of (**a**) a normal convolution and (**b**) a deformable convolution. From the bottom to the top, the sampling points are fixed in the normal convolution, while the sampling points in the deformable convolution adjust according to the shape of the object.

A normal two-dimensional convolution includes two steps: (1) sampling on the input feature mapping x with a normal convolution kernel K and (2) summing over the w-weighted sampled values. For each position p on the output feature mapping y,

$$y(p) = \sum_{k=1}^{K} w_k \cdot x(p + p_k),$$ (5)

where k lists the positions in K.

In a deformable convolution, the position offsets are first obtained through a normal convolution layer, and then the offsets and magnitudes of the features learned from each sampling point are modulated. Finally, a more complex geometric transform feature learning process is performed, which is calculated as,

$$y(p) = \sum_{k=1}^{K} w_k \cdot x(p + p_k + \Delta p_k) \cdot \Delta m_k,$$ (6)

where Δp_k and Δm_k represent the learnable offset and weight scalar at position k, respectively. Δm_k represents the modulation scalar at position k in the range $[0, 1]$, and the calculation of the offset value is executed using bilinear interpolation. Δp_k and Δm_k can be obtained by applying a convolution to the same input feature map layer. The number of output channels is $3K$, and K represents the convolutional kernel size of the backbone. The first $2K$ dimensions represent the x and y offsets (Δp_k) at each position, and the subsequent K dimensions are used to obtain the weight (Δm_k) of each position according to the sigmoid layer.

We add a deformable convolution layer to enhance the feature extraction ability of our model for geometric shapes before the final pixel classification output of the network.

2.5. Warmup and Cosine Annealing Learning Rate Policy

During training, gradient descent is usually adopted to optimize models. The learning rate (LR) is one of the hyperparameters that affect the model optimization process; it plays a guiding role in how to use the loss function gradient to adjust the network weights in the gradient descent step. When model training starts, a large initial LR is generally set to rapidly decrease the loss value of the network, and the LR decreases in a certain way as the number of iterations increases to ensure small model fluctuations during the later training stages as it gradually approaches the global optimal solution. The LR usually decreases via exponential decay, piecewise constant decay, or cosine annealing [48].

Since the network is relatively unstable in the early training stage, a large initial LR causes the gradient of the weights to fluctuate back and forth, and a small initial LR decelerates network convergence; thus, we employ the "warmup and cosine annealing" LR policy to ensure the performance of the model. As shown in Figure 6b, in the first 10 epochs, the LR linearly increases from 0 to the base LR and then gradually decreases from the base LR to 0 as the number of epochs increases. The LR is calculated by:

$$\eta_t = \begin{cases} \frac{\eta_{max} T_{cur}}{T_{wu}}, & T_{cur} \leq T_{wu} \\ \frac{\eta_{max}}{2}\left(1 + \cos\left(\frac{T_{cur} - T_{wu}}{T_{max}}\pi\right)\right), & T_{wu} < T_{cur} \leq T_{max} \end{cases}$$ (7)

where η_t is the LR of the current training epoch, η_{max} is the base LR, T_{cur} is the current number of training epochs, T_{wu} is the total number of warmup epochs, and T_{max} is the maximum number of training epochs.

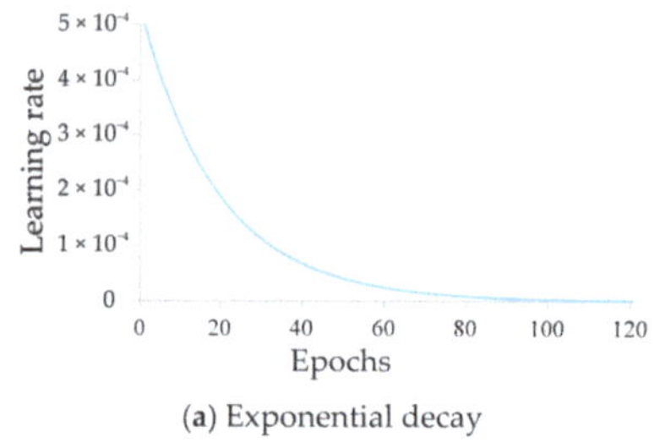
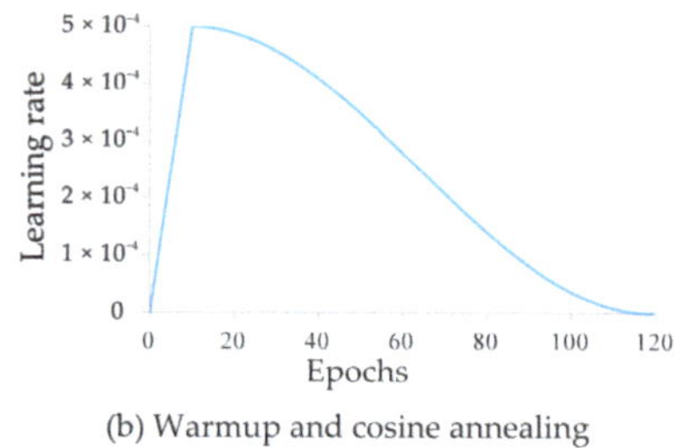

(**a**) Exponential decay   (**b**) Warmup and cosine annealing

Figure 6. Comparison between two LR policies. The left figure (**a**) shows the exponential decay LR policy, while the right figure (**b**) shows the warmup and cosine annealing LR policy.

3. Experiments and Results

3.1. Descriptions of the Datasets

In this study, we use the WHU Building Aerial Imagery dataset (WHU dataset) [3], Inria Aerial Image Labeling dataset (Inria dataset) [49], and Massachusetts Buildings dataset (Massachusetts dataset) [50], with spatial resolutions ranging from 0.3 m to 1.0 m; thus, we can fully test the performance of the developed model. The details of each dataset are listed in Table 1.

Table 1. Details of each dataset.

Dataset	Spatial Resolution (m)	Pixels	Area (km²)	Tiles
WHU dataset	0.3	512 × 512	450	8189
Inria dataset	0.3	5000 × 5000	810	180
Massachusetts dataset	1.0	1500 × 1500	240	151

For both the Massachusetts dataset and the WHU dataset, the training and validation steps are performed directly using the training, validation, and test set ratios that have been partitioned by default in these datasets. For the Inria dataset, the training, validation, and test sets are divided at a ratio of 8:1:1. The numbers of images in the training, validation and test sets used for each dataset in the experiments are shown in Table 2.

Table 2. Division of each dataset.

Dataset	Training Set	Validation Set	Test Set
WHU dataset	4737	1036	2416
Inria dataset	14418	1782	1800
Massachusetts dataset	1233	36	90

3.2. Experimental Settings

The main software and hardware used in our study are listed in Table 3.

Table 3. Details of the employed hardware and software.

Item	Details
CPU	Intel i7-12700K @ 3.61 GHz
GPU	GeForce RTX 3090 (24 GB)
OS	Windows 10 x64
Language	Python 3.8
Framework	PyTorch 1.8.1

Since buildings are the only experimental objects, the pixel value range of the labeled binary map is adjusted from [0, 255] to [0, 1] before training, where pixels with values of 1 represent the buildings and pixels with values of 0 represent the background. The widely used and high-performing U-Net, PSPNet, and DeepLabv3+ are selected for comparison, and all four models are tested on the above three datasets using the same training, validation, and test sets. The input images are one-hot coded, the batch size is set to 12, the loss function is a direct summation of the focal loss [51] and dice loss [52], the Adam optimizer is used in the training process, the base LR is set to 0.0005, and each comparison model is trained for 120 epochs using an exponential decay LR policy with a gamma value of 0.0005. MSL-Net is first warmed up for 10 epochs and then trained for 110 epochs using the cosine annealing LR policy.

In the training stage, data augmentation strategies are applied to preprocess the input images to obtain more feature information from the limited data. Due to the rich geometric features of the buildings in the dataset, we first apply spatial data augmentation strategies, including random horizontally mirror flipping with a 50% probability, random rotation with 50% probabilities at different angles ($-10°$ to $10°$), and random scaling with ratios between 0.25 and 2. Then, spectral data augmentation strategies are applied to reduce the impact caused by imaging condition differences to improve the generalization ability of the network. The spectral data augmentation techniques include hue augmentation, saturation augmentation, value augmentation and random Gaussian blurring.

3.3. Evaluation Metrics

To quantitatively evaluate the reliability and accuracy of each model, we use six metrics, the Intersection-Over-Union (IoU), F1-score, Accuracy, Recall, Precision and Kappa, to evaluate the segmentation results. The building segmentation results are compared with the corresponding building labels at the pixel level, and the instances are classified as positive or negative, with true indicating a correct prediction and false indicating an incorrect prediction; thus, true positives (TPs) represent the correctly predicted building pixels, false positives (FPs) represent the pixels that predict the background as buildings, false negatives (FNs) represent the pixels that predict buildings as the background, and true negatives (TNs) represent the correctly predicted background pixels. The confusion matrix is shown in Table 4.

Table 4. Confusion matrix.

Ground Truth	Prediction	Building	Background
Building		TP	FN
Background		FP	TN

Among these six metrics, Recall denotes the proportion of correctly predicted building pixels among all real building pixels, Precision denotes the proportion of correctly predicted building pixels among all predicted building pixels, and the IoU, which is currently the most commonly used evaluation metric in semantic segmentation tasks, denotes the ratio of the intersection to the union of the predicted building and real building pixels. Accuracy indicates the proportion of correctly predicted pixels among all pixels. The F1-score is the harmonic mean of the Recall and Precision. Kappa is a metric that considers both the target and background accuracies. The Recall, Precision, IoU, Accuracy, F1-score and Kappa are defined as:

$$\text{Recall} = \text{TP}/(\text{FN} + \text{TP}), \tag{8}$$

$$\text{Precision} = \text{TP}/(\text{FP} + \text{TP}), \tag{9}$$

$$\text{IoU} = \text{TP}/(\text{FN} + \text{FP} + \text{TP}), \tag{10}$$

$$\text{Accuracy} = (\text{TP} + \text{TN})/(\text{FP} + \text{FN} + \text{TP} + \text{TN}), \tag{11}$$

$$\text{F1-score} = 2\text{TP}/(2\text{TP} + \text{FN} + \text{FP}), \tag{12}$$

$$P_0 = (\text{TP} + \text{FP})/(\text{FP} + \text{FN} + \text{TP} + \text{TN}), \tag{13}$$

$$P_e = ((TP + FN)(TP + FP) + (FN + TN)(FP + TN))/((FP + FN + TP + TN)^2), \quad (14)$$

$$Kappa = (P_0 - P_e)/(1 - P_e). \quad (15)$$

4. Discussion

4.1. Comparisons on Each Dataset

4.1.1. Comparison on the WHU Dataset

The WHU dataset is the most accurate building dataset available to date [53]; it consists of aerial images of Christchurch, New Zealand, with extraordinarily high image resolutions. Building images with obvious spectral features, geometric features, and spatial distribution features are selected from the dataset for display, as shown in Figure 7, where white pixels indicate the correctly detected parts of a building, red pixels indicate incorrectly detected parts of a building, blue pixels indicate the missed parts of a building, and black pixels indicate the correctly detected background.

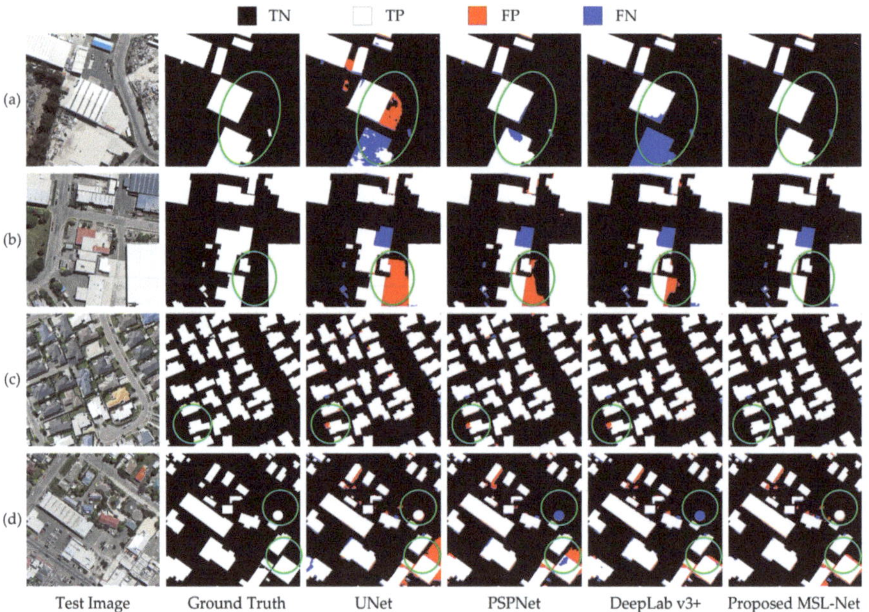

Figure 7. Examples of the segmentation results obtained using different methods. (**a–d**) Results selected from the WHU dataset.

In Figure 7a, U-Net, PSPNet, and DeepLabv3+ all fail to detect large bungalows to some degree, and some results have rough building edges with some discrete pixels, whereas MSL-Net decreases the salt and pepper noise by fusing multiscale and multilevel features through its ASPP module and long skip connections. In Figure 7b, the other three methods have varying degrees of false detection. This is because the building roofs are similar to the ground in terms of color and texture, which signifies "intraclass spectral heterogeneity". The results of Figure 7b demonstrate that MSL-Net effectively reduces the negative impact of "intraclass spectral heterogeneity" on the extraction of buildings and achieves enhanced recognition accuracy. Figure 7c depicts a region with a significant number of densely distributed small-scale buildings, and MSL-Net efficiently mitigates false detection. Buildings of various sizes and shapes can be found in Figure 7d. According to the extraction results, U-Net and PSPNet are unable to effectively discriminate between the ground and buildings. Buildings with irregular shapes, such as round buildings, are ineffectively extracted by PSPNet and DeepLabv3+. MSL-Net not only eliminates the negative impact of "intraclass spectral heterogeneity" but also completely extracts round buildings and maintains their continuity, revealing that the deformable convolutional layer is involved in the detection of target features with irregular shapes.

To quantitatively evaluate the extraction effect of each method, the results of each metric are calculated, as shown in Table 5.

Table 5. Metrics produced on the WHU dataset. The highest scores are bolded; the second-highest scores are underlined.

Method	IoU (%)	Accuracy (%)	F1-Score (%)	Kappa (%)	Precision (%)	Recall (%)
U-Net	84.9	98.2	91.9	90.8	90.0	93.8
PSPNet	87.6	98.5	93.4	92.6	92.6	<u>94.3</u>
DeepLabv3+	<u>88.0</u>	<u>98.6</u>	<u>93.6</u>	<u>92.8</u>	<u>94.4</u>	92.9
MSL-Net	**90.4**	**98.9**	**95.0**	**94.3**	**95.1**	**94.8**

In Table 5, MSL-Net exceeds 90% in every metric, with IoU, F1-score and Kappa values that are about 2.4%, 1.4% and 1.5% higher than those of the second-best model, respectively. Our proposed model outperforms the widely used models in accuracy of recognition outcomes.

4.1.2. Comparison on the Inria Dataset

The Inria dataset comprises a variety of urban landscapes, covering an area of 810 km^2 in 10 different cities. Various places have diverse architectural types, and the spectral features and shadow features of the images also vary depending on their imaging times and meteorological conditions, so this dataset might be a good indicator of a model's robustness in different scenarios. Some of the original images and labels and the corresponding results extracted by each method are shown in Figure 8.

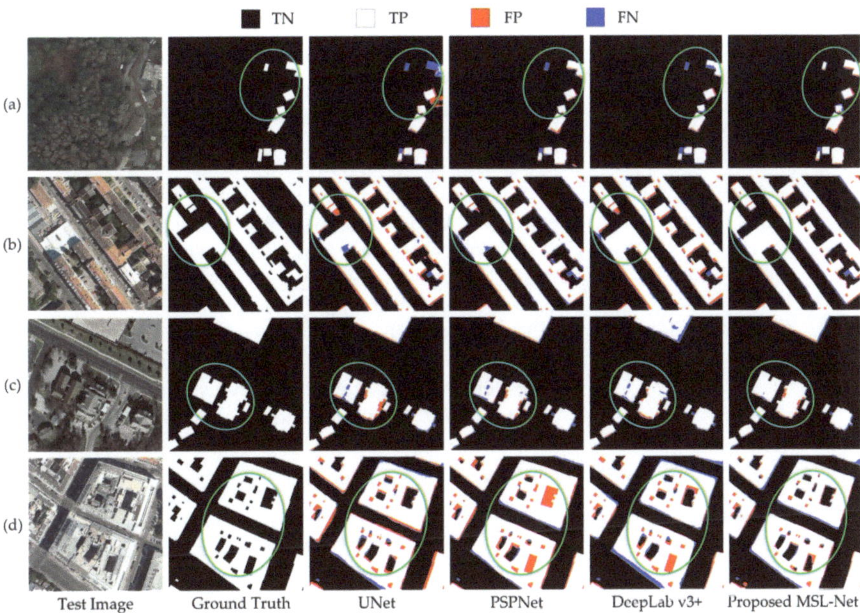

Figure 8. Examples of the segmentation results obtained by the different methods. (**a–d**) Results selected from the Inria dataset.

The spatial distribution of the buildings in Figure 8a is uneven, and the materials and spectral features of the roofs of the scattered buildings vary. The buildings in Figure 8b are large in scale and are connected by several buildings with different spectral features. The building structures are complex, and shadows obscure the buildings. Figure 8c contains two types of buildings, villas and large bungalows, and the spectral features of the buildings and the ground are relatively similar. The buildings in Figure 8d are rectangular ambulatory planes in terms of shape, and the spectral features of the roofs are complex.

MSL-Net effectively reduces the amount of missed detection and effectually suppresses the occurrence of false detection, as shown in Figure 8a,c. Figure 8b shows how MSL-Net successfully distinguishes rooftops and road surfaces with similar spectral features, weakening the influence of the "interclass spectral homogeneity". MSL-Net also effectively distinguishes between white and brown rooftops with different spectral features and successfully extracts white buildings shaded by trees, indicating that MSL-Net can reduce the impact of shaded buildings to some extent. We can observe in Figure 8b,d that MSL-Net can recognize buildings with complicated structures.

Table 6 lists the computed evaluation metric results. The IoU, F1-score and Kappa coefficient of MSL-Net are 0.2%, 0.2% and 0.2% higher than those of the second-best model PSPNet, respectively; these results are not much higher than those of the PSPNet, but they still indicate that MSL-Net has good competitiveness in the building recognition task in various situations.

Table 6. Metrics produced on the Inria dataset. The highest scores are bolded; the second-highest scores are underlined.

Method	IoU (%)	Accuracy (%)	F1-Score (%)	Kappa (%)	Precision (%)	Recall (%)
U-Net	78.2	96.2	87.8	85.5	88.5	87.0
PSPNet	80.9	96.7	89.4	87.5	88.9	89.9
DeepLabv3+	78.1	96.2	87.7	85.4	88.2	87.1
MSL-Net	81.1	96.8	89.6	87.7	89.3	89.9

4.1.3. Comparison on the Massachusetts Dataset

The Massachusetts dataset includes Boston aerial images with 1-m spatial resolutions, which are significantly lower than the 0.3-m resolutions of the WHU dataset and Inria dataset. With these lower spatial resolutions, the feature information of buildings is rough and more difficult to extract. Some of the original images and labels and the corresponding results extracted by each method are shown in Figure 9.

In general, most of the buildings in the segmentation results are displayed in fragmented patchy distributions, which greatly test each model's ability to extract small targets. Figure 9a,b demonstrate that MSL-Net can alleviate the occurrences of missed and false detections to a certain extent, and Figure 9c,d demonstrate that MSL-Net is also able to effectively identify irregular buildings.

The results of the calculated evaluation metrics are shown in Table 7. The IoU, F1-score and Kappa coefficient values of MSL-Net are 3.3%, 2.3% and 2.9% higher than those of the second-best model U-Net, respectively, and all other metrics are also better. The results show that MSL-Net still has strong robustness even when working with images possessing poor spatial resolutions.

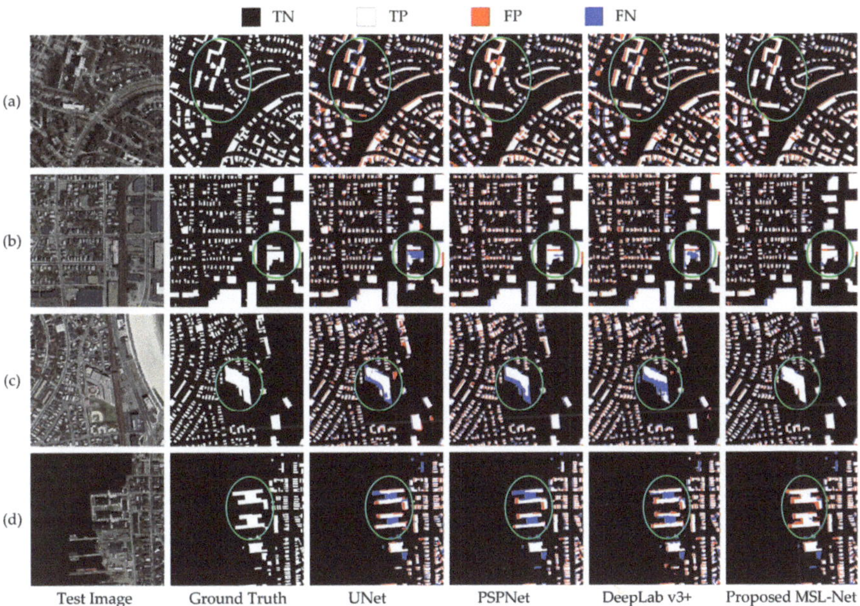

Figure 9. Examples of the segmentation results obtained by different methods. (**a–d**) Results selected from the Massachusetts dataset.

Table 7. Metrics produced on the Massachusetts dataset. The highest scores are bolded; the second-highest scores are underlined.

Method	IoU (%)	Accuracy (%)	F1-Score (%)	Kappa (%)	Precision (%)	Recall (%)
U-Net	<u>67.6</u>	<u>92.6</u>	<u>80.7</u>	<u>76.1</u>	78.5	<u>83.0</u>
PSPNet	67.2	92.6	80.4	75.8	<u>79.3</u>	81.5
DeepLabv3+	63.3	91.3	77.5	74.5	74.9	80.4
MSL-Net	**70.9**	**93.6**	**83.0**	**79.0**	**81.9**	**84.1**

4.2. Complexity Comparison

Complexity is a critical factor that affects the practical application of a model. In the building extraction task based on the CNN method, lower numbers of parameters and floating-point operations (FLOPs) often result in faster training and inference speeds. A model with lower complexity is more convenient for practical applications. To objectively evaluate the complexity of each model, the number of parameters, the number of FLOPs, the training speed and the inference speed are calculated separately for each model. On an NVIDIA RTX 3090 GPU, the training speed is expressed as the number of frames per second (FPS) required for an input image of size $3 \times 512 \times 512$, and the inference speed is expressed as the number of FPS required for 2 input images of size $3 \times 512 \times 512$. The quantitative comparison results are shown in Figure 10.

In Figure 10, we can easily find that MSL-Net obviously achieves the fastest training speed and inference speed with very small numbers of parameters and FLOPs. In detail, the numbers of parameters and FLOPs required by MSL-Net are much lower than those of DeepLabv3+ and PSPNet, which are approximately 14% and 37% of the numbers required by the suboptimal U-Net model, respectively. Our proposed MSL-Net reaches a competitive training speed of 53.1 FPS, which is 65% faster than that of the second-best model, while the inference speed surpasses those of other models by more than 57.1% with an FPS of 101.4.

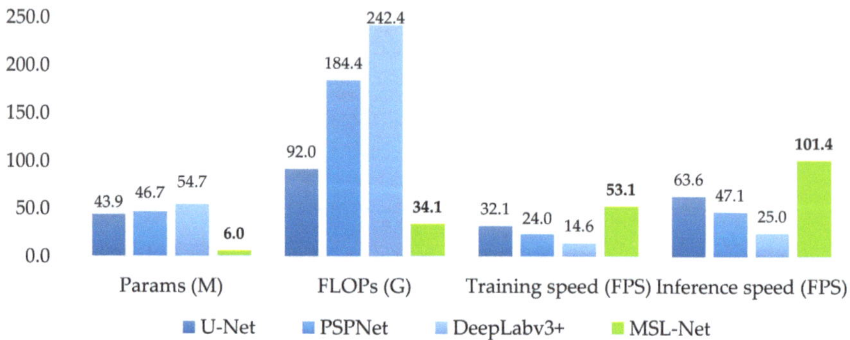

Figure 10. Complexity comparison.

4.3. Comparison with State-of-the-Art Methods

To verify the effectiveness of the proposed network, MSL-Net is compared with recent state-of-the-art building extraction methods, including AGs-Unet [54], PISANet [55], DR-Net [56], RSR-Net [57], BRRNet [58], and SRI-Net [59], on the accurate WHU dataset. As WHU dataset is a publicly available dataset, we use the reported model performances for our comparisons. The quantitative comparison results are shown in Table 8.

Table 8. Comparison with state-of-the-art methods on the WHU dataset. Here, '–' denotes that the paper did not provide relevant data. The best results are bolded; the second-best results are underlined.

Method	IoU (%)	F1-Score (%)	Params (M)
AGs-Unet	85.5	–	34.9
PISANet	88.0	93.6	–
DR-Net	88.3	93.8	9.0
RSR-Net	88.3	–	**2.9**
BRRNet	89.0	94.1	17.3
SRI-Net	89.1	94.2	–
MSL-Net	**90.4**	**95.0**	6.0

As seen in Table 8, the IoU and F1-score of MSL-Net are superior to those of all the tested methods that have been developed in recent studies, demonstrating the state-of-the-art performance of our proposed method. Compared with RSR-Net, BRRNet, and SRI-Net, our proposed MSL-Net achieves IoU improvements of 2.1%, 1.4%, and 1.3% on the WHU dataset, respectively. MSL-Net also achieves competitive number of parameters, demonstrating its great tradeoff between complexity and accuracy.

4.4. Ablation Experiments

To verify the effectiveness of each improvement, we perform ablation experiments on the WHU dataset, Inria dataset and Massachusetts dataset. Based on the baseline (MSL-Net with the unimproved MobileNetV2), we first change the LR policy from exponential decay (ED) to warmup and cosine annealing (WCA), and then we add a DCN layer at the end of the network. Finally, we replace the second BN layer with a GN layer in the inverse residual structure. The results of the six metrics are calculated, as shown in Table 9 and Figures 11–13.

The metrics of the model demonstrate constant trends toward superiority with the adjustment of the LR policy, the addition of the DCN, and the embedding of GN in the inverse residual module. For instance, on the WHU dataset, the WCA increases the IoU by 0.8% and the F1-score by 0.7%, the DCN module increases the IoU by 0.3% and the F1-score by 0.3%, and the improvement of the inverse residual structure increases the IoU by 1.0% and the F1-score by 0.1%. The ablation experimental results strongly prove the effectiveness of our introduced improvements.

Table 9. Ablation experimental results obtained on the three datasets. The highest scores are bolded; the second-highest scores are underlined.

Method	WHU		Inria		Massachusetts	
	IoU	F1-Score	IoU	F1-Score	IoU	F1-Score
Baseline + ED	88.3%	93.9%	77.3%	87.2%	68.7%	81.4%
Baseline + WCA	89.1%	94.6%	80.5%	89.2%	70.1%	82.4%
Baseline + WCA + DCN	<u>89.4%</u>	<u>94.9%</u>	<u>80.7%</u>	<u>89.3%</u>	<u>70.7%</u>	<u>82.8%</u>
Baseline + WCA + DCN + GN	**90.4%**	**95.0%**	**81.1%**	**89.6%**	**70.9%**	**83.0%**

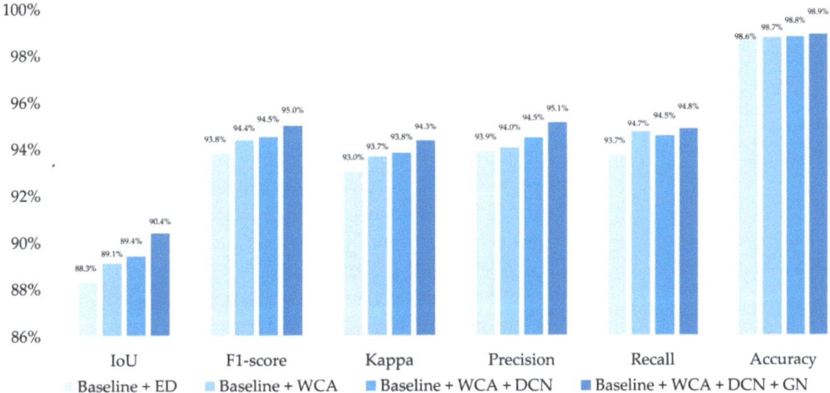

Figure 11. Ablation experimental results obtained on the WHU dataset.

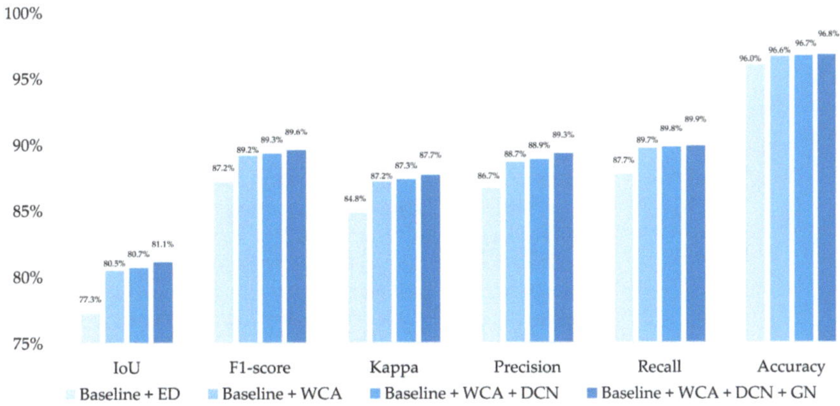

Figure 12. Ablation experimental results obtained on the Inria dataset.

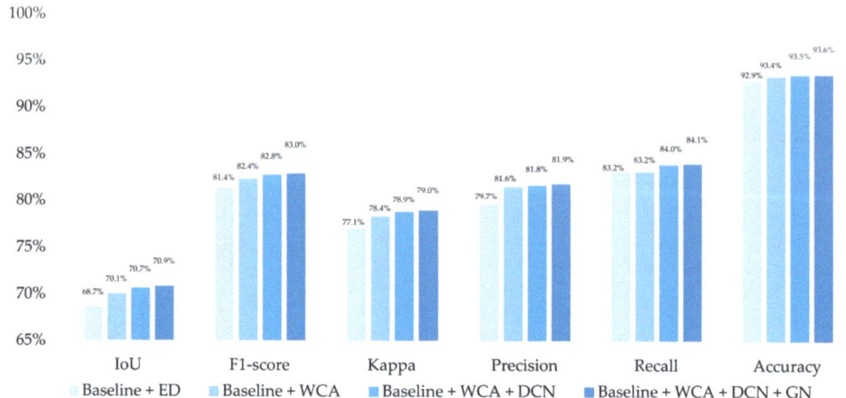

Figure 13. Ablation experimental results obtained on the Massachusetts dataset.

4.5. Limitations and Future Work

Despite superior performance achieved by the proposed MSL-Net in accuracy and complexity, MSL-Net still has limitations that need to be addressed. The experimental results of MSL-Net on the Inria dataset with more complex scenes have insignificant advantages over those of PSPNet, and it is still necessary to strengthen the robustness of lightweight MSL-Net in complex scenes. Additionally, the number of parameters of the ASPP module accounts for a large proportion of that of the whole network. In future work, the improvements or replacements of the ASPP module can be considered to address this limitation.

5. Conclusions

In this paper, we propose MSL-Net, an efficient neural network for building extraction. In terms of its network structure, MSL-Net adopts an ASPP module and skip connections to obtain the multiscale features of buildings and the multilevel features of images. The numbers of network parameters and computations are reduced by a DSC, GN is embedded in the inverted residual structure to alleviate network degradation, and the extraction capability of the model for irregularly shaped buildings is ensured by a DCN layer. Experiments are conducted on three publicly available datasets with varying spatial resolutions and building styles, and MSL-Net outperforms the comparison methods in model accuracy, demonstrating its superiority and robustness. Complexity evaluation experiments reveal that MSL-Net surpasses other models by more than 57.1% with an inference speed of 101.4 FPS; it requires only 14% of the parameters and 37% of the FLOPs required by the second-best method, manifesting the efficiency of MSL-Net. Ablation experiments indicate the effectiveness of each improvement. However, we find that the experimental results obtained by MSL-Net on the Inria dataset with more complex scenes are not sufficiently superior, and we will focus on enhancing the robustness of MSL-Net to complex scenes in our future work so that MSL-Net can perform well when monitoring urbanization in different cities.

Author Contributions: Conceptualization, Y.Q.; methodology, Y.Q. and F.W.; software, Y.Q.; validation, Y.Q. and J.Y.; formal analysis, C.L.; investigation, X.G.; resources, A.W.; data curation, Y.Q.; writing—original draft preparation, Y.Q.; writing—review and editing, Y.Q.; visualization, Y.Q.; supervision, F.W.; project administration, F.W.; funding acquisition, F.W. All authors have read and agreed to the published version of the manuscript.

Funding: This research was funded by the Natural Science Foundation for Distinguished Young Scholars of Henan Province under grant number 212300410014; the Research and Practice Projects of Higher Education Reform in Henan Province under grant number 2021SJGLX299.

Data Availability Statement: The datasets and code presented in this study are available at https://github.com/ParkourX/MSLNet, accessed on 3 August 2022.

Acknowledgments: We thank the editors and reviewers for their constructive and helpful comments that led to the substantial improvement of this paper.

Conflicts of Interest: The authors declare no conflict of interest.

Abbreviations

ASPP	Atrous spatial pyramid pooling
BN	Batch normalization
CPU	Central processing unit
DCN	Deformable convolution network
DSC	Depthwise separable convolution
FCN	Fully convolutional network
FLOPs	Floating-point operations
FN	False negative
FP	False positive
FPS	Frames per second
GN	Group normalization
GPU	Graphics processing unit
IoU	Intersection over union
OS	Operating system
TP	True positive
TN	True negative

References

1. Zeng, Y.; Guo, Y.; Li, J. Recognition and Extraction of High-Resolution Satellite Remote Sensing Image Buildings Based on Deep Learning. *Neural. Comput. Appl.* **2022**, *34*, 2691–2706. [CrossRef]
2. Ghanea, M.; Moallem, P.; Momeni, M. Building Extraction from High-Resolution Satellite Images in Urban Areas: Recent Methods and Strategies Against Significant Challenges. *Int. J. Remote Sens.* **2016**, *37*, 5234–5248. [CrossRef]
3. Ji, S.; Wei, S.; Lu, M. Fully Convolutional Networks for Multisource Building Extraction from an Open Aerial and Satellite Imagery Data Set. *IEEE Trans. Geosci. Remote Sens.* **2019**, *57*, 574–586. [CrossRef]
4. Chen, Q.; Wang, L.; Waslander, S.L.; Liu, X. An End-to-End Shape Modeling Framework for Vectorized Building Outline Generation from Aerial Images. *ISPRS J. Photogramm. Remote Sens.* **2020**, *170*, 114–126. [CrossRef]
5. Katartzis, A.; Sahli, H.; Nyssen, E.; Cornelis, J. Detection of Buildings from a Single Airborne Image Using a Markov Random Field Model. In Proceedings of the IGARSS 2001, Scanning the Present and Resolving the Future, IEEE 2001 International Geoscience and Remote Sensing Symposium (Cat. No.01CH37217), Sydney, Australia, 9–13 July 2001; Volume 6, pp. 2832–2834.
6. Simonetto, E.; Oriot, H.; Garello, R. Rectangular Building Extraction from Stereoscopic Airborne Radar Images. *IEEE Trans. Geosci. Remote Sens.* **2005**, *43*, 2386–2395. [CrossRef]
7. Jung, C.R.; Schramm, R. Rectangle Detection Based on a Windowed Hough Transform. In Proceedings of the 17th Brazilian Symposium on Computer Graphics and Image Processing, Curitiba, Brazil, 20–20 October 2004; pp. 113–120.
8. Li, L. Research on Shadow-Based Building Extraction from High Resolution Remote Sensing Images. Master's Thesis, Hunan University of Science and Technology, Xiangtan, China, 2011.
9. Zhao, Z.; Zhang, Y. Building Extraction from Airborne Laser Point Cloud Using NDVI Constrained Watershed Algorithm. *Acta Optica Sin.* **2016**, *36*, 503–511.
10. Zhou, S.; Liang, D.; Wang, H.; Kong, J. Remote Sensing Image Segmentation Approach Based on Quarter-Tree and Graph Cut. *Comput. Eng.* **2010**, *36*, 224–226.
11. Wei, D. Research on Buildings Extraction Technology on High Resolution Remote Sensing Images. Master's Thesis, Information Engineering University, Zhengzhou, China, 2013.
12. Tournaire, O.; Brédif, M.; Boldo, D.; Durupt, M. An Efficient Stochastic Approach for Building Footprint Extraction from Digital Elevation Models. *ISPRS J. Photogramm. Remote Sens.* **2010**, *65*, 317–327. [CrossRef]
13. Parsian, S.; Amani, M. Building Extraction from Fused LiDAR and Hyperspectral Data Using Random Forest Algorithm. *Geomatica* **2017**, *71*, 185–193. [CrossRef]
14. Ferro, A.; Brunner, D.; Bruzzone, L. Automatic Detection and Reconstruction of Building Radar Footprints from Single VHR SAR Images. *IEEE Trans. Geosci. Remote Sens.* **2013**, *51*, 935–952. [CrossRef]
15. Wei, Y.; Zhao, Z.; Song, J. Urban Building Extraction from High-Resolution Satellite Panchromatic Image Using Clustering and Edge Detection. In Proceedings of the IEEE International Geoscience and Remote Sensing Symposium, Anchorage, AK, USA, 20–24 September 2004; IEEE: Anchorage, AK, USA, 2004; Volume 3, pp. 2008–2010.

16. Huang, X.; Zhang, L. Morphological Building/Shadow Index for Building Extraction from High-Resolution Imagery Over Urban Areas. *IEEE J. Sel. Top. Appl. Earth Obs. Remote Sens.* **2012**, *5*, 161–172. [CrossRef]
17. Gao, X.; Wang, M.; Yang, Y.; Li, G. Building Extraction from RGB VHR Images Using Shifted Shadow Algorithm. *IEEE Access* **2018**, *6*, 22034–22045. [CrossRef]
18. Maruyama, Y.; Tashiro, A.; Yamazaki, F. Use of Digital Surface Model Constructed from Digital Aerial Images to Detect Collapsed Buildings during Earthquake. *Procedia Eng.* **2011**, *14*, 552–558. [CrossRef]
19. Guo, H.; Du, B.; Zhang, L.; Su, X. A Coarse-to-Fine Boundary Refinement Network for Building Footprint Extraction from Remote Sensing Imagery. *ISPRS J. Photogramm. Remote Sens.* **2022**, *183*, 240–252. [CrossRef]
20. Long, J.; Shelhamer, E.; Darrell, T. Fully Convolutional Networks for Semantic Segmentation. In Proceedings of the IEEE Conference on Computer Vision and Pattern Recognition, Boston, MA, USA, 7–12 June 2015; pp. 3431–3440.
21. Yuan, J. Learning Building Extraction in Aerial Scenes with Convolutional Networks. *IEEE Trans. Pattern Anal. Mach. Intell.* **2018**, *40*, 2793–2798. [CrossRef] [PubMed]
22. Maggiori, E.; Tarabalka, Y.; Charpiat, G.; Alliez, P. High-Resolution Aerial Image Labeling with Convolutional Neural Networks. *IEEE Trans. Geosci. Remote Sens.* **2017**, *55*, 7092–7103. [CrossRef]
23. Zhao, H.; Shi, J.; Qi, X.; Wang, X.; Jia, J. Pyramid Scene Parsing Network. In Proceedings of the IEEE Conference on Computer Vision and Pattern Recognition, Honolulu, HI, USA, 21–26 July 2017; pp. 2881–2890.
24. Chen, L.-C.; Papandreou, G.; Schroff, F.; Adam, H. Rethinking Atrous Convolution for Semantic Image Segmentation. *arXiv* **2017**, arXiv:1706.05587.
25. Xu, Z.; Shen, Z.; Li, Y.; Zhao, L.; Ke, Y.; Li, L.; Wen, Q. Classification of High-Resolution Remote Sensing Images Based on Enhanced DeepLab Algorithm and Adaptive Loss Function. *Nat. Remote Sens. Bull.* **2022**, *26*, 406–415.
26. Chen, L.-C.; Zhu, Y.; Papandreou, G.; Schroff, F.; Adam, H. Encoder-Decoder with Atrous Separable Convolution for Semantic Image Segmentation. In Proceedings of the European Conference on Computer Vision (ECCV), Munich, Germany, 8–14 September 2018; pp. 833–851.
27. Ronneberger, O.; Fischer, P.; Brox, T. U-Net: Convolutional Networks for Biomedical Image Segmentation. In Proceedings of the Medical Image Computing and Computer-Assisted Intervention—MICCAI 2015, Munich, Germany, 5–9 October 2015; pp. 234–241.
28. Wang, Z.; Zhou, Y.; Wang, S.; Wang, F.; Xu, Z. House Building Extraction from High-Resolution Remote Sensing Images based on IEU-Net. *Nat. Remote Sens. Bull.* **2021**, *25*, 2245–2254.
29. Xu, L.; Liu, Y.; Yang, P.; Chen, H.; Zhang, H.; Wang, D.; Zhang, X. HA U-Net: Improved Model for Building Extraction from High Resolution Remote Sensing Imagery. *IEEE Access* **2021**, *9*, 101972–101984. [CrossRef]
30. Liu, Y.; Chen, D.; Ma, A.; Zhong, Y.; Fang, F.; Xu, K. Multiscale U-Shaped CNN Building Instance Extraction Framework with Edge Constraint for High-Spatial-Resolution Remote Sensing Imagery. *IEEE Trans. Geosci. Remote Sens.* **2021**, *59*, 6106–6120. [CrossRef]
31. Zhang, Y.; Yan, Q.; Deng, F. Multi-Path RSU Network Method for High-Resolution Remote Sensing Image Building Extraction. *Acta Geod. Cartogr. Sin.* **2022**, *51*, 135–144.
32. Xu, J.; Liu, W.; Shan, H.; Shi, J.; Li, E.; Zhang, L.; Li, H. High-Resolution Remote Sensing Image Building Extraction Based on PRCUnet. *J. Geo-inf. Sci.* **2021**, *23*, 1838–1849.
33. He, Z.; Ding, H.; An, B. E-Unet: A Atrous Convolution-Based Neural Network for Building Extraction from High-Resolution Remote Sensing Images. *Acta Geod. Cartogr. Sin.* **2022**, *51*, 457–467.
34. Zhang, C.; Liu, H.; Ge, Y.; Shi, S.; Zhang, M. Multi-Scale Dilated Convolutional Pyramid Network for Building Extraction. *J. Xi'an Univ. Sci. Technol.* **2021**, *41*, 490–497, 574.
35. Rashidian, V.; Baise, L.G.; Koch, M. Detecting Collapsed Buildings After a Natural Hazard on VHR Optical Satellite Imagery Using U-Net Convolutional Neural Networks. In Proceedings of the IGARSS 2019—2019 IEEE International Geoscience and Remote Sensing Symposium, Yokohama, Japan, 28 July–2 August 2019; pp. 9394–9397.
36. Xiong, C.; Li, Q.; Lu, X. Automated Regional Seismic Damage Assessment of Buildings Using an Unmanned Aerial Vehicle and a Convolutional Neural Network. *Autom. Constr.* **2020**, *109*, 102994. [CrossRef]
37. Cooner, A.J.; Shao, Y.; Campbell, J.B. Detection of Urban Damage Using Remote Sensing and Machine Learning Algorithms: Revisiting the 2010 Haiti Earthquake. *Remote Sens.* **2016**, *8*, 868. [CrossRef]
38. Shimoni, M.; Haelterman, R.; Perneel, C. Hypersectral Imaging for Military and Security Applications: Combining Myriad Processing and Sensing Techniques. *IEEE Geosci. Remote Sens. Mag.* **2019**, *7*, 101–117. [CrossRef]
39. Sandler, M.; Howard, A.; Zhu, M.; Zhmoginov, A.; Chen, L.-C. MobileNetV2: Inverted Residuals and Linear Bottlenecks. In Proceedings of the 2018 IEEE/CVF Conference on Computer Vision and Pattern Recognition, Salt Lake City, UT, USA, 18–23 June 2018; pp. 4510–4520.
40. Sifre, L. Rigid-Motion Scattering for Image Classification. Ph.D. Thesis, École Polytechnique, Paris, France, 2014.
41. Wu, Y.; He, K. Group Normalization. In Proceedings of the European Conference on Computer Vision (ECCV), Munich, Germany, 8–14 September 2018; pp. 3–19.
42. Srivastava, R.K.; Greff, K.; Schmidhuber, J. Highway Networks. *arXiv* **2015**, arXiv:1505.00387.
43. Zhu, X.; Hu, H.; Lin, S.; Dai, J. Deformable ConvNets V2: More Deformable, Better Results. In Proceedings of the 2019 IEEE/CVF Conference on Computer Vision and Pattern Recognition, Long Beach, CA, USA, 15–20 June 2019; pp. 9300–9308.

44. He, K.; Zhang, X.; Ren, S.; Sun, J. Deep Residual Learning for Image Recognition. In Proceedings of the IEEE Conference on Computer Vision and Pattern Recognition, Las Vegas, NV, USA, 27–30 June 2016; pp. 770–778.
45. Ioffe, S.; Szegedy, C. Batch Normalization: Accelerating Deep Network Training by Reducing Internal Covariate Shift. In Proceedings of the 32nd International Conference on International Conference on Machine Learning, Lille, France, 6–11 July 2015; pp. 448–456.
46. Huang, L.; Zhou, Y.; Wang, T.; Luo, J.; Liu, X. Delving into the Estimation Shift of Batch Normalization in a Network. *arXiv* **2022**, arXiv:2203.10778.
47. Yu, F.; Koltun, V. Multi-Scale Context Aggregation by Dilated Convolutions. *arXiv* **2015**, arXiv:1511.07122.
48. Loshchilov, I.; Hutter, F. SGDR: Stochastic Gradient Descent with Warm Restarts. *arXiv* **2017**, arXiv:1608.03983.
49. Maggiori, E.; Tarabalka, Y.; Charpiat, G.; Alliez, P. Can Semantic Labeling Methods Generalize to Any City? The Inria Aerial Image Labeling Benchmark. In Proceedings of the 2017 IEEE International Geoscience and Remote Sensing Symposium (IGARSS), Fort Worth, TX, USA, 23–28 July 2017; pp. 3226–3229.
50. Mnih, V. Machine Learning for Aerial Image Labeling. Ph.D. Thesis, University of Toronto, Toronto, ON, Canada, 2013.
51. Lin, T.Y.; Goyal, P.; Girshick, R.; He, K.M.; Dollar, P. Focal Loss for Dense Object Detection. *IEEE Trans. Pattern Anal. Mach. Intell.* **2020**, *42*, 318–327. [CrossRef] [PubMed]
52. Li, X.; Sun, X.; Meng, Y.; Liang, J.; Wu, F.; Li, J. Dice Loss for Data-imbalanced NLP Tasks. *arXiv* **2020**, arXiv:1911.02855.
53. Ji, S.; Wei, S. Building Extraction via Convolutional Neural Networks from an Open Remote Sensing Building Dataset. *Acta Geod. Cartogr. Sin.* **2019**, *48*, 448–459.
54. Yu, M.; Chen, X.; Zhang, W.; Liu, Y. AGs-Unet: Building Extraction Model for High Resolution Remote Sensing Images Based on Attention Gates U Network. *Sensors* **2022**, *22*, 2932. [CrossRef]
55. Zhou, D.; Wang, G.; He, G.; Long, T.; Yin, R.; Zhang, Z.; Chen, S.; Luo, B. Robust Building Extraction for High Spatial Resolution Remote Sensing Images with Self-Attention Network. *Sensors* **2020**, *20*, 7241. [CrossRef]
56. Chen, M.; Wu, J.; Liu, L.; Zhao, W.; Tian, F.; Shen, Q.; Zhao, B.; Du, R. DR-Net: An Improved Network for Building Extraction from High Resolution Remote Sensing Image. *Remote Sens.* **2021**, *13*, 294. [CrossRef]
57. Huang, H.; Chen, Y.; Wang, R. A Lightweight Network for Building Extraction from Remote Sensing Images. *IEEE Trans. Geosci. Remote Sens.* **2022**, *60*, 1–12. [CrossRef]
58. Shao, Z.; Tang, P.; Wang, Z.; Saleem, N.; Yam, S.; Sommai, C. BRRNet: A Fully Convolutional Neural Network for Automatic Building Extraction from High-Resolution Remote Sensing Images. *Remote Sens.* **2020**, *12*, 1050. [CrossRef]
59. Liu, P.; Liu, X.; Liu, M.; Shi, Q.; Yang, J.; Xu, X.; Zhang, Y. Building Footprint Extraction from High-Resolution Images via Spatial Residual Inception Convolutional Neural Network. *Remote Sens.* **2019**, *11*, 830. [CrossRef]

Communication

Time Ring Data: Definition and Application in Spatio-Temporal Analysis of Urban Expansion and Forest Loss

Xin Liu, Xinhu Li * and Haijun Bao

School of Spatial Planning and Design, Hangzhou City University, Hangzhou 310015, China
* Correspondence: lixh@zucc.edu.cn

Abstract: Remote sensing can provide spatio-temporal continuous Earth observation data and is becoming the main data source for spatial and temporal analysis. Remote sensing data have been widely used in applications such as meteorological monitoring, forest investigation, environmental health, urban planning, and water conservancy. While long-time-series remote sensing data are used for spatio-temporal analysis, this analysis is usually limited because of the large data volumes and complex models used. This study intends to develop an innovative and simple approach to reveal the spatio-temporal characteristics of geographic features from the perspective of remote sensing data themselves. We defined an efficient remote sensing data structure, namely time ring (TR) data, to depict the spatio-temporal dynamics of two common geographic features. One is spatially expansive features. Taking nighttime light (NTL) as an example, we generated a NTL TR map to exhibit urban expansion with spatial and temporal information. The speed and acceleration maps of NTL TR data indicated extraordinary expansion in the last 10 years, especially in coastal cities and provincial capitals. Beijing, Tianjin, Hebei Province, Shandong Province, and Jiangsu Province exhibited fast acceleration of urbanization. The other is spatially contractive features. We took forest loss in the Amazon basin as an example and produced a forest cover TR map. The speed and acceleration were mapped in two 10-year periods (2000–2010 and 2010–2020) in order to observe the changes in Amazon forest cover. Then, combining cropland TR data, we determined the consistency of the spatio-temporal variations and used a linear regression model to detect the association between the acceleration of cropland and forest. The forest TR map showed that, spatially, there was an apparent phenomenon of forest loss occurring in the southern and eastern Amazon basin. Temporally, the speed of forest loss was more drastic between 2000 and 2010 than that in 2010–2020. In addition, the acceleration of forest loss showed a dispersed distribution, except for in Bolivia, which demonstrated a concentrated regional acceleration. The R-squared value of the linear regression between forest and cropland acceleration reached 0.75, indicating that forest loss was closely linked to the expansion of cropland. The TR data defined in this study not only optimized the use of remote sensing data, but also facilitated their application in spatio-temporal integrative analysis. More importantly, multi-field TR data could be jointly applied to explore the driving force at spatial and temporal scales.

Keywords: time ring data; long time series; spatio-temporal characteristics; Amazon forest cover; urban expansion; nighttime light

Citation: Liu, X.; Li, X.; Bao, H. Time Ring Data: Definition and Application in Spatio-Temporal Analysis of Urban Expansion and Forest Loss. *Remote Sens.* **2023**, *15*, 972. https://doi.org/10.3390/rs15040972

Academic Editors: Ioannis Z. Gitas and Gregory Giuliani

Received: 11 November 2022
Revised: 13 January 2023
Accepted: 8 February 2023
Published: 10 February 2023

1. Introduction

Spatial and temporal variation analysis of geographic features is an important part of geographic research that helps us to understand the characteristics of geographic features [1]. Furthermore, we can predict the potential variations in such characteristics so as to assess the rationality of the variations and the possible risks, and put forward corresponding countermeasures. At present, the data sources for spatio-temporal variation analysis are usually divided into ground observation data and remote sensing (RS) data [2]. With the development of RS techniques, satellite RS images have gradually become the main data source because they can provide spatio-temporal continuous Earth observations [3].

RS data, to date, have been widely used in applications such as meteorological monitoring, forest investigation, environmental health, urban planning, water conservancy, and global climate change [4–9].

However, remote sensing data may be a burden when performing long-time-series and large-geographical-scale research [10,11]. These studies involve massive amounts of data, forcing researchers to perform additional data organization work to conduct spatio-temporal change detection analysis [12]. For specific geographic features with the characteristics of a large covered range and long time series, such as flood expansion [13], urban expansion [14], and forest cover change monitoring [15], remote sensing data need to be pre-processed to fit the model inputs, which is time-consuming and labor-intensive work. A large body of work has been conducted due to the proliferation of RS data. For example, Ma built a data-intensive index to decrease the computing complexity for long-time-span and large-range RS applications [16]. Xu put forward a spatial featured data cube analysis tool conducting time-series RS data processing and analysis based on multidimensional spatio-temporal data models, took Poyang Lake data cube datasets for 20 years as an example, and successfully mapped the percentage change in the water area of Poyang Lake over a 20-year time series [10]. Such methods solved the problem of tedious data processing, but research about RS data processing for spatio-temporal distribution analysis of features is still insufficient.

It Is also essential to take into account appropriate approaches in the spatio-temporal analysis of geographic features [17]. Most studies conducting spatiotemporal analysis of RS data simply involved the cognition of the spatial distribution of features at different times [18–21]. These studies inherently spatially and temporally separated out difficult-to-mine spatio-temporal integrative information. Furthermore, when involving long-time-series or large-scale analysis, these approaches are unwieldy for understanding spatio-temporal characteristics. At present, there are no studies on how to present spatio-temporal integrative characteristics in practical applications based on long-time-series RS data.

Several studies have utilized an effective data form to store long-term changing features and present the spatial variation with time on one map [22–24]. For example, Gong produced one global artificial impervious area (GAIA) map from 1985 to 2018, and this data storage was convenient and the spatial variation in GAIA with time could be clearly understood. However, the GAIA data were used only for mapping, lacking further spatio-temporal dynamic analysis based on the data structure. In addition, similar studies failed to explore the general specification and potential practical value of the data form. Therefore, this study proposed time ring (TR) data from the perspective of remote sensing data organization on the basis of previous studies. The TR data are intended to exhibit the spatial variation in geographic features over time, which can intuitively show the spatio-temporal characteristics of geographic features. The TR data structure that we defined is efficient for two types of features. One is space-expansive features over time, such as urban buildings [14] and nighttime light [25]. The other is space-contractive features over time, such as forest cover [26] and glacier cover [27]. Different processing methods will be performed when producing TR data according to the different types of features.

This study provides a detailed definition of TR data and illustrates the steps involved in transforming multi-temporal RS data into TR data. Then, we introduce related calculations and applications of TR data. Two examples of nighttime light TR data and forest cover TR data are provided to explain the spatio-temporal characteristics. Additionally, we perform an ordinary linear regression analysis combining forest and cropland TR data to reveal the driving mechanism spatially and temporally. Finally, this study discusses the strengths and weaknesses of TR data and states the possible application prospects in the future.

2. Details of Time Ring (TR) Data

2.1. Definition of TR Data

Time ring (TR) data are a new form based on RS data in which geographic features are exhibited like tree rings, which can clearly present the spatial variation characteristics over time (Figure 1). These data rely on long-time-series remote sensing images because a long time series can better show the spatio-temporal variation characteristics of geographic features. Too few time slices may be detrimental to their subsequent exploitation in space and time. The primary trait of TR data for geographic features is either expansive or contractive (Figure 1). For expansive features, the spatial distribution area of these features will gradually increase with temporal change, while for contractive features, the spatial distribution area of these features will gradually decrease with temporal change. The determination of expansion and contraction can be found in Section 2.2. Each grid unit in the TR data is marked by a time label in order to represent integrative temporal and spatial information. The time label denotes when the feature appears or disappears in this grid. For example, for expansive features, such as urban areas, there are different urban areas in t1, t2, and t3. Then, when forming TR data, t2 in each grid indicates that the urban area was emerging in time t2. This expression of data assures that information can be obtained regarding the time at which the spatial changes happened. Furthermore, researchers can perform analysis using a regression model combining multiple TR variables in various fields of application.

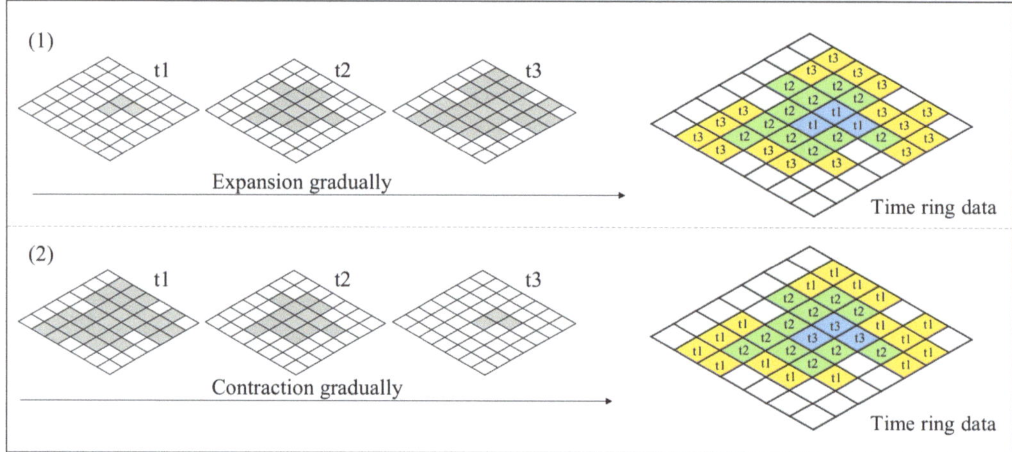

Figure 1. The general form of time ring (TR) data: (**1**) for spatially expansive feature; and (**2**) for spatially contractive feature.

2.2. Generation of TR Data

The generation of TR data involves four steps (Figure 2). The first step is to obtain a relevant remote sensing dataset with notable geographic features. Some datasets can be downloaded based on open-source satellite sensor data, such as normalized differential vegetation index (NDVI) data released by the Moderate-Resolution Imaging Spectroradiometer (MODIS) and nighttime light data produced by the Visible Infrared Imaging Radiometer Suite (VIIRS). Others are subject to further interpretation according to the original image, such as land cover data and urban built-up area data. The second step is to reclassify all images by assigning each pixel of the feature a time label and setting the other pixels to null. The third step is to determine whether the geographic feature is spatially expansive or contractive. We sum up the number of time labels on each image and establish whether the total increased or decreased over time. If the count is increasing, we consider it an expanding feature; otherwise, it is a contracting feature. The last step is to fuse all images in chronological order and produce the TR data. If a feature is expanding outward, the central region represents an older time and the peripheral regions' time is newer. If a feature is contracting inward, then the central region represents a newer time and the peripheral region time is older.

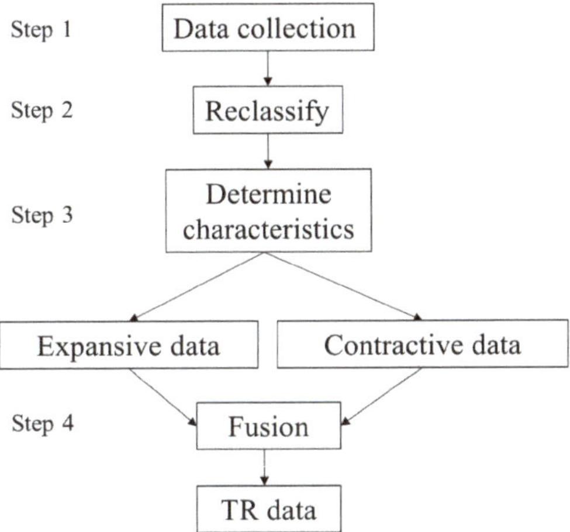

Figure 2. The generation steps of time ring (TR) data.

2.3. Calculation of TR Data

Producing TR data is a key step, and the subsequent related analysis will be performed with TR data. In this study, we take into account speed and acceleration as a method of calculating TR data. Speed is used to describe the state of change in features in different periods regarding, specifically, a changing extent of expansion or contraction. Additionally, acceleration is a further calculation based on speed denoting the magnitude of the impact of the driving force [14]. The calculated expression was improved on the basis of our previously published article regarding the analysis of built-up area expansion [14]. Concretely, we first set the pixel with time labels to 1 using produced TR data (Figure 3). Then, we define a fishing net with a coarser resolution (5 × 5 km) than the pixel. The 5 km grid is suitable for larger-scale studies. In practice, we can flexibly set the grid size in terms of spatial scale in the study. The sum of all pixel values in each grid of the fishing net is counted as the grid value (Figure 3). In the end, the grid value is used to compute the speed and acceleration.

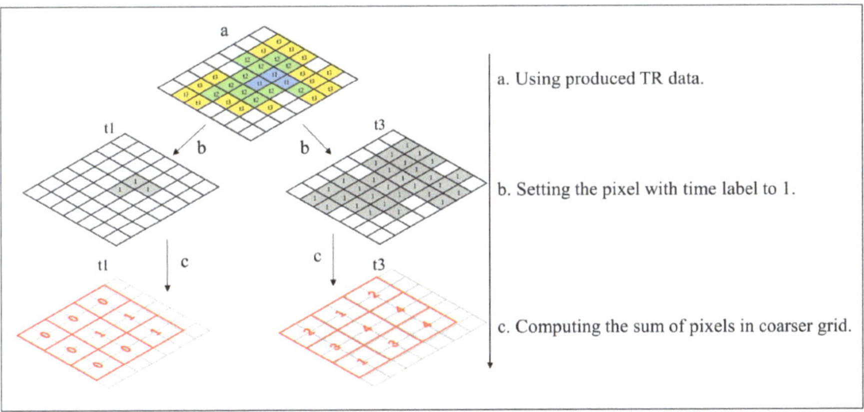

Figure 3. The data preparation for computing speed and acceleration. The data take the TR data with expansive features as an example. One grid can contain four pixels in this figure.

In this calculation, speed is defined as the difference in grid value between two time points multiplied by the pixel size and divided by the time interval. Acceleration is defined as the difference in speed between two periods divided by the time interval. The formulae for speed and acceleration are as follows:

$$Speed_{i,t0 \to t1} = (area_{i,t1} - area_{i,t0})/(t1 - t0) \tag{1}$$

$$Speed_{i,t1 \to t2} = (area_{i,t2} - area_{i,t1})/(t2 - t1) \tag{2}$$

$$Acceleration_{i,T} = (Speed_{i,t1 \to t2} - Speed_{i,t0 \to t1})/T \tag{3}$$

where $area_{t0}$, $area_{t1}$, and $area_{t2}$ represent the object areas at time t_1, t_2, and t_3, respectively. i represents each grid defined during data preparation (Figure 3). $t0 \to t1$ and $t1 \to t2$ represent two periods for calculating their speed. T represents the interval between the two periods for calculating their acceleration. The speed and acceleration are just one form of the calculation of TR data. Subsequent studies will further explore the analysis of TR data based on other methods, such as Moran's I.

3. Application to Urban Expansion

3.1. Nighttime Light Data

The TR data, on one hand, can be clearly used to recognize the time label when there is a spatial variation. On the other hand, they are able to characterize the driving force by combining other TR data. It is appropriate for use in the analysis of urban expansion under the context of global urbanization. Many studies have discussed the temporal variation and driving force of urban expansion by extracting the build-up area from remote sensing images, such as Landsat and Sentinel images [14,22]. In this study, we attempt to map the nighttime light (NTL) data to exhibit the urban expansion in China in 2000–2020. NTL data with a long time series can be obtained from the website https://dataverse.harvard.edu, accessed on 1 September 2022 [28]. Urban expansion represents a type of geographic feature with spatial expansion. As many studies on the spatial and temporal characteristics of urban expansion have been conducted, further influential mechanisms will not be discussed in this study. However, we produced the speed and acceleration map of NTL TR data to analyze the potential spatio-temporal characteristics.

3.2. Results of Urban Expansion

3.2.1. Spatio-Temporal Characteristics of NTL TR Data

According to the generated NTL TR map (Figure 4), comparing the three cities of Chengdu, Wuhan, and Beijing, the result shows that the largest NTL area was in Beijing in 2000, indicating the rapid development of urbanization in this period [29,30], while Chengdu and Wuhan were still in the early stages of development during this period. The NTL TR map also shows that the eastern part of Chengdu has maintained a state of rapid development in recent years. The application of urban expansion may be used to explain some phenomena caused by urban development, such as ecological environmental change.

3.2.2. Analysis of Speed and Acceleration of NTL in China

The increase in the NTL area can reflect the urban expansion, while the changing speed and acceleration of the NTL can also indirectly show the expansion rate and acceleration of urban areas. During the two periods (2000–2010 and 2010–2020), from the distribution map of speed and acceleration of NTL in China (Figure 5), we can establish that the overall speed in eastern China has been relatively high since 2000, indicating that these cities have experienced a rapid expansion. In addition, some provincial capitals in the central and western regions, such as Chengdu, Xi'an, and Wuhan, have also experienced significant expansion. In contrast, some cities developed more rapidly in period 2. The acceleration map shows that Beijing, Tianjin, Hebei Province, Shandong Province, and Jiangsu Province exhibited a high acceleration in urbanization. However, there are negative variations in acceleration in the central area of some provincial capitals, indicating that the NTL in the center of these cities is decreasing, which is especially significant around Shanghai.

NTL TR Map

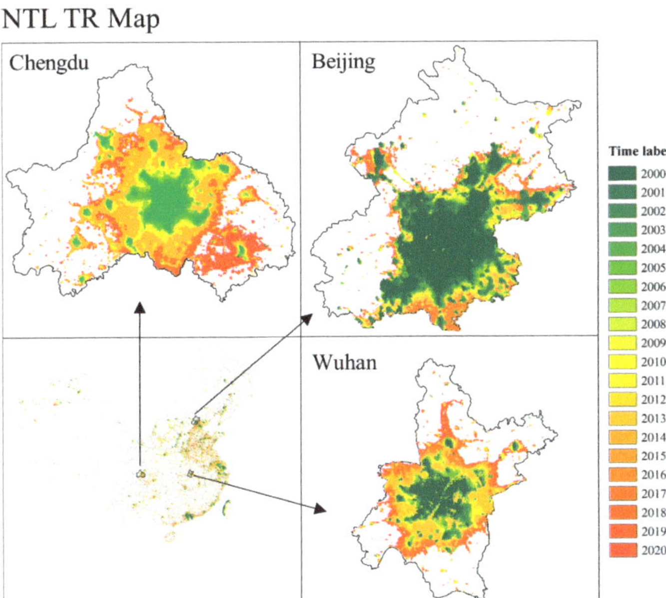

Figure 4. The urban NTL time ring (TR) map from 2000 to 2020 in China.

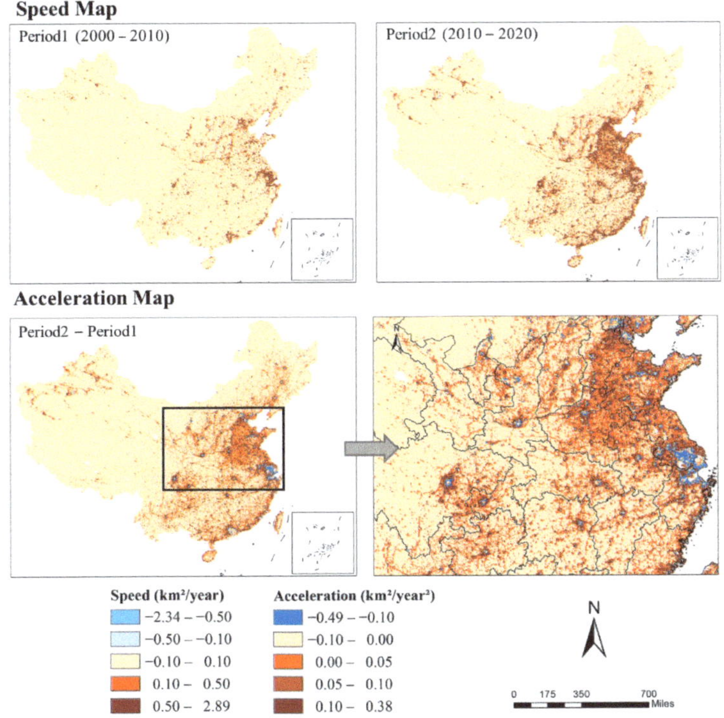

Figure 5. The speed and acceleration maps of NTL in China. The negative value in the speed legend represents the speed of NTL reduction, and the negative value of acceleration represents the acceleration of NTL reduction.

4. Application to Forest Cover Change

4.1. Study Area and Data

This study takes forest cover data in the Amazon basin as an example to check the utility of the TR data. The Amazon basin covers 7.05 million square kilometers, accounting for 40% of the total area of the South American continent. The region consists of parts of Brazil, Bolivia, Peru, Ecuador, Colombia, Venezuela, Guyana, Suriname, and French Guiana (Figure 6). The region has the largest tropical rainforest in the world, benefitting humankind by purifying the atmosphere and adjusting the climate. However, due to the impact of human activities, such as agricultural expansion, logging, and urban expansion, the land cover in the Amazon basin has undergone dramatic changes, especially with the massive loss of forest [31].

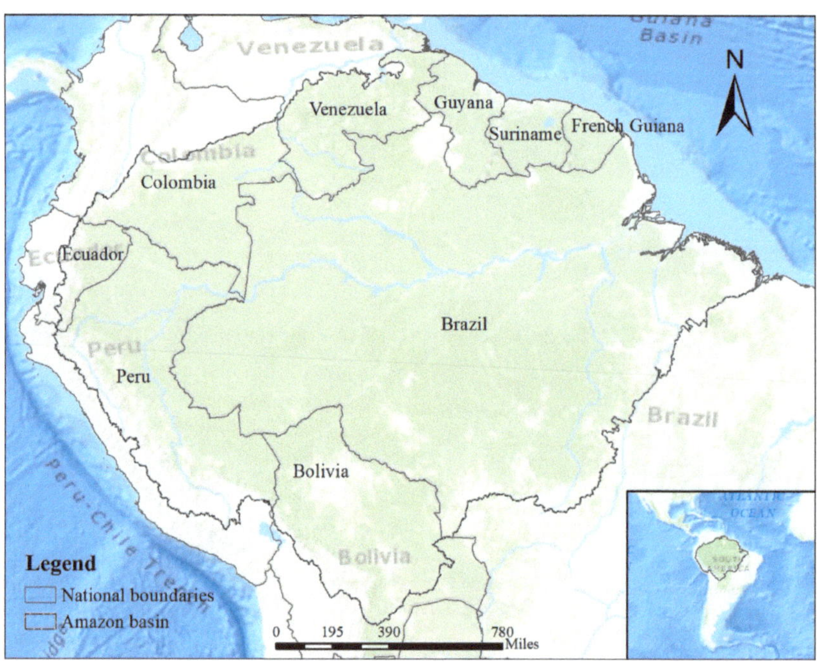

Figure 6. The study area of the Amazon basin in South America.

In this case, we screened out the forest and cropland datasets in the Amazon basin from 2000 to 2020 based on global land cover (LC) classification data. The global LC classification data can be obtained from the website https://cds.climate.copernicus.eu (accessed on 7 February 2023), which includes two sources. One is the global annual LC maps from 1992 to 2015 produced by the European Space Agency (ESA) Climate Change Initiative (CCI) LC project [32,33]. The other is the Copernicus Climate Change Service (C3S), which generated global LC maps from 2016 to 2020 that are consistent with the ESA-CCI LC maps. The LC classification data were defined using the United Nations Food and Agriculture Organization's (UN FAO) Land Cover Classification System (LCCS). The long-time-series LC classification data with a spatial resolution of 300 m are well suited to fine spatio-temporal analysis.

We produced the forest cover TR data in the Amazon basin based on the proposed generation method. Then, we extracted forest cover data from the TR data in 2000, 2010, and 2020 in order to create speed and acceleration maps. Furthermore, we also created a cropland TR map and an acceleration map in the Amazon basin to analyze the driving force of forest cover change, and observed their spatio-temporal variation characteristics based on these two TR maps. At the same time, ordinary linear regression was also utilized to establish the relationship between acceleration changes in forest and cropland.

4.2. Results of Forest Cover Change

4.2.1. Spatio-Temporal Characteristics of Forest TR Data

We produced TR data based on forest cover from 2000 to 2020 in the Amazon basin (Figure 7). Despite the high level of forest cover in the Amazon basin, there has been significant loss since the beginning of the 21st century. Forest loss is usually a process occurring from the edge to the center at the local scale, so the display of TR data was obvious. Overall, forest loss mostly occurred at the edge of the Amazon basin and near the main river and its tributaries (Figure 7). From a regional perspective, an apparent phenomenon of forest loss occurring in the southern and eastern Amazon basin was observed. According to the mapping result, forest loss in the Amazon basin was most significant between 2000 and 2010.

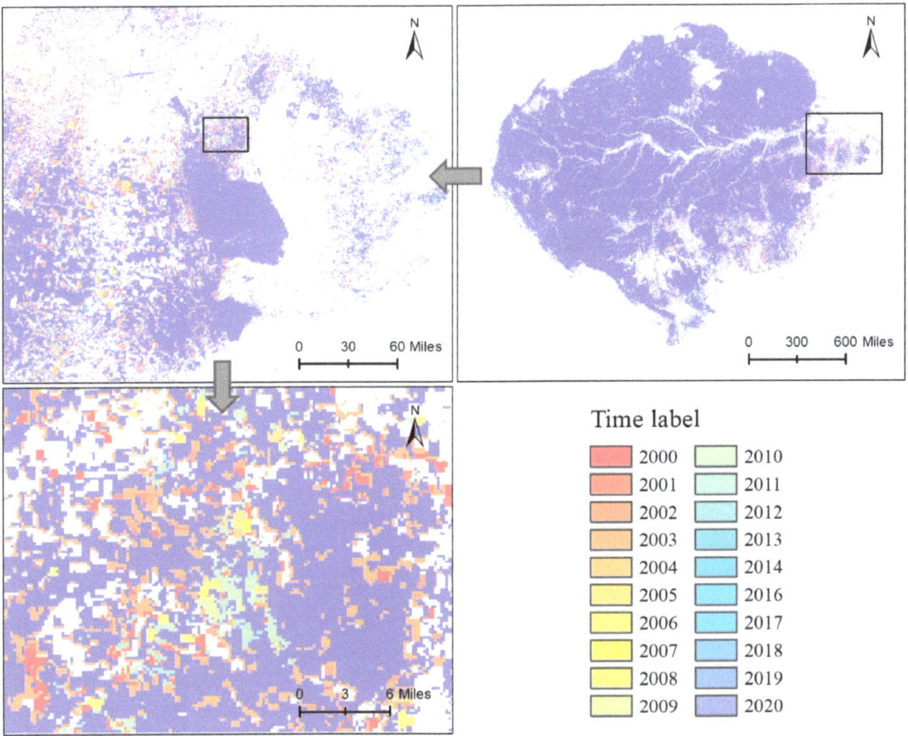

Figure 7. The forest time ring (TR) map from 2000 to 2020 in the Amazon basin.

4.2.2. Analysis of Speed and Acceleration

We mapped the speed and acceleration of forest loss for two periods (2000–2010 and 2010–2020) in the Amazon basin (Figure 8). By comparing the variation in speed between these two periods, we found that the speed of forest loss was more drastic in the first period. Spatially, the speed of forest loss was higher in the southeastern Amazon basin, mainly in Brazil, than in other regions. There was also a high speed of forest loss in some local areas, such as Caquetá Province in Colombia and central Bolivia. In the period of 2010–2020, the speed of forest loss slowed down overall, with higher speed only in the southern Amazon basin. From the acceleration map, the regional characteristics of forest loss were not obvious. Only in Bolivia was there a concentrated area of acceleration of forest loss. However, there was an acceleration of forest loss occurring in the whole region of the Amazon basin.

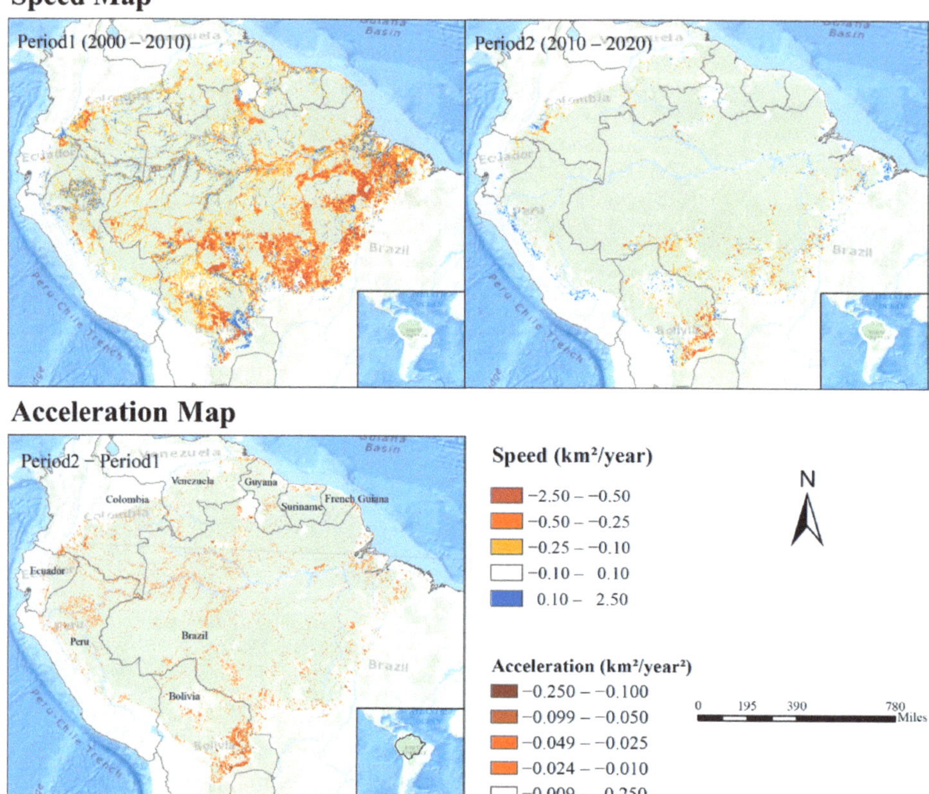

Figure 8. The speed and acceleration maps of forest loss in the Amazon basin. The negative value in the speed legend represents the speed of forest loss, and the negative value of acceleration represents the acceleration of forest loss.

4.2.3. Driving Force of Forest Cover Change

We compared forest and cropland variation characteristics from three perspectives. Firstly, the annual temporal variation in forest and cropland in the Amazon basin showed a symmetric change (Figure 9). The forest area in the Amazon basin demonstrated a downward trend from 2000 to 2008, while the cropland area, on the contrary, displayed an upward trend. The overall forest area in the Amazon basin displayed a slight decrease between 2010 and 2020, while the cropland area displayed a slight increase during the same period. Further, we produced a cropland TR map to compare the spatio-temporal change with that of the forest TR map (Figure 10). The cropland and forest TR maps showed a high similarity of spatial variation characteristics over time. That is, areas of forest loss were usually converted into cropland in the same period. Further, the acceleration of cropland was mapped and linear regression analysis was conducted (Figure 11). The result showed that there was a negative relationship between the acceleration of cropland and forest. The R-squared value reached 0.7484, indicating a strong correlation between them.

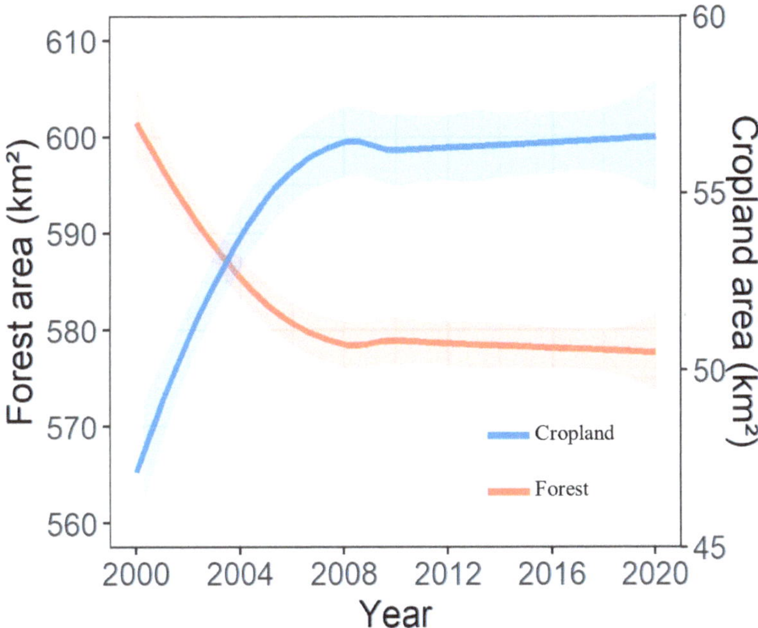

Figure 9. The temporal variation of forest and cropland in the period of 2000–2020 in the Amazon basin.

Cropland TR Map

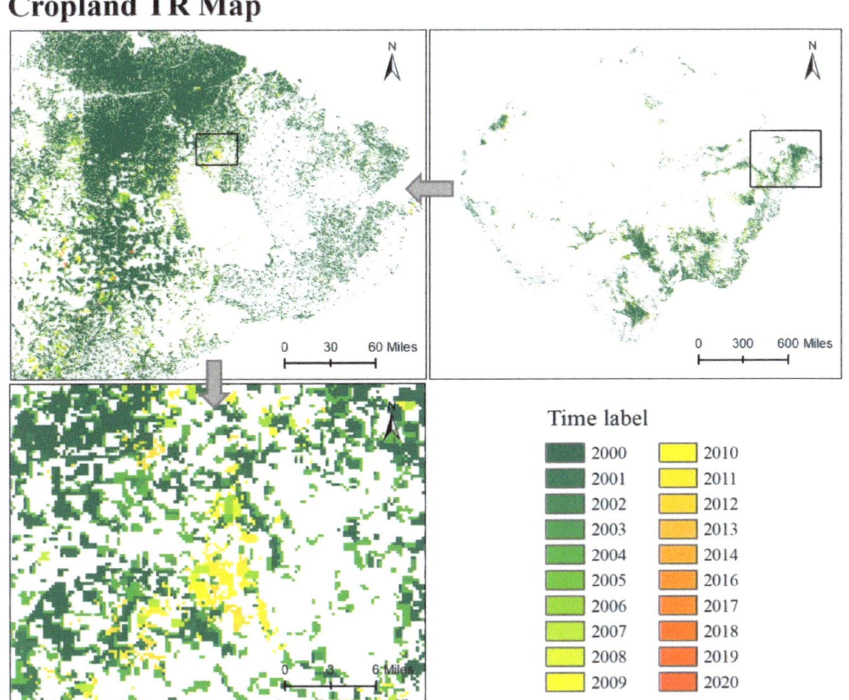

Figure 10. The cropland time ring (TR) map from 2000 to 2020 in the Amazon basin.

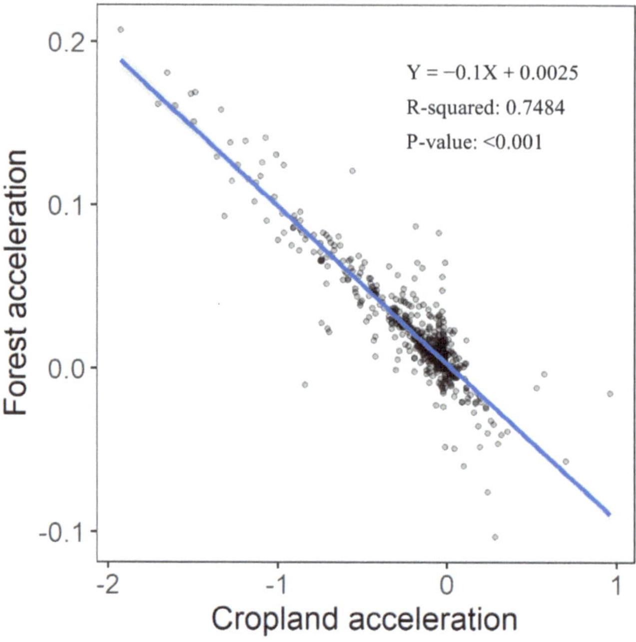

Figure 11. The linear regression result between the acceleration of cropland and forest.

5. Discussion

In this study, TR data were proposed to optimize the organizational approach of long-time-series remote sensing data, which was convenient for spatio-temporal analysis and modeling, as well as compensating for the defects of the separate analyses of time and space. Specifically, we used two applications to demonstrate the advantages of TR data. One is NTL in China, representing expansive geographic features. We created the NTL TR map to show the change in urban expansion. Previous studies have shown that Chinese cities and islands have undergone rapid urban expansion [34–36], which is consistent with our study mapping NTL TR data. However, we further found that this characteristic was particularly pronounced in coastal cities and some provincial capitals, based on the speed and acceleration maps. In addition, superior to other spatio-temporal analysis methods on urban expansion, such as the expansion contribution rate and expansion intensity index [24,37,38], we found that there was a high speed of expansion on the edge of developed cities in China, while there was a negative acceleration at the center of these cities. This may be due to the limiting of city light by the government and a shift in health awareness in these large urban centers, leading to a reduction in the use of NTL [39,40].

In addition, taking Amazon forest cover as another example, this study identified the spatio-temporal characteristics after mapping the forest TR data, providing explicit information about when and where the forest changed. We observed significant losses in forest cover occurring in the southern and eastern Amazon basin in the period of 2000–2010. The annual rates of deforestation in the Amazon maintained a high level in this decade [41]. Some behaviors, including the development of agriculture and mining, and constructing infrastructure, especially in Brazil, caused large-area deforestation until the related government's deforestation-control program was enacted [42,43]. The massive forest loss had a great impact on climate change, such as reducing dry season rainfall, which may have directly caused the local temperature to rise [44], even leading to destructive effects on regional forest resilience because of changing climatic conditions [45].

Through the speed and acceleration maps of forest cover and cropland, we found that forest loss was linked to the expansion of cropland in the Amazon basin. Nearly 75% of the forest loss could be attributed to cropland expansion between 2000 and 2020. In addition, according to a survey of soybean expansion in South America, the Brazilian Amazon also experienced the most rapid expansion, where the land that was originally natural vegetation was converted into soybean plantations [46]. Land speculators deforested these areas and usually sold off the timber and converted the land into pasture to further sell it to a soy producer [47]. These deleterious land use changes could contribute to environmental degradation by modifying seasonal water balance and increasing the risk of flooding [48].

Urban expansion and forest loss are both the results of land use change. This transformation is linked to the economic benefits behind it. People, in terms of their own development, tend to seek out land with the greatest economic value. However, this may be wrong when looking at the long term. For example, although the deforestation for cropland in the Amazon basin may promote the economic development of the countries in the region in the short term, the loss may be enormous due to the ecological imbalance caused by forest loss. Therefore, the short-term and long-term benefits should be considered together when planning land use.

As TR data mainly rely on long-time-series data, the temporal integrity and noise of RS data may affect the quality of the production of TR data. Many remote sensing reconstruction methods have been developed [49–51], while TR data may help to construct missing and low-quality data from the perspective of spatial–temporal consistency. If geographic features demonstrate evident contraction and expansion, we think that this is predictable for adjacent missing time labels. However, research needs to further explore the prediction technique for TR data.

The TR data have a concise data organization form and, through mapping, these data allow researchers to intuitively understand the spatio-temporal variation characteristics of features. The generation of TR data comprises a simple data-production process, including data preparation, reclassification, feature identification, and data fusion. Similar maps for the urban expansion process have been drawn based on remote sensing technology [35,36,52], but our study systematically puts forward the concept of TR data to visualize the spatiotemporal process. Furthermore, we present the calculation methods of speed and acceleration for TR data analysis. Speed calculation provides a basis for researchers to understand the changing state of features in space and time. Additionally, acceleration helps us to explore the driving mechanism of features. In addition, it is possible to develop other approaches that address TR data to mine more spatial–temporal information.

Time intervals are important in the analysis of time ring data. When we produced TR data using long-time-series RS data, we used the acquisition time of RS data as the time label of TR data, but we also needed to determine the time span of speed and acceleration analysis. Different time spans may affect the mapping result. For that, we, taking NTL TR data in China as an example, reproduced five-year speed and acceleration maps (Supplement Figures S1 and S2). Comparing the 10-year and 5-year maps, we concluded that the longer the time span, the more obvious the characteristics in space and time. However, a shorter time span can provide finer temporal information. Therefore, we advise researchers to balance the temporal resolution and observability of results when choosing the time span to map speed and acceleration.

From our perspective, time ring (TR) data have great application prospects because they can be used to obtain information relating to spatial expansion or contraction over time and can be suitably linked to other data to analyze the change mechanism, which has practical significance in many applications. For instance, lumpy skin disease (LSD), an infectious disease of cattle transmitted by arthropods, has undergone a rapid expansion in its geographic distribution since 2012 [53]. The spatial expansion of LSD has caused substantial economic losses [54] and is closely linked to environmental factors, so we could easily identify the spatio-temporal influencing pattern by mapping the TR data of related factors. Many vector-borne diseases, such as malaria [55,56] and dengue [57,58], as a result of climate change, may spatially expand or contract over time. Additionally, the variational speed regarding space could be different when affected by environmental conditions at different times, meaning that it is usually difficult to determine the influencing relationship [59]. However, the analysis based on TR data may help us to clearly recognize the spatial and temporal characteristics of environmental risk factors associated with vector-borne diseases.

In our study, we emphasized the advantages of using RS data as the data source of TR data, but TR data are also inclusive of other data sources, such as vector features, including point features, line features, or polygon features, in geographic information science (GIS). For instance, we can use yearly OpenStreetMap road data to produce road TR data. Roads can be seen as an expansive feature, usually extending on the basis of the previous roads. This expansion is related to human activity and may cause a series of social environmental problems, so road TR data potentially have great analytic value.

There are some limitations of TR data. Firstly, TR data cannot be applied to all geographic features, as they can only be used for features that undergo spatial expansion or contraction over time. Of course, many geographic features are both expansive and contractive. One TR map struggles to display both of these states simultaneously. However, it is possible to separately exhibit the expansion and contraction of one feature based on TR data. For example, we tend to produce contractive TR maps of forests in the Amazon basin because forest loss is more dominant than forest growth in the region. However, we can also generate a TR map for forest expansion in order to find out if there are areas with a growing forest over time. The difficulty may lie in how to combine forest increases and decreases for interactive analysis, which is expected to be solved in the following study. Secondly, the produced TR data were used to reveal the distribution change in geographic features in space and time, but changes in other feature attributes have been not considered, such as the NDVI value or the strength of nighttime light. Thirdly, TR data are used to display the spatial change in features with time, inevitably leaving out information due to data fusion, but this can be ignored in the analysis of spatio-temporal distribution variation, unless researchers focus on the other attributes of geographic features.

6. Conclusions

This study defined an efficient data structure, namely time ring (TR) data, to reveal the spatio-temporal changes in geographic features based on long-time-series remote sensing data. Common geographic features are either spatially expanding or spatially contracting, so we list two applications to illustrate the practicality of TR data. Firstly, we used NTL remote sensing data as an example to exhibit urban expansion over time by producing a NTL TR map. Cities in eastern China have experienced a rapid expansion since 2000. Beijing, Tianjin, Hebei Province, Shandong Province, and Jiangsu Province exhibited a high acceleration of urbanization. However, in the center of many developed cities, the NTL decreased, revealing laws related to urban development. In addition to revealing the spatio-temporal characteristics of NTL data, a significant aspect is that we can use NTL TR data to perform an exploration of spatio-temporal dynamics, such as the relationships between NTL expansion and population distribution. In addition, taking Amazon forest cover as another example, we illustrated its spatio-temporal variation by mapping forest TR data and demonstrated an applied method of TR data by conducting the calculation of speed and acceleration and using ordinary regression modeling. The results indicate that forest loss in the Amazon basin was more significant between 2000 and 2010 in the southern and eastern regions. Forest loss is closely linked to the expansion of cropland in the Amazon basin, so it is essential to pay more attention to the causes of cropland expansion in order to take measures to avoid continuous forest loss. TR data have the advantages of small data volumes and the easy analysis of spatially and temporally changing characteristics, providing this process with great application prospects. As far as we are concerned, TR data can be applied to environmental health investigation, land use/cover change, disaster monitoring, urban development, and so on, where more analytical methods related to TR data are expected to be discussed.

Supplementary Materials: The following supporting information can be downloaded at: https://www.mdpi.com/article/10.3390/rs15040972/s1.

Author Contributions: Conceptualization, X.L. (Xinhu Li); methodology, X.L. (Xinhu Li); Validation, X.L. (Xinhu Li) and H.B.; formal analysis, X.L. (Xin Liu); data curation, X.L. (Xin Liu); writing—original draft preparation, X.L. (Xin Liu); Writing—review and editing, X.L. (Xin Liu) and X.L. (Xinhu Li); visualization, X.L. (Xin Liu); supervision, X.L. (Xinhu Li) and H.B.; project administration, X.L. (Xinhu Li) and H.B.; funding acquisition, X.L. (Xinhu Li). All authors have read and agreed to the published version of the manuscript.

Funding: This research was funded by the National Natural Science Foundation of China, grant numbers 41671444.

Data Availability Statement: The data presented in this study are available on request from the corresponding author.

Conflicts of Interest: The authors declare no conflict of interest.

References

1. Wang, J.; Han, P.; Zhang, Y.; Li, J.; Xu, L.; Shen, X.; Yang, Z.; Xu, S.; Li, G.; Chen, F. Analysis on Ecological Status and Spatial–Temporal Variation of Tamarix Chinensis Forest Based on Spectral Characteristics and Remote Sensing Vegetation Indices. *Environ. Sci. Pollut. Res.* **2022**, *29*, 37315–37326. [CrossRef] [PubMed]
2. Despini, F.; Ferrari, C.; Bigi, A.; Libbra, A.; Teggi, S.; Muscio, A.; Ghermandi, G. Correlation between Remote Sensing Data and Ground Based Measurements for Solar Reflectance Retrieving. *Energy Build.* **2016**, *114*, 227–233. [CrossRef]
3. Militino, A.F.; Ugarte, M.D.; Pérez-Goya, U. An Introduction to the Spatio-Temporal Analysis of Satellite Remote Sensing Data for Geostatisticians. In *Handbook of Mathematical Geosciences: Fifty Years of IAMG*; Daya Sagar, B.S., Cheng, Q., Agterberg, F., Eds.; Springer International Publishing: Cham, Switzerland, 2018; pp. 239–253. ISBN 9783319789996.
4. Liu, L.; Huang, R.; Cheng, J.; Liu, W.; Chen, Y.; Shao, Q.; Duan, D.; Wei, P.; Chen, Y.; Huang, J. Monitoring Meteorological Drought in Southern China Using Remote Sensing Data. *Remote Sens.* **2021**, *13*, 3858. [CrossRef]
5. Gao, Y.; Skutsch, M.; Paneque-Gálvez, J.; Ghilardi, A. Remote Sensing of Forest Degradation: A Review. *Environ. Res. Lett.* **2020**, *15*, 103001. [CrossRef]
6. Lin, L.; Di, L.; Zhang, C.; Guo, L.; Di, Y. Remote Sensing of Urban Poverty and Gentrification. *Remote Sens.* **2021**, *13*, 4022. [CrossRef]
7. Viana, J.; Santos, J.V.; Neiva, R.M.; Souza, J.; Duarte, L.; Teodoro, A.C.; Freitas, A. Remote Sensing in Human Health: A 10-Year Bibliometric Analysis. *Remote Sens.* **2017**, *9*, 1225. [CrossRef]
8. Yang, H.; Kong, J.; Hu, H.; Du, Y.; Gao, M.; Chen, F. A Review of Remote Sensing for Water Quality Retrieval: Progress and Challenges. *Remote Sens.* **2022**, *14*, 1770. [CrossRef]
9. He, M.; Xu, Y.; Li, N. Population Spatialization in Beijing City Based on Machine Learning and Multisource Remote Sensing Data. *Remote Sens.* **2020**, *12*, 1910. [CrossRef]
10. Xu, D.; Ma, Y.; Yan, J.; Liu, P.; Chen, L. Spatial-Feature Data Cube for Spatiotemporal Remote Sensing Data Processing and Analysis. *Computing* **2020**, *102*, 1447–1461. [CrossRef]
11. Xu, C.; Du, X.; Fan, X.; Yan, Z.; Kang, X.; Zhu, J.; Hu, Z. A Modular Remote Sensing Big Data Framework. *IEEE Trans. Geosci. Remote Sens.* **2022**, *60*, 1–11. [CrossRef]
12. Koyama, C.N.; Watanabe, M.; Hayashi, M.; Ogawa, T.; Shimada, M. Mapping the Spatial-Temporal Variability of Tropical Forests by ALOS-2 L-Band SAR Big Data Analysis. *Remote Sens. Environ.* **2019**, *233*, 111372. [CrossRef]
13. Liu, H.; Hitchcock, D.B.; Samadi, S.Z. Spatio-Temporal Analysis of Flood Data from South Carolina. *J. Stat. Distrib. Appl.* **2020**, *7*, 11. [CrossRef]
14. Wang, L.; Jia, Y.; Li, X.; Gong, P. Analysing the Driving Forces and Environmental Effects of Urban Expansion by Mapping the Speed and Acceleration of Built-up Areas in China between 1978 and 2017. *Remote Sens.* **2020**, *12*, 3929. [CrossRef]
15. Suleiman, M.S.; Wasonga, O.V.; Mbau, J.S.; Elhadi, Y.A. Spatial and Temporal Analysis of Forest Cover Change in Falgore Game Reserve in Kano, Nigeria. *Ecol. Process.* **2017**, *6*, 11. [CrossRef]
16. Ma, Y.; Wang, L.; Liu, P.; Ranjan, R. Towards Building a Data-Intensive Index for Big Data Computing—A Case Study of Remote Sensing Data Processing. *Inf. Sci.* **2015**, *319*, 171–188. [CrossRef]
17. Chi, M.; Plaza, A.; Benediktsson, J.A.; Sun, Z.; Shen, J.; Zhu, Y. Big Data for Remote Sensing: Challenges and Opportunities. *Proc. IEEE* **2016**, *104*, 2207–2219. [CrossRef]
18. Zhang, Y.; Cheng, J. Spatio-Temporal Analysis of Urban Heat Island Using Multisource Remote Sensing Data: A Case Study in Hangzhou, China. *IEEE J. Sel. Top. Appl. Earth Obs. Remote Sens.* **2019**, *12*, 3317–3326. [CrossRef]
19. Ma, J.; Jin, S.; Li, J.; He, Y.; Shang, W. Spatio-Temporal Variations and Driving Forces of Harmful Algal Blooms in Chaohu Lake: A Multi-Source Remote Sensing Approach. *Remote Sens.* **2021**, *13*, 427. [CrossRef]
20. Silva, D.; Galvanin, E.A.S.; Menezes, R. Spatio-Temporal Analysis of Land Use/Land Cover Change Dynamics in Paraguai/Jauquara Basin, Brazil. *Environ. Monit. Assess.* **2022**, *194*, 400. [CrossRef]

21. Wu, J.; Zhang, Z.; He, Q.; Ma, G. Spatio-Temporal Analysis of Ecological Vulnerability and Driving Factor Analysis in the Dongjiang River Basin, China, in the Recent 20 Years. *Remote Sens.* **2021**, *13*, 4636. [CrossRef]
22. Gong, P.; Li, X.; Wang, J.; Bai, Y.; Chen, B.; Hu, T.; Liu, X.; Xu, B.; Yang, J.; Zhang, W.; et al. Annual Maps of Global Artificial Impervious Area (GAIA) between 1985 and 2018. *Remote Sens. Environ.* **2020**, *236*, 111510. [CrossRef]
23. Gong, P.; Li, X.; Zhang, W. 40-Year (1978–2017) Human Settlement Changes in China Reflected by Impervious Surfaces from Satellite Remote Sensing. *Sci. Bull.* **2019**, *64*, 756–763. [CrossRef]
24. Zhong, Y.; Lin, A.; He, L.; Zhou, Z.; Yuan, M. Spatiotemporal Dynamics and Driving Forces of Urban Land-Use Expansion: A Case Study of the Yangtze River Economic Belt, China. *Remote Sens.* **2020**, *12*, 287. [CrossRef]
25. Yin, Z.; Li, X.; Tong, F.; Li, Z.; Jendryke, M. Mapping Urban Expansion Using Night-Time Light Images from Luojia1-01 and International Space Station. *Int. J. Remote Sens.* **2020**, *41*, 2603–2623. [CrossRef]
26. Ram, A.K.; Yadav, N.K.; Kandel, P.N.; Mondol, S.; Pandav, B.; Natarajan, L.; Subedi, N.; Naha, D.; Reddy, C.S.; Lamichhane, B.R. Tracking Forest Loss and Fragmentation between 1930 and 2020 in Asian Elephant (Elephas Maximus) Range in Nepal. *Sci. Rep.* **2021**, *11*, 19514. [CrossRef]
27. Hugonnet, R.; McNabb, R.; Berthier, E.; Menounos, B.; Nuth, C.; Girod, L.; Farinotti, D.; Huss, M.; Dussaillant, I.; Brun, F.; et al. Accelerated Global Glacier Mass Loss in the Early Twenty-First Century. *Nature* **2021**, *592*, 726–731. [CrossRef] [PubMed]
28. Chen, Z.; Yu, B.; Yang, C.; Zhou, Y.; Yao, S.; Qian, X.; Wang, C.; Wu, B.; Wu, J. An Extended Time Series (2000-2018) of Global NPP-VIIRS-like Nighttime Light Data from a Cross-Sensor Calibration. *Earth Syst. Sci. Data* **2021**, *13*, 889–906. [CrossRef]
29. Li, C.; Wu, K.; Wu, J. Urban Land Use Change and Its Socio-Economic Driving Forces in China: A Case Study in Beijing, Tianjin and Hebei Region. *Environ. Dev. Sustain.* **2018**, *20*, 1405–1419. [CrossRef]
30. Tian, Y.; Yin, K.; Lu, D.; Hua, L.; Zhao, Q.; Wen, M. Examining Land Use and Land Cover Spatiotemporal Change and Driving Forces in Beijing from 1978 to 2010. *Remote Sens.* **2014**, *6*, 10593–10611. [CrossRef]
31. Qin, Y.; Xiao, X.; Dong, J.; Zhang, Y.; Wu, X.; Shimabukuro, Y.; Arai, E.; Biradar, C.; Wang, J.; Zou, Z.; et al. Improved Estimates of Forest Cover and Loss in the Brazilian Amazon in 2000–2017. *Nat. Sustain.* **2019**, *2*, 764–772. [CrossRef]
32. Defourny, P.; Schouten, L.; Bartalev, S.; Bontemps, S.; Caccetta, P.; De Wit, A.J.W.; Di Bella, C.; Gérard, B.; Giri, C.; Gond, V.; et al. Accuracy Assessment of a 300 m Global Land Cover Map: The GlobCover Experience. In Proceedings of the 33rd International Symposium on Remote Sensing of Environment (ISRSE), Stresa, Italy, 4–8 May 2009; pp. 400–403.
33. Bontemps, S.; Defourny, P.; Radoux, J.; Van Bogaert, E.; Lamarche, C.; Achard, F.; Mayaux, P.; Boettcher, M.; Brockmann, C.; Kirches, G.; et al. Consistent Global Land Cover Maps for Climate Modeling Communities: Current Achievements of the ESA's Land Cover CCI. In Proceedings of the ESA Living Planet Symposium, Edinburgh, UK, 9–13 September 2013; Volume 2013.
34. Huang, X.; Xia, J.; Xiao, R.; He, T. Urban Expansion Patterns of 291 Chinese Cities, 1990–2015. *Int. J. Digit. Earth* **2019**, *12*, 62–77. [CrossRef]
35. Cao, X.; Gao, X.; Shen, Z.; Li, R. Expansion of Urban Impervious Surfaces in Xining City Based on GEE and Landsat Time Series Data. *IEEE Access* **2020**, *8*, 147097–147111. [CrossRef]
36. Cao, W.; Zhou, Y.; Li, R.; Li, X.; Zhang, H. Monitoring Long-Term Annual Urban Expansion (1986–2017) in the Largest Archipelago of China. *Sci. Total Environ.* **2021**, *776*, 146015. [CrossRef]
37. Hu, Z.-L.; DU, P.-J.; Guo, D.-Z. Analysis of Urban Expansion and Driving Forces in Xuzhou City Based on Remote Sensing. *J. China Univ. Min. Technol.* **2007**, *17*, 267–271. [CrossRef]
38. Meng, H.; Zhang, K.; Ba, M.; Wen, D. Spatial Characteristics Analysis of Urban Expansion in Luoyang, China. *J. Geogr. Inf. Syst.* **2022**, *14*, 153–174. [CrossRef]
39. Du, X.; Shen, L.; Wong, S.W.; Meng, C.; Yang, Z. Night-Time Light Data Based Decoupling Relationship Analysis between Economic Growth and Carbon Emission in 289 Chinese Cities. *Sustain. Cities Soc.* **2021**, *73*, 103119. [CrossRef]
40. Xu, D.; Gao, J. The Night Light Development and Public Health in China. *Sustain. Cities Soc.* **2017**, *35*, 57–68. [CrossRef]
41. PRODES 2019 Amazon Deforestation Database. Available online: www.obt.inpe.br/prodes (accessed on 1 September 2022).
42. Kröger, M. Inter-Sectoral Determinants of Forest Policy: The Power of Deforesting Actors in Post-2012 Brazil. *For. Policy Econ.* **2017**, *77*, 24–32. [CrossRef]
43. Fearnside, P.M. The Roles and Movements of Actors in the Deforestation of Brazilian Amazonia. *Ecol. Soc.* **2008**, *13*, 23. [CrossRef]
44. Cohn, A.S.; Bhattarai, N.; Campolo, J.; Crompton, O.; Dralle, D.; Duncan, J.; Thompson, S. Forest Loss in Brazil Increases Maximum Temperatures within 50 Km. *Environ. Res. Lett.* **2019**, *14*, 84047. [CrossRef]
45. Zemp, D.C.; Schleussner, C.F.; Barbosa, H.M.J.; Rammig, A. Deforestation Effects on Amazon Forest Resilience. *Geophys. Res. Lett.* **2017**, *44*, 6182–6190. [CrossRef]
46. Song, X.P.; Hansen, M.C.; Potapov, P.; Adusei, B.; Pickering, J.; Adami, M.; Lima, A.; Zalles, V.; Stehman, S.V.; Di Bella, C.M.; et al. Massive Soybean Expansion in South America since 2000 and Implications for Conservation. *Nat. Sustain.* **2021**, *4*, 784–792. [CrossRef] [PubMed]
47. Zalles, V.; Hansen, M.C.; Potapov, P.V.; Stehman, S.V.; Tyukavina, A.; Pickens, A.; Song, X.P.; Adusei, B.; Okpa, C.; Aguilar, R.; et al. Near Doubling of Brazil's Intensive Row Crop Area since 2000. *Proc. Natl. Acad. Sci. USA* **2019**, *116*, 428–435. [CrossRef]
48. Nosetto, M.D.; Paez, R.A.; Ballesteros, S.I.; Jobbágy, E.G. Higher Water-Table Levels and Flooding Risk under Grain vs. Livestock Production Systems in the Subhumid Plains of the Pampas. *Agric. Ecosyst. Environ.* **2015**, *206*, 60–70. [CrossRef]
49. Li, S.; Xu, L.; Jing, Y.; Yin, H.; Li, X.; Guan, X. High-Quality Vegetation Index Product Generation: A Review of NDVI Time Series Reconstruction Techniques. *Int. J. Appl. Earth Obs. Geoinf.* **2021**, *105*, 102640. [CrossRef]

50. Chu, D.; Shen, H.; Guan, X.; Chen, J.M.; Li, X.; Li, J.; Zhang, L. Long Time-Series NDVI Reconstruction in Cloud-Prone Regions via Spatio-Temporal Tensor Completion. *Remote Sens. Environ.* **2021**, *264*, 112632. [CrossRef]
51. Shen, H.; Li, X.; Cheng, Q.; Zeng, C.; Yang, G.; Li, H.; Zhang, L. Missing Information Reconstruction of Remote Sensing Data: A Technical Review. *IEEE Geosci. Remote Sens. Mag.* **2015**, *3*, 61–85. [CrossRef]
52. Liu, F.; Zhang, Z.; Zhao, X.; Liu, B.; Wang, X.; Yi, L.; Zuo, L.; Xu, J.; Hu, S.; Sun, F.; et al. Urban Expansion of China from the 1970s to 2020 Based on Remote Sensing Technology. *Chinese Geogr. Sci.* **2021**, *31*, 765–781. [CrossRef]
53. Şevik, M.; Doğan, M. Epidemiological and Molecular Studies on Lumpy Skin Disease Outbreaks in Turkey during 2014–2015. *Transbound. Emerg. Dis.* **2017**, *64*, 1268–1279. [CrossRef] [PubMed]
54. Machado, G.; Korennoy, F.; Alvarez, J.; Picasso-Risso, C.; Perez, A.; VanderWaal, K. Mapping Changes in the Spatiotemporal Distribution of Lumpy Skin Disease Virus. *Transbound. Emerg. Dis.* **2019**, *66*, 2045–2057. [CrossRef]
55. Liu, X.; Song, C.; Ren, Z.; Wang, S. Predicting the Geographical Distribution of Malaria-Associated Anopheles Dirus in the South-East Asia and Western Pacific Regions Under Climate Change Scenarios. *Front. Environ. Sci.* **2022**, *10*, 841966. [CrossRef]
56. Gething, P.W.; Smith, D.L.; Patil, A.P.; Tatem, A.J.; Snow, R.W.; Hay, S.I. Climate Change and the Global Malaria Recession. *Nature* **2010**, *465*, 342–345. [CrossRef] [PubMed]
57. Ren, Z.; Wang, D.; Ma, A.; Hwang, J.; Bennett, A.; Sturrock, H.J.W.; Fan, J.; Zhang, W.; Yang, D.; Feng, X.; et al. Predicting Malaria Vector Distribution under Climate Change Scenarios in China: Challenges for Malaria Elimination. *Sci. Rep.* **2016**, *6*, 20604. [CrossRef] [PubMed]
58. Messina, J.P.; Brady, O.J.; Golding, N.; Kraemer, M.U.G.; Wint, G.R.W.; Ray, S.E.; Pigott, D.M.; Shearer, F.M.; Johnson, K.; Earl, L.; et al. The Current and Future Global Distribution and Population at Risk of Dengue. *Nat. Microbiol.* **2019**, *4*, 1508–1515. [CrossRef] [PubMed]
59. Rocklöv, J.; Dubrow, R. Climate Change: An Enduring Challenge for Vector-Borne Disease Prevention and Control. *Nat. Immunol.* **2020**, *21*, 479–483. [CrossRef]

Article

Coupling Random Forest, Allometric Scaling, and Cellular Automata to Predict the Evolution of LULC under Various Shared Socioeconomic Pathways

Jiangfu Liao [1], **Lina Tang** [2,*] and **Guofan Shao** [3]

[1] Computer Engineering College, Jimei University, Xiamen 361021, China
[2] Key Lab of Urban Environment and Health, Institute of Urban Environment, Chinese Academy of Sciences, Xiamen 361021, China
[3] Department of Forestry and Natural Resources, Purdue University, West Lafayette, IN 47907, USA
[*] Correspondence: lntang@iue.ac.cn

Abstract: Accurately estimating land-use demand is essential for urban models to predict the evolution of urban spatial morphology. Due to the uncertainties inherent in socioeconomic development, the accurate forecasting of urban land-use demand remains a daunting challenge. The present study proposes a modeling framework to determine the scaling relationship between the population and urban area and simulates the spatiotemporal dynamics of land use and land cover (LULC). An allometric scaling (AS) law and a Markov (MK) chain are used to predict variations in LULC. Random forest (RF) and cellular automata (CA) serve to calibrate the transition rules of change in LULC and realize its micro-spatial allocation ($MKCA_{RF-AS}$). Furthermore, this research uses several shared socioeconomic pathways (SSPs) as scenario storylines. The $MKCA_{RF-AS}$ model is used to predict changes in LULC under various SSP scenarios in Jinjiang City, China, from 2020 to 2065. The results show that the figure of merit (FoM) and the urban FoM of the $MKCA_{RF-AS}$ model improve by 3.72% and 4.06%, respectively, compared with the $MKCA_{ANN}$ model during the 2005–2010 simulation period. For a 6.28% discrepancy between the predicted urban land-use demand and the actual urban land-use demand over the period 2005–2010, the urban FoM degrades by 21.42%. The growth of the permanent urban population and urban area in Jinjiang City follows an allometric scaling law with an exponent of 0.933 for the period 2005–2020, and the relative residual and R^2 are 0.0076 and 0.9994, respectively. From 2020 to 2065, the urban land demand estimated by the Markov model is 19.4% greater than the urban area predicted under scenario SSP5. At the township scale, the different SSP scenarios produce significantly different spatial distributions of urban expansion rates. By coupling random forest and allometric scaling, the $MKCA_{RF-AS}$ model substantially improves the simulation of urban land use.

Keywords: cellular automata; urban growth; shared socioeconomic pathways; allometric scaling; land-use change

Citation: Liao, J.; Tang, L.; Shao, G. Coupling Random Forest, Allometric Scaling, and Cellular Automata to Predict the Evolution of LULC under Various Shared Socioeconomic Pathways. *Remote Sens.* **2023**, *15*, 2142. https://doi.org/10.3390/rs15082142

Academic Editor: Georgios Mallinis

Received: 4 March 2023
Revised: 10 April 2023
Accepted: 16 April 2023
Published: 18 April 2023

1. Introduction

In the context of global climate change, change in land use and land cover (LULC) has become one of the areas where natural and human processes most intimately intersect. On a global scale, change in LULC is affected by slow long-term effects caused by natural factors. On a regional scale, change in LULC is driven by human activities related to land use and management [1,2]. Advances in technology and medicine have led to a rapidly growing population that has created a huge demand for natural resources. This demand, which is mainly driven by human activity, appears in the form of ever-increasing changes in land use, especially urban built-up land [3]. At the same time, the high degree of homogeneity and heterogeneity of the LULC landscape dominated by human activity produces an irreversible negative impact on regional ecosystem services, such as the water cycle, biodiversity, and climate regulation [4]. Once a highly urbanized area forms, its spatial form and development pattern are difficult to change for a long time [5]. Therefore, the double effect of change in LULC due to population growth and the negative impact of this process on the population has attracted significant research attention to the study of how changes in LULC can contribute to global environmental sustainable development [6].

Process modeling and scenario simulations of changes in LULC can reproduce the historical development dynamics of urbanized regions and can therefore be used to investigate their future spatial development patterns [7]. The emergence of GIS-based spatial data triggered the emergence of numerous techniques to simulate and analyze changes in LULC. Examples of these techniques include the intelligent cloud learning model, cellular automata (CAs), agent-based modeling, machine learning, and hybrid methods [4,8,9]. Cellular state transition and neighborhood interactions between CAs are ideally suited for expressing the process by which LULC changes, so CAs are often used to simulate and analyze changes in LULC. A CA is a lattice dynamics model with discrete time, space, and state, in which a complex system output is generated from simple interactions at the level of individual cells [10]. The characteristics of a CA model are similar to those of the complex urban system in terms of self-organization, self-replication, and the storage and transmission of information, so numerous studies have used such models to simulate urban growth [11,12].

The mining of conversion rules and the associated parametric calibration are key components of CA models used to simulate changes in LULC. Compared with traditional statistical learning techniques, machine learning algorithms and deep learning networks are better for mining the transition rules of urban CA models [13,14]. These intelligent technologies include particle swarm optimization, artificial neural networks, the random forest (RF) algorithm, fully convolutional neural networks, U-Net, and long short-term memory networks [15,16]. The deep learning network that extends the neural network is essentially a black box and is limited by computational cost, exploding and disappearing gradients, and overfitting. In contrast, the random forest algorithm better balances the performance and clearly explains the driving force contribution [17].

Numerous CA models such as CLUE-S, IDRISI, FLUS, LUSD, and PLUS divide the cell-based model of land-use change into two modules, one involving macro-scale land-use demand and one involving micro-scale changes in land use [18–22]. Multiple models or techniques were integrated into the constrained cellular-based land-use model and now serve as the basis for discerning land demand [23–25]. When overall land use is static in a particular area, growth in one type of land use must be accompanied by a decrease in another type (or types) of land use. Land-use and land cover dynamics can be interpreted as the conversion of several types of land use into a single type of land use or vice versa, thus presenting a zero-sum game of changing land-use types. For example, in areas dominated by farmland and wetlands, urban development implies the change in land use from farmland or wetland to built-up land. LULC at specific locations can change from one state to another or maintain its current state. Its state space, state sequence, change interval, and transition probability are consistent with a Markov (MK) discrete-event stochastic process.

Based on mathematical theories and observed data, a Markov chain can be used to predict the probability of changes in LULC, starting from the initial state while avoiding the data-independence assumption required by statistical methods [26,27]. Therefore, numerous studies have coupled Markov models with CAs to predict land conversion demand and simulate changes in LULC [28]. Were the analysis period restricted to several years or decades, the dynamics of land use would have their own inertia and correlation between different land-use types [29], which means that a Markov transition probability matrix based on a historical empirical analysis should accurately predict the dynamics of land-use conversion over a relatively short period [30]. However, few studies have investigated the possible biases of Markov models when predicting land demand over longer periods. Given the uncertainties faced by social and economic development, the use of Markov models in long-term land-demand prediction over multiple cycles has yet to be evaluated [31].

Among all the socioeconomic factors that lead to changes in LULC, scaling relationships connecting urban populations and urban areas have received the most research attention [32]. The urban population is one of the most critical determinants of city size and is molded by complex regional policy factors [33]. For example, China's birth control policies and household registration system significantly affected the birth rate and regional migration of the population, which ultimately affected the population size of a region [34,35]. A clear and strong correlation exists between urban population size and urban area, although determining the exact functional form of this relationship requires further study [36]. Similar to the law of relative growth observed in the development of organisms in nature, allometric phenomena involving various components of urban systems have been evinced by numerous studies [37]. For example, Abdulrasheed et al. [38] investigated the quantitative scaling relationship between clear-sky upwelling energy and the population size in 35 cities in the United Kingdom (UK). The "economy of scale" implied by the allometry slope reflects the relationship between urban form and urban size and inspires planning responses. Based on a survey of 500 cities in China, Lei et al. [39] reported that a robust scaling law exists in eight types of urban land use and population size, including residential and industrial land.

The definition of urban scale or urban boundary affects whether an allometric scaling (AS) law exists between urban population size and city size [40]. One vital factor for judging this type of human–land relationship based on geographic geometry is the shape of urban settlements. Longley, Batty, and Shepherd [36] established a theoretical urban growth model similar to the diffusion-limited aggregation (DLA) model, which is widely applied in the theoretical physics of particle clusters. This model was used to analyze empirically the urban settlement system in Norfolk, UK. The results reveal a clear allometric relationship between population size and urban settlement area, which supports the allometric scaling hypothesis proposed by Woldenberg [41] and Dutton [42]. Describing urban growth in terms of the DLA model provides a perspective for understanding why urban forms have evolved in such a "clear and unambiguous" way. This analogy suggests that the seemingly irregular evolution of urban morphology is ruled by laws similar to those that produce the DLA model or the theoretical model proposed by Longley, Batty, and Shepherd [36], as verified by empirical calculations [36,41]. The allometric relationship between population size and settlement area is an important component of these laws, so using the urban allometric scaling law to determine how urban land development reacts to changes in urban population size and to calibrate the Markov transition probability is a feasible way to model long-term patterns in LULC.

Storylines have been widely used to predict the likely dynamics of future urban growth and have proven to be an effective way to impose specific scenarios based on historical trends. Shared socioeconomic pathways (SSPs) are the core of the ScenarioMIP in the Sixth International Coupling Model Intercomparison Project. Compared with the self-defined-scenario storyline, the SSP scenario fully considers the socioeconomic factors that affect the future population size and the evolution of the urban spatial form [43]. SSP scenarios are a series of socioeconomic scenarios and emission scenarios proposed by the Intergovernmental Panel on Climate Change in 2010 to consider the challenges of mitigating and adapting to climate change. SSPs cover five scenarios: sustainable path (SSP1), middle-of-the-road path (SSP2), regional rivalry path (SSP3), inequality path (SSP4), and fossil-fuel development path (SSP5). Because it assists climate change research and facilitates comparisons between different studies, a multi-scenario analysis based on SSPs has been widely used in various fields [44]. However, few studies have explored the evolution of the urban spatial form under different SSP scenarios from the perspective of allometric scaling laws of urban systems. Constructing a new framework for deriving and predicting the future spatial development pattern of urbanized regions through the storyline of SSP population scenarios thus has potential value.

The goal of this study is to explore the allometric relationship between population size and urban area and to investigate via SSP scenarios the long-term urban land-use dynamics spanning multiple simulation periods. First, by integrating CAs into the Markov model, we construct a spatiotemporal evolution model of urban change in LULC involving various land-use types. Second, the random forest optimization algorithm is used to mine the contribution of the natural, social, and economic driving factors of land-use change and intelligently extract the occurrence probability of land-use conversion. Subsequently, the allometric scaling hypothesis is integrated into the SSP scenario framework to predict the demand for urban construction land under the different SSP population projections. Finally, the demand for urban land is used to calibrate the Markov transition probability matrix and the proposed MKCA$_{\text{RF-AS}}$ model is applied to predict, under different SSP scenarios, the evolution of urban spatial morphology in Jinjiang City, China. The contribution of this work is thus to tightly couple the observed scaling invariance of urban systems into a highly self-calibrating urban CA model via land-use transition probabilities. This novel modeling framework is a good way to simulate the changing dynamics of urban LULC when facing large uncertainties in socioeconomic factors.

2. Study Area and Data Sources

2.1. Study Area

Jinjiang is an industrial city on the southeast coast of Fujian Province at 24°30′44″–24°54′21″N latitude and 118°24′56″–118°41′10″E longitude (Figure 1). With a land surface area of 649 km^2, it faces the sea on three sides: the Quanzhou bay to the northeast, the Taiwan Strait to the southeast, and Kinmen Island to the south. The city has a subtropical marine monsoon climate, with a warm and hot climate and an average annual precipitation of 1147 mm. The terrain in Jinjiang is relatively flat, and the landform types consist mainly of terraces, plains, and hills (terraces account for 67.3% of the city area). Jinjiang is a county-level city with 6 subdistricts and 13 towns under its jurisdiction. As of 1 November 2020, the city's resident population was 2.062 million, and its population density of 3177 people/km^2 ranked second among districts and counties in China [45]. Figure 1 shows the geographical location, topographic features, land use, and population density of Jinjiang City in 2020. Areas A, B, and C are located in the northeast, midwest, and south of Jinjiang City, respectively. The land use of area A is dominated by urban land, and its average population density exceeds 3877 people/km^2. Land use in areas B and C mainly involves urban land and agricultural land. Areas B and C have lower population densities than area A. Jinjiang City is an industrial city dominated by advanced manufacturing and has many industrial clusters with a scale of over 100 billion or 10 billion CNY, such as shoes and clothing, textiles, building materials, equipment, and medical care. Its gross domestic product in 2021 is 298.64 billion CNY, and it is the district and county that ranks first in the output value of the secondary industry in Fujian Province [46]. The rapid industrialization of the past few decades has led to the continuous and rapid expansion of urban land, so a large amount of cultivated land has been converted into construction land. According to the 14th Five-Year Plan, the per capita gross domestic product of Jinjiang City in 2035 is expected to be twice that of 2020, and the rapid expansion of urban land in the coming period will still be the main feature of regional changes in land use. However, in the mid-to-long term (i.e., beyond 2050), China's overall population will slowly decline [47]. This makes the highly industrialized area of Jinjiang City an excellent region to explore how the medium- and long-term land-use dynamics will evolve under SSP scenarios.

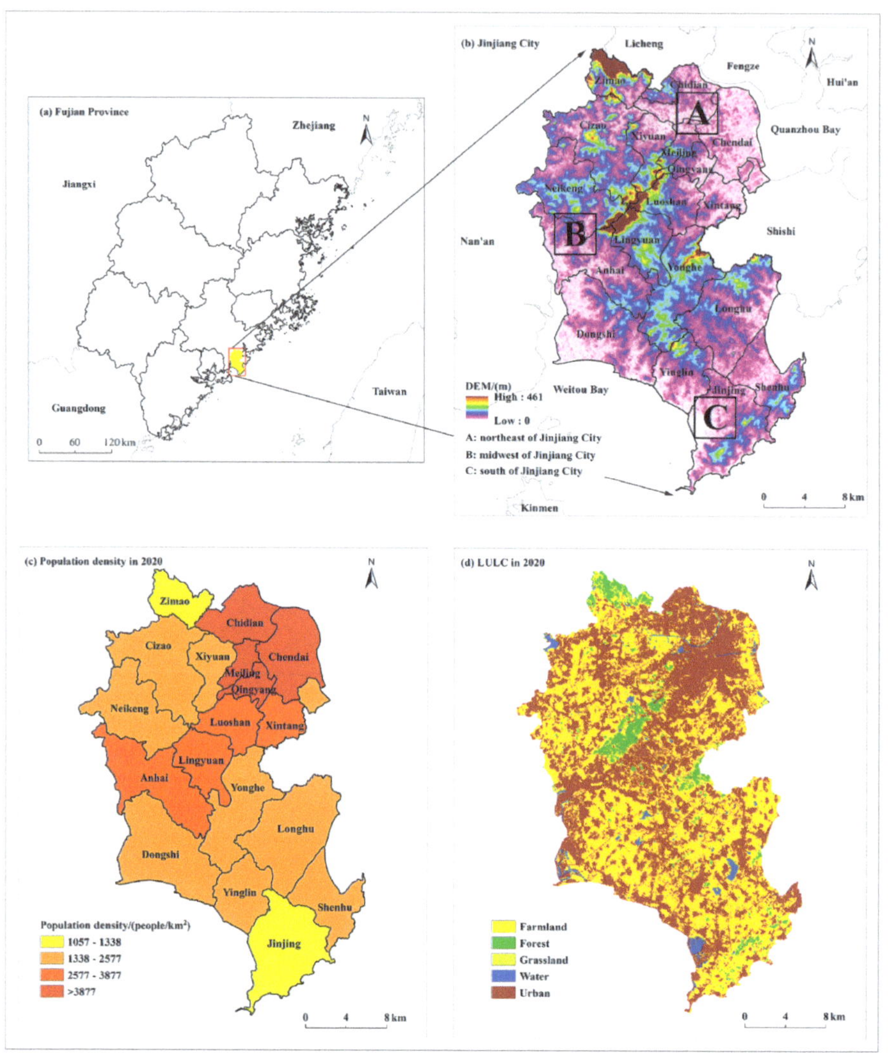

Figure 1. Case study area: Jinjiang, Fujian Province, China.

2.2. Data Source

The basic data used in this study include LULC data, traffic and cadastral data, socioeconomic data, urban development planning data, and population data for SSP scenarios. LULC data (2005–2020) were obtained by interpreting and processing Landsat TM/ETM + remote sensing images based on the Google Earth Engine (GEE) platform. This dataset combines data from Landsat TM and ETM+ sensors before 2013, and only uses Landsat 8 Operational Land Imager (OLI) data after 2013 (available at http://code.earthengine.google.com, last accessed 6 August 2021). Yang and Huang [48] interpreted the data. This set of LULC year-by-year datasets covers 5 types of land use, including agricultural land, forest land, grassland, built-up land, and water, and has a spatial resolution of 30 m × 30 m. Traffic and cadastral data mainly include the location of administrative centers at various levels: expressways, railways, main roads, and township administrative division maps. Because Jinjiang City is located in the Quanzhou Plain and occupies a flat terrain, the elevation and slope are excluded from the factors driving the change in LULC. The selected driving factors include distance to the city center, distance to the town center, distance to a train station, distance to the airport, distance to a highway, distance to the railway, distance to the main road, and distance to the coastline. The population socioeconomic data come from the statistical yearbook of Jinjiang City from 2005 to 2020 and mainly involve data from permanent urban residents. Population data include the sixth and seventh national household census data from 2010 and 2020. The standard time points of the two censuses are 0:00 on 1 November 2010 and 2020, respectively. The permanent resident population in the census includes the following three situations: (1) population residing in the given township with household registration in the township, (2) population living in the township and leaving the registered township for more than half a year, and (3) population with household registration in the township and less than half a year away from the township. The urban development planning data come from the 14th Five-Year Plan outline of Jinjiang national economic and social development (2020–2035). The SSP scenario data come from the gridded datasets for China's urban and rural population from 2020 to 2100 under SSP1–5, which was released by Jiang et al. from Nanjing University of Information Science and Technology and with a resolution of $0.5° \times 0.5°$. The dataset was downscaled to 1 km resolution through Kriging interpolation to obtain the urban population size of Jinjiang City from 2020 to 2100.

3. Methods

3.1. Calibrating Transition Rules Using Random Forest

Random forest is a dual-state classifier composed of multiple decision trees, and its output state is determined by the mode of the state vectors output by all subtrees [49]. Random forest adopts a random selection strategy with replacement from the original dataset when constructing the dataset for the sub-decision tree and uses the undrawn samples as the test set to evaluate its prediction error. At the same time, random forest subtrees randomly select feature subsets for each split node from all features and on this basis determine the optimal classification feature for the given node [50]. These random strategies endow the sub-decision trees with sufficient diversity, thereby ensuring the optimal performance of the random forest classification algorithm. When using random forest to mine the transition rules of the urban CA model, the original dataset consists of cells in the cellular grid, and the features consist of elements in the driving force vector. In this study, the random forest module in the PLUS model is used to mine the contribution of different factors driving change in LULC [17]. In the urban land-use evolution model, the formula for obtaining the probability of land-use conversion using the random forest classification algorithm is

$$P^r_{ij,k} = \frac{\sum_{n=1}^{M} I(h_n(v) = r)}{M},$$ (1)

where $P^r_{ij,k}$ is the probability that the state of cell (i, j) transforms into land-use type k, v is the vector of the factors driving change in LULC, $h_n(v)$ is the prediction type of the nth decision tree after inputting the independent variable $v(v_1, v_2, \ldots, v_i, \ldots, v_n)$, and r is a binary value (0 or 1), where 1 means that other land-use types change to type k, and 0 means otherwise. M is the number of sub-decision trees and $I(\cdot)$ is the indicative function of the decision tree set and returns 1 (0) when the $h_n(v) = r$ (otherwise). In the urban CA model, the driving factor v_i is the independent variable or feature in the random forest classifier. Therefore, a significant advantage of mining cellular transition rules with random forests is the ability to assess the importance of each factor in the driving-factor vector to the dependent variable [17]. By adding random noise to the corresponding out of bag at the sub-decision tree feature v_i, calculating the change in out-of-bag error helps to accurately estimate the importance of feature v_i [51].

3.2. Coupling an Allometric Scaling Law and the Markov Chain to Predict Demand for Change in LULC

A Markov chain describes a random motion process in a discrete index set and state space. The conditional probability of its state transition has no aftereffect [52]. Under a specific spatiotemporal background, change in LULC in urbanized areas shows a mutual transformation relationship, and the average transfer state of land-use structure remains relatively stable. Therefore, for the spatiotemporal evolution of urban land-use systems with multiple land-use types, the Markov model is appropriate for predicting the demand for change in LULC [28]. The key to using a Markov chain to predict the demand for urban land-use conversion is to construct the transition probability matrix [27]

$$P = (P_{ij}) = \begin{vmatrix} P_{11} & P_{12} & \cdots & P_{1M} \\ P_{21} & P_{22} & \cdots & P_{2M} \\ \cdots & & P_{ii} & \\ P_{M1} & P_{M2} & \cdots & P_{MM} \end{vmatrix}, \tag{2}$$

where P_{ij} is the probability that the ith land-use type converts to the jth land-use type between time T and $T+1$. The value of P_{ij} is obtained from the occurrence frequency of the transition from land-use type i to land-use type j during the period (i.e., the conversion area counted in units of cells divided by the total area of land-use type i at time T). M is the number of land-use types, and each element P_{ij} of the transition probability matrix satisfies the conditions [53]

$$\begin{cases} 0 \leq P_{ij} \leq 1, & i,j = 1,2,\ldots,M \\ \sum_{j=1}^{M} P_{ij} = 1, & i = 1,2,\ldots,M. \end{cases} \tag{3}$$

Most studies calculate the area of converted land-use types by using the historical land-use data from two points in time and then calculate the transition probability matrix for this period.

The transition probability matrix of the Markov model objectively reflects the status of land-use conversion between different land-use types and is suitable for predicting the evolution of land use and the corresponding demand for land use. However, the Markov transition probability matrix is usually calculated based on historical land-use data, and less consideration is given to socioeconomic or other driving factors. However, socioeconomic models or land-use planning on the regional macro-scale should be used to obtain the development demand for urban land and to revise the transition probability matrix output by the Markov model.

The evolution of urban spatial patterns generally follows the allometric scaling law [54], which condemns each part of the urban system to a different speed of development (i.e., the relative growth rate of one element of urban development such as population size and that of another element such as urban area are proportional) [55]. The longitudinal allometric growth model is generally used to analyze the expansion of urban land use. Longitudinal allometries describe the evolution of a dynamic process over time by using time series or corresponding sample paths based on observational data. The mathematical expression of the longitudinal allometric growth model is generally based on a power-law relationship [32],

$$A_t = \alpha(P_t)^b, \tag{4}$$

where t is the period of the historical land-use data or statistical information, A_t is the urban area at time t, P_t is the urban population size at time t, α is a proportionality coefficient, and b is a constant scaling exponent, generally representing the ratio of the relative growth rate of urban built-up land to that of population size. Thus, if the parameters α and b are determined from historical data, the area of urban built-up land can be estimated based on demographic predictions of population size.

If A_a and A_m are, respectively, used to represent the area of new urban construction land predicted by the allometric model and the Markov model, and the ratio of A_a to A_m is denoted ψ_j, the Markov transition probability matrix can be revised as follows: (1) Multiplying ψ_j by the probability of converting other land-use types into urban land, the probability of non-conversion of urban land remains unchanged. (2) By using ψ_i multiplied by the probability that land use i is converted into a non-urban land use ($i \neq j$), the revised Markov transition probability matrix using the allometric growth model takes the form

$$
P' = (P_{ij}') = \begin{vmatrix}
\psi_1 P_{11} & \psi_1 P_{12} & \cdots & \psi_j P_{1j} & \cdots & \psi_1 P_{1M} \\
\psi_2 P_{21} & \psi_2 P_{21} & \cdots & \psi_j P_{2j} & \cdots & \psi_2 P_{2M} \\
\cdots \\
\psi_i P_{i1} & \psi_i P_{i2} & \cdots & \psi_j P_{ij} & \cdots & \psi_i P_{iM} \\
\cdots \\
P_{j1} & P_{j2} & \cdots & P_{jj} & \cdots & P_{jM} \\
\cdots \\
\psi_M P_{M1} & \psi_M P_{M2} & \cdots & \psi_j P_{Mj} & \cdots & \psi_M P_{MM}
\end{vmatrix}, \tag{5}
$$

where ψ_i and ψ_j are the correction coefficients of the above transition probability matrix, and subscript j means urban land. The correction factor ψ_i is calculated by the following formula:

$$
\psi_i = \frac{1 - \psi_j P_{ij}}{\sum_{l=1}^{M} P_{il} - P_{ij}}, \qquad i = 1, 2, \ldots, M; \ i \neq j, \tag{6}
$$

where l is the land-use subscript in the transition probability matrix, i is the subscript of land-use types other than urban land, j is the subscript of urban land use, and P_{il} is the probability that land use i will be transformed into land use l.

3.3. Using Cellular Automata to Allocate Micro-Spatial Changes in LULC

The suitability of various land uses obtained by the random forest algorithm and the demand for land-use conversion output by the Markov model were input into the MKCA$_{RF-AS}$ model to simulate the evolution of LULC. In this study, the Markov model and CA model were designed and implemented by using Matlab (MathWorks, Natick, MA, USA). To simulate the micro-spatial allocation based on CAs, the transition rules are critical to obtaining suitable land use at each location. The transition rules generally use cell state, neighborhood state, and land conversion potential to estimate the probability of land-use changes in cells. The state of a cell at time t is a function of the cell state at $t - 1$, the neighborhood state at time $t - 1$, and many other factors. Its general form is as follows [56]:

$$
S_{ij}{}^t = f\left(S_{ij}{}^{t-1}, P_{c,ij}, \Omega_{ij}{}^{t-1}, Cons, R \right) \tag{7}
$$

where $S_{ij}{}^t$ and $S_{ij}{}^{t-1}$ are the states of cell (i, j) at times t and $t - 1$, respectively; $P_{c,ij}$ is the cell's land conversion potential (i.e., the developmental suitability of cell (i, j)); $\Omega_{ij}{}^{t-1}$ is the distribution of specific LULC in the neighboring space of cell (i, j); $Cons$ is a constraint mechanism that restricts land-use conversion; R is a random factor that simulates the uncertainty of land-use dynamics; and f is a series of transition rules that determines changes in cell state.

The neighborhood interaction is one of the core components of a cellular-based land-use evolution model. The number of cells of various LULCs in the neighborhood changes as the cellular model evolves. The state of a cell is directly affected by the distribution of land-use types in the neighborhood. The most abundant land-use types in the neighborhood often determine the future transition state of the central cell. The neighborhood function based on a square neighborhood window can be constructed as follows [57]:

$$
\Omega_{ij,k}{}^{t-1} = \frac{\sum_q^{m \times m} con\left(S_{ij}{}^{t-1} = k \right)\left(q \neq 1 \right)}{m \times m - 1}, \tag{8}
$$

where $\Omega_{ij,k}^{t-1}$ is the value of the neighborhood function that affects the conversion of the central cell (i, j) to land use k, and $con\left(S_{ij}^{t-1} = k\right)$ is the number of cells of land use k in the $m \times m$ moving window neighborhood with cell (i, j) as the center.

If a local probability expresses development suitability, then the probability is generally affected by a series of spatial distance variables. The values of each spatial variable at sample points are extracted as independent variables and input into the random forest model to obtain the local probability that the cell be converted to a specific land use. After adding factors such as random perturbation and constraints to the cellular model, the final total conversion probability of the central cell is [58]

$$P_{total,k}^{t} = \left[1 + (-\ln \gamma)^{\alpha}\right] P_{c,ij} \Omega_{ij,k}^{t-1} \times Cons\left(S_{ij,k}^{t-1}\right), \tag{9}$$

where t is the iterative sequence of cellular automaton evolution, $P_{total,k}^{t}$ is the probability of conversion to land use k at time t, $1 + (-\ln \gamma)^{\alpha}$ is a random disturbance term reflecting the uncertainty of an urban system [59], γ is a random number between 0 and 1, α is an integer between 1 and 10, $\Omega_{ij,k}^{t-1}$ is the neighborhood influence, and $Cons(\cdot)$ is a constraining function constrains the change in cell states.

3.4. Scenario Setting

China has implemented a universal two-child population policy since 2015 [60]. We use the scenario settings based on this policy and the global SSPs framework to carry out urban population estimates for China's 5 SSP scenarios from 2020 to 2100 [47]. Based on the allometric relationship between population and area in the urban system, this research predicts various future scenarios of urban land demand as a function of different population scenario storylines. The evaluation considers the basic factors affecting population development, such as fertility rate, mortality rate, migration rate, and education level in historical periods. It also calibrates parameters and verifies the population-development-environment model.

The SSP scenario depicts the current status of social and economic development and the following possible future development trends: (1) In the SSP1 scenario, regional development gradually shifts toward a relatively sustainable green development path, which emphasizes human well-being while reducing inequalities within countries. (2) The SSP2 scenario is an intermediate path between SSP1 and SSP3. Under this scenario, the region's population, economy, and urbanization will not deviate significantly from the historical development path. (3) In the SSP3 scenario, the revival of nationalism prompts countries to focus on domestic or regional problems. The region experiences slow economic development and urbanization, and the decline in education and investment worsens social inequality. (4) In the SSP4 scenario, income inequality within and between countries increases and becomes significantly stratified. Human capital investment and economic development opportunities are highly unequal, resulting in high levels of development in high-tech industries but low income and educational opportunities in labor-intensive sectors. (5) The SSP5 scenario adopts a resource- and energy-intensive lifestyle, and each country's technology, economy, and population develop rapidly.

3.5. Method for Evaluating Accuracy

In this work, the overall accuracy (OA), Kappa coefficient, and figure of merit (FoM) are used to qualify the simulation of the proposed model. The overall accuracy and Kappa coefficient provide a comparative analysis at the pixel level. The calculation superimposes the real and simulated maps and constructs a confusion matrix based on a point-to-point comparison. The FoM quantifies the consistency of land-use transition patterns and is obtained by overlaying and analyzing the initial, actual, and simulated maps. The FoM is the fraction of correctly predicted land use to the overall land use in the simulation results [61]:

$$FoM = \frac{B}{A + B + C + D},$$ (10)

where B is the number of cells where the land use changes and the simulated scenario also produces the correct transition; A is the number of cells whose land use actually changes, whereas the simulated land use remains unchanged; C is the number of cells in which both the observed data and the simulated results change, but the resulting land uses differ; and D is the number of cells where the actual land use does not change, but the simulated land use changes.

4. Simulation Results

4.1. Land-Use Suitability

The 2005–2010 land-use data and driving factors were input into the land-use expansion analysis strategy (LEAS) module of the PLUS model to intelligently mine the land development suitability of the MKCA$_{RF-AS}$ model [17]. The LEAS module is driven by the random forest algorithm. The sampling method in its parameter configuration is random sampling; there are 20 decision trees, the sampling rate is 0.05, and 8 features participate in the training. The results output by the LEAS module of the PLUS model reflect the suitability for the development of land use on a scale of 1 to 255. In Jinjiang City, the change in land use from 2005 to 2010 mainly involves the expansion of urban land. Figure 2 shows the suitability of urban land development extracted by the random forest algorithm and the importance of the driving factors. In general, the main urban area in the northeast, the satellite towns in the south, and the central region and the west are all highly suitable for development from 2005 to 2010. The top 3 drivers contributing to urban development are the distance to main roads, distance to railway stations, and distance to railways, and the relative importance is 0.2164, 0.1270, and 0.1238, respectively. As a typical industrial city, the urbanization of Jinjiang depends strongly on the transportation infrastructure.

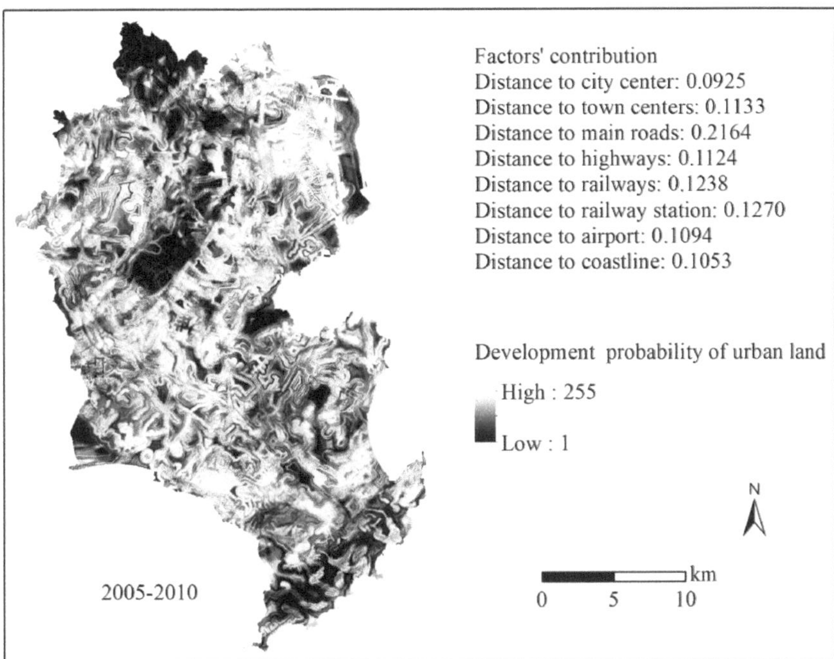

Figure 2. Suitability of urban land development mined by random forest algorithm in Jinjiang City from 2005 to 2010.

4.2. Simulation of Spatial Distribution of LULC

We input the obtained suitability of land-use development into the $MKCA_{RF-AS}$ model and simulated the spatial distribution of land during 2005–2010 (see Figure 3). Analyzing and comparing the spatial details, Figure 3a,b,e show that forest land, grassland, and water body are the smallest three land-use types. Woodland and grassland are mainly distributed in the northern and central parts, and water bodies are mainly distributed in lakes, rivers, and depressions around Quanzhou Bay and Weitou Bay. Cultivated land and construction land are the two most widely distributed land-use types in the study area. The construction land is mainly distributed in the coastal plain area with flat terrain and well-developed transportation, and its expansion mainly comes from the occupation of cultivated land. The expansion of construction land mainly occurs in the central urban area in the northeast, the satellite cities in the central and western regions, and the industrial technology innovation corridor between them. The land-use spatial pattern simulated in Figure 3e is similar to that in Figure 3b. Based on the point-to-point overlay analysis, the evaluation metrics of the FoM, urban FoM, overall accuracy, and Kappa coefficient are 25.5%, 35.0%, 84.2%, and 0.709, respectively. Studies show that the FoM is more accurate than the Kappa coefficient for evaluating the simulation of land-use change models, and the FoM and urban FoM of the $MKCA_{RF-AS}$ model both have high values in the literature [62].

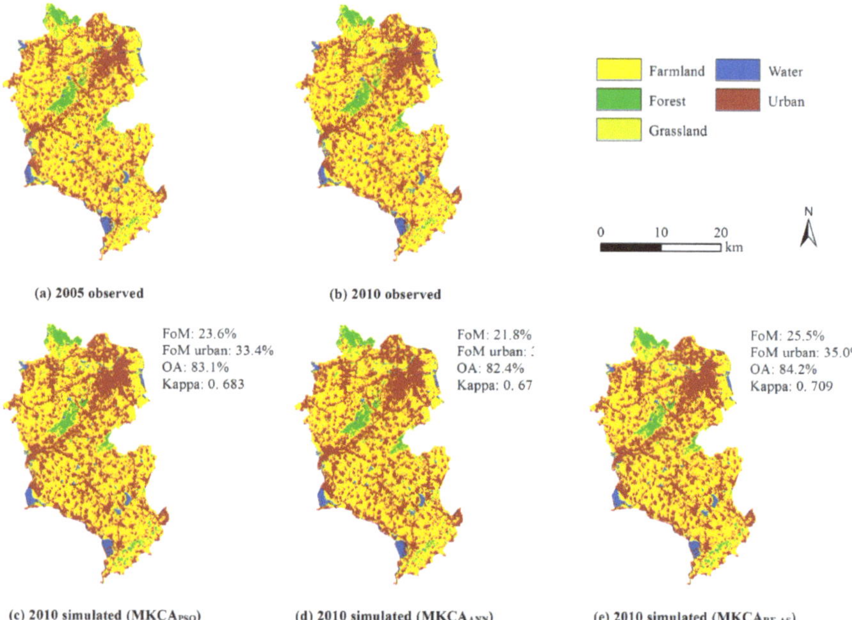

Figure 3. Comparison of spatial distribution of LULC simulated using different cellular automaton models.

Particle swarm optimization (PSO) and the artificial neural network (ANN) are typical artificial intelligence algorithms and perform well in the automatic calibration of the transition rules of urban CA models [63,64]. To verify the ability of the random forest algorithm to mine the transition rules of urban LULC change models, this work selects PSO and ANN as references for comparative research. To test the simulation of $MKCA_{RF-AS}$, $MKCA_{PSO}$ and $MKCA_{ANN}$ were used to simulate the spatiotemporal evolution of the LULC over the same period. The results show that the above four accuracy indicators are significantly greater for the $MKCA_{RF-AS}$ model than for the $MKCA_{PSO}$ or $MKCA_{ANN}$ models. For example, the FoM and urban FoM produced by the $MKCA_{RF-AS}$ model are 3.72% and 4.06% greater than that produced by the MKCA$_{ANN}$ model. The MKCA$_{RF-AS}$ model is thus an effective tool for simulating land-use change and explaining the driving factors.

4.3. Sensitivity Analysis

To explore how different configurations of land-use demand affect the simulation results of the MKCA$_{RF-AS}$ model, we carried out a sensitivity analysis of the model based on the land-use demand predicted by the Markov model spanning multiple periods. The simulation covers the three periods of 2005–2010, 2005–2015, and 2005–2020, and the land-use conversion demand is calculated by using the Markov transition probability matrix extracted from the period 2000–2005. Considering the difference between the actual urban land-use demand in the above 3 periods and the values predicted by the Markov model, 2% was selected as the minimum change for the sensitivity analysis (see Table 1). Compared with the FoM, the urban FoM is more sensitive to changes in urban land-use demand. During the period 2005–2010, when the urban land-use demand changed from 261,381 cells (actual value in 2010) to 244,954 cells (predicted value of Markov model), the urban FoM decreased from 35.02% to 27.52%. In other words, an error of 6.28% in the predicted urban land-use demand changes the urban FoM by 21.42%. When the predicted demand for urban land use in the two periods 2005–2015 and 2005–2020 produces similar errors, the urban FoM decreases by 17.43% and 17.57%, respectively. At the same time, when the error in urban land-use demand reaches 2%, the decrease in the urban FoM differs for different simulation periods. This sensitivity analysis illustrates that the error of predicted land-use demand amplifies the spatial error in the simulation of the change in LULC.

Table 1. Sensitivity of the MKCA$_{RF-AS}$ model to demand for urban land use.

Changes in Demand for Urban Land Use	2005–2010			2005–2015			2005–2020		
	Urban Land (Cells)	FoM (%)	Urban FoM (%)	Urban Land (Cells)	FoM (%)	Urban FoM (%)	Urban Land (Cells)	FoM (%)	Urban FoM (%)
Predicted change in land use by Markov model	244,954	23.45	27.52	263,221	27.21	32.40	277,949	21.37	27.76
+2%	249,853	24.31	29.95	268,485	27.46	33.74	283,508	21.48	28.55
+2%	254,752	24.78	32.05	273,750	27.75	35.11	289,067	21.57	29.31
+2%				279,014	28.00	36.43	294,626	21.76	30.18
+2%				284,279	28.19	37.63	300,185	21.96	31.13
+2%							305,744	21.97	31.85
+2%							311,303	22.21	32.79
Actual demand for urban land use	261,381	25.48	35.02	292,024	28.27	39.24	318,210	22.21	33.68

4.4. Allometric Relations between Population Size and Urban Area

The least squares method was used to fit the allometric growth equation to the permanent urban population as a function of the urbanized land area during the period 2005–2020, and the results are shown in Table 2. The urban area uses the classification data of the urban land use from the LULC yearly dataset obtained from remote sensing images from 2005 to 2020, and the permanent urban population comes from the statistical yearbook and national census data. The results show that the coefficient of proportionality $\alpha = 2.800$, the scale index $b = 0.933$, the relative residual is 0.0076, and the $R^2 = 0.9994$. The 2005–2020 urban area predicted by the calibrated allometric model is very close to the observed result. The urban area predicted by the calibrated allometric model closely matches the observed data for the period 2005–2020.

Table 2. Allometric scaling analysis of population size and urban area in Jinjiang from 2005 to 2020.

Year	2005	2006	2007	2008	2009	2010	2011	2012	2013	2014	2015	2016	2017	2018	2019	2020
POP	67.7	72.2	74.0	76.7	80.9	117.3	120.7	124.9	128.4	132.2	133.5	135.8	139.0	141.4	142.8	141.6
POP-SSNPC	94.8	101.1	103.5	107.4	113.3	117.3	118.5	122.7	126.1	129.8	131.1	133.4	136.5	138.9	140.3	141.6
UBA	200.4	207.7	213.7	218.7	228.4	235.6	241.4	246.8	253.4	259.6	263.2	267.9	274.9	281.2	285.6	286.4
UBA-AS	195.7	207.7	212.4	219.8	231.1	238.6	241.0	249.0	255.4	262.4	264.9	269.0	275.0	279.5	282.0	284.6

Note: POP: resident population (urban) (10^4), POP-SSNPC: revised population data from the Sixth and Seventh National Population Censuses (10^4), UBA: urban area (km^2), UBA-AS: urban area predicted by the allometric scaling model (km^2).

This research revises the statistical data for 2005–2020 based on the results of the Sixth and Seventh National Population Censuses conducted in 2010 and 2020. The data from the Sixth National Population Census (2010) differ significantly from that in 2009. For example, the permanent urban population in 2009 and 2010 was 809,200 and 1,172,800, respectively, an increase of 44.9%. Considering that the average annual growth rate of the urban resident population in 2005–2009 was between 2.49% and 6.65%, the statistics for 2005–2009 are revised upward by 40%. Similarly, according to the Seventh National Population Census data (2020), the statistical data for 2011–2019 are revised downward by 1.78%. The population data for 2005–2020 were revised by two national household censuses and are highly accurate. This greatly enhances the reliability of the analysis of the allometric relationship between urban population size and urban area.

4.5. Predicting Urban Land-Use Demand for 2020–2100

Figure 4a shows the storyline of Jinjiang's population growth from 2020 to 2100 under the various SSP scenarios after the localization calibration. The calibration first sets the population size in 2020 under the SSP scenarios to the actual population in 2020, and the correction coefficient obtained is then used to proportionally adjust the population data in future years under each SSP scenario. Figure 4b uses the 2005–2020 Markov transition probability matrix to predict the urban area of Jinjiang City in 2035, 2050, 2065, 2080, 2095, and 2110. Furthermore, we linearly interpolate to obtain year-by-year population forecasts for 2020–2100. At the same time, the parameters obtained for the allometric growth model are used to predict the urban area from 2020 to 2100 under the various SSP scenarios. The results show that the urban area predicted by the Markov model in 2020–2100 even exceeds that of the SSP5 scenario. When using a Markov model to predict long-term urban land-use demand, the results may deviate significantly from the baseline or even in the opposite direction.

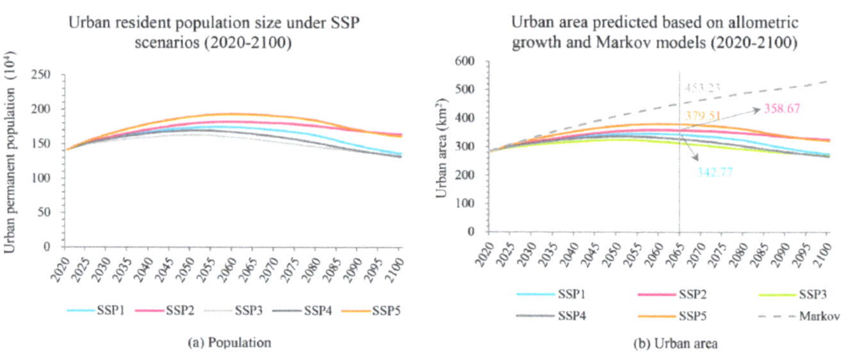

Figure 4. Predicted population size and urban area in Jinjiang City from 2020 to 2100 under the different SSP scenarios.

4.6. Spatiotemporal Evolution of LULC from 2020 to 2065

4.6.1. Correction of Transition Probability Matrix

To measure the demand for land-use conversion from 2020 to 2065, the Markov transition probability matrix from 2005 to 2020 is revised based on the urban area obtained under each SSP scenario in 2065. Table 3 summarizes the transition probability under typical SSP scenarios of the initial land-use type to the target land-use type (including no change in land-use type). The results show that, under each SSP scenario and compared with 2005–2020, the probability of transforming agricultural land, grassland, and water bodies into built-up land decreases significantly, while forest land and built-up land do not change significantly. For example, under the SSP5 scenario, the probabilities of transforming agricultural land, grassland, and water into built-up land decreases from 0.1948, 0.6031, and 0.2561 during 2005–2020 to 0.0994, 0.2826, and 0.1260, respectively. Figure 4b shows the projection of urban area for the period 2020–2100 in the Markov model, calculated based on the transition probability matrix from 2005 to 2020. Therefore, the comparison confirms the negative impact of the slower growth trend of urban land use on the Markov transition probability under the SSP scenarios. Conversely, the probability that cultivated land remains unchanged increases significantly under each SSP scenario. This probability increases from the original values of 0.7829 to 0.9132, 0.8966, and 0.8740 under the scenarios SSP1, SSP2, and SSP5, respectively. The probability of forest land, grassland, and water bodies remaining unchanged also increases to a certain extent.

Table 3. Corrected Markov transition probability matrices for three SSPs scenarios during 2020–2065.

SPP Scenarios	Probability of Shifting to the Following Land-Use Type				
	Agricultural	Woodland	Grassland	Water	Built-Up
2005–2020					
Agricultural	0.7829	0.0166	0.0018	0.0039	0.1948
Woodland	0.3092	0.6653	0.0005	0.0001	0.0249
Grassland	0.3487	0.0068	0.0327	0.0088	0.6031
Water	0.2218	0.0004	0.0024	0.5193	0.2561
Built-up	0.0004	0.0000	0.0000	0.0025	0.9971
2020–2065 (SSP1)					
Agricultural	0.9132	0.0211	0.0021	0.0051	0.0585
Woodland	0.2895	0.6989	0.0004	0.0000	0.0114
Grassland	0.7454	0.0161	0.0476	0.0306	0.1606
Water	0.2870	0.0002	0.0030	0.6366	0.0733
Built-up	0.0002	0.0000	0.0000	0.0022	0.9976
2020–2065 (SSP2)					
Agricultural	0.8966	0.0206	0.0021	0.0050	0.0757
Woodland	0.2913	0.6943	0.0004	0.0000	0.0142
Grassland	0.7043	0.0150	0.0409	0.0287	0.2114
Water	0.2793	0.0003	0.0029	0.6223	0.0953
Built-up	0.0002	0.0000	0.0000	0.0022	0.9975
2020–2065 (SSP5)					
Agricultural	0.8740	0.0198	0.0021	0.0048	0.0994
Woodland	0.2941	0.6881	0.0004	0.0000	0.0175
Grassland	0.6402	0.0133	0.0385	0.0257	0.2826
Water	0.2685	0.0003	0.0027	0.6025	0.1260
Built-up	0.0003	0.0000	0.0000	0.0023	0.9975

4.6.2. Scenario Simulation

We used the revised Markov transition probability matrix and the area of various land-use types in 2020 to predict the area where the land-use type changes from 2020 to 2065. The resulting land-use demand was then input into the MKCA$_{\text{RF-AS}}$ model to simulate the 2065 spatial distribution of land use in Jinjiang City under various SSP scenarios (Figure 5). The simulation results show that agricultural land and built-up land are the most important land-use types in the study area. Although the development trend is slow, the main change in the LULC under each SSP scenario during 2020–2065 remains the continuous sprawl of urban land. Over time, the development of urban built-up land is particularly significant in the northeastern, central, western, and southern regions. By 2065, due to the continuous expansion of the main urban area and the implementation of the strategy of integrating Quanzhou Bay, a large-scale high-density urban area should appear to the northeast of Jinjiang City (see Figure 5, enlarged area A). At the same time, affected by the radiation of Xiamen City, the west and south should develop satellite-city clusters that would play an important role in supporting the aggregation and development of the central urban area. Conversely, the comparison of enlarged areas B and C in Figure 5 under different SSP scenarios shows that construction land expansion under SSP2 and SSP5 is significantly faster than under the other SSP scenarios. Driven by the dense transportation network and the integration of urban and rural areas, Jinjiang should develop a medium-to-high-density urbanization area within the city under the scenarios SSP2 and SSP5.

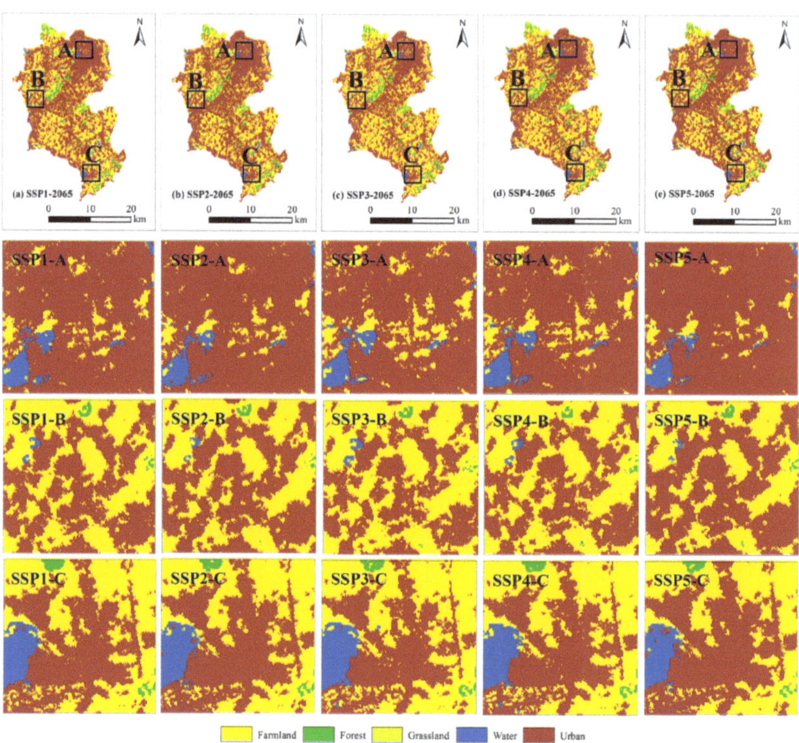

Figure 5. Simulation of LULC changes in Jinjiang City from 2020 to 2065 under five SSP scenarios.

Table 4 lists the percent change in the land-use area under the different SSP scenarios over 2020–2065. The percent change in the LULC in 2065 depends strongly on the SSP scenario. For example, the percent change in the urban land (agricultural land) area under SSP5 is 3.49 (3.73) times that under SSP3. In all SSP scenarios, the expansion of built-up land is counter-balanced by a reduction in the area of other land-use types. Agricultural land changes relatively little under some SSP scenarios, such as −6.7% and −10.7% under scenarios SSP3 and SSP4. However, the reduction in agricultural land area is larger under each SSP scenario than that of forest land, grassland, and water bodies. In particular, under scenarios SSP1, SSP2, and SSP5, the area of agricultural land reduction is 49.88, 64.63, and 83.58 km^2, respectively.

Table 4. Areas assigned to different land uses and the percent change in land-use area over the period 2020–2065 under the different SSP scenarios.

Land-Use Types	2020	2065 (SSP1)	2020–2065 (SSP1)	2065 (SSP2)	2020–2065 (SSP2)	2065 (SSP3)	2020–2065 (SSP3)	2065 (SSP4)	2020–2065 (SSP4)	2065 (SSP5)	2020–2065 (SSP5)
	Area (km²)	Area (km²)	%	Area (km²)	%	Area (km²)	%	Area (km²)	%	Area (km²)	%
Agricultural	331.36	281.47	−15.05	266.72	−19.51	308.92	−6.77	296.15	−10.62	247.78	−25.22
Woodland	25.44	23.23	−8.67	22.39	−11.99	24.83	−2.40	24.03	−5.54	21.51	−15.44
Grassland	0.92	0.70	−23.90	0.68	−26.63	0.80	−12.78	0.74	−19.71	0.65	−29.56
Water	11.84	8.36	−29.37	8.18	−30.87	9.10	−23.12	8.70	−26.49	7.84	−33.80
Built-up	286.39	342.63	19.64	358.43	25.16	312.94	9.27	326.89	14.14	378.92	32.31

4.6.3. Analysis of Urban Expansion on the Township Scale

Figure 6 depicts the spatial distribution of the urbanization rate on the township scale based on the calculated growth rate of the urban area over the period 2020–2065. Under scenario SSP1, which continues the historical development trend, most regions show a medium rate of development. Under scenario SSP4 (unbalanced development pathway), the entire region develops at a medium-low rate. The spatial distribution of the urban development rate of scenarios SSP2, SSP3, and SSP5 differs significantly. Under scenario SSP2, the urbanization rate of most towns in the west, central, and eastern regions is medium, whereas under scenarios SSP3 and SSP5, most of these towns develop at a low or high rate. The development rate of streets or townships in the northeast and central urban areas is relatively low, which is closely related to the high urbanization rate in this area at the beginning of the simulation. The analysis of the township-scale urbanization rate under scenarios such as SSP2 and SSP5 provides an objective reference for the urban hierarchy in the spatial planning of land use.

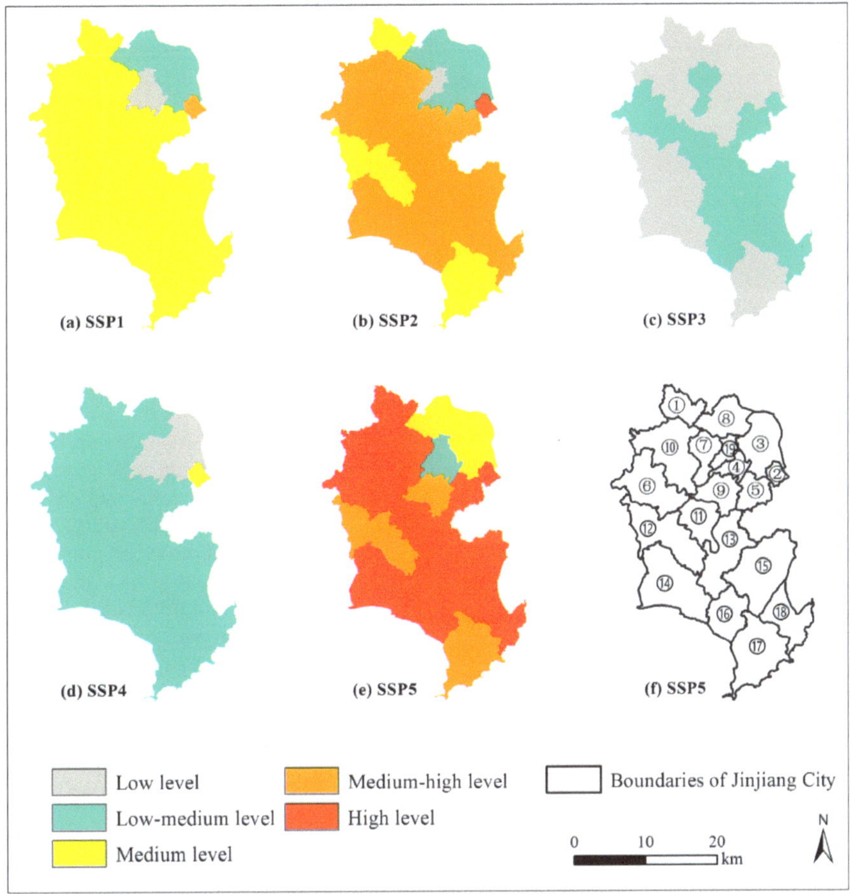

Figure 6. Spatial distribution of rate of expansion of Jinjiang City under the different SSP scenarios from 2020 to 2065. The town numbers in panel (f) are assigned as follows: ① Zimao, ② Xibin, ③ Chendai, ④ Qingyang, ⑤ Xintang, ⑥ Neikeng, ⑦ Xiyuan, ⑧ Chidian, ⑨ Luoshan, ⑩ Cizao, ⑪ Lingyuan, ⑫ Anhai, ⑬ Yonghe, ⑭ Dongshi, ⑮ Longhu, ⑯ Yinglin, ⑰ Jinjing, ⑱ Shenhu, and ⑲ Meiling.

5. Discussion

The scaling exponent of the allometric growth model has a clear geographical meaning. The two most important characteristics of cities are the concentration of population and the more intense use of urban material resources [54]. This study investigates the scale invariance associated with complex urban development. In other words, we are interested in the range and ideal value of the scaling exponent in the scale invariance of an urban system. Bettencourt [54] analyzed the urban infrastructure and socioeconomic metrics of metropolitan areas in the United States in 2006 and calculated the scaling exponents governing the scaling relationship between (i) the gross domestic product and population and (ii) the roads and population to be 0.849 and 1.126, respectively. At the same time, Bettencourt [54] claims that the superlinear scaling exponent for wealth, innovation, and crime in urban systems is about 1.15 (7/6), and the sublinear scaling exponent for infrastructure networks such as roads and cables is about 0.85 (5/6). Bettencourt [54] concludes that sublinear scaling means the material infrastructure saved per capita as cities grow. Moreover, the scaling exponent of the urban infrastructure network with respect to population size is more sublinear than that of the urban area. Nordbeck [65] considered that, because people are active in three-dimensional space and urban expansion occurs on a two-dimensional plane, the allometric exponent should be 2/3. Recently, Bettencourt [54] proposed a microscopic urban growth model that considers the hierarchical organization of infrastructure networks, which theoretically confirms the 2/3 scaling exponent between population size and urban area.

Lee [66] concluded that the allometric exponent of urban growth should fluctuate between 2/3 and 1. If the dimensions of population and urban area are assumed to be 2 and 1.7, respectively, then the theoretical value of the scaling exponent should be close to 0.85 [67]. Based on an empirical analysis of typical cities around the world, Lee [66] concluded that the scaling exponents between population size and urban area in the United States, Japan, and Canada in 1960 were 0.8757, 0.9140, and 0.874, respectively. Cao et al. [32] used the population density flow diagram to describe the 6 development levels of China's urban agglomerations, and the scaling exponents between the population size and urban area fall in the range [0.79, 0.95]. Based on the data input during 2005–2020, the scaling exponent of the allometric relationship between population size and urban area in Jinjiang is 0.933. Strictly speaking, when the scaling exponent is greater than 1, the expansion rate of urban construction land is greater than the growth rate of the urban population. When the scaling exponent is equal to 1, the relative rate of urban land area expansion equals the relative growth of the urban population. When the scaling exponent is less than 1, the relative rate of expansion of urban land is less than the relative growth of the urban population. With the expansion of urban land area, the land area in the urban system of Jinjiang City is efficiently used, on average. If compared with the theoretical value of the scaling exponent, the urban land use of Jinjiang City can be planned and designed in a more compact form. This means that Jinjiang can explore an intensive development path with a greater economic value per unit land area [68].

Natural biological organisms have undergone long-term adaptation and evolution, so allometric scaling phenomena (weight/volume) are widely observed [69]. The spatial expansion of urban areas is often driven or constrained by natural conditions, historical culture, and socioeconomic factors. Scaling relations thus appear in the context of individual cities and urban agglomerations between population size and urban area or between other components of urban systems [32,70]. The large differences in factors that influence urban regions and their selection or adaptation produce different scaling relations. The existing literature divides the values of scaling exponents into positive or negative allometry based on a threshold of 1 or 0.85 [71,72]. In contrast with this method of categorization, we assign positive allometric growth to the case of urban expansion and negative allometric growth to urban contraction. This categorization scheme makes it vital to simulate urban shrinkage from the perspective of the negative allometric growth of population size and urban area.

Conversely, a new research focus involves combining remote sensing measurement technology and CA to deduce the three-dimensional growth process of urbanized areas [73]. Based on accurate measurements of the area and volume of built-up urbanized areas, the scaling relationship between population size and urban volume and its scaling exponents can be further clarified. In this way, population size can be used to predict the building volumes of an urban area. Considering urban volume as a macro demand and combining CA and allometric scaling to simulate the horizontal and vertical growth of three-dimensional urban space is also a new field to be explored.

This research revises the Markov transition probability matrix for long time series to span multiple periods. We use the Markov transition probability matrix from 2000 to 2005 to predict the demand for land use from 2005 to 2020, which helps to test the advantages and limitations of the Markov chain for estimating land-use demand across multiple cycles. The results in Table 1 show that the error ratio between the predicted urban land use and the observed urban land use in 2020 reaches 12.7% when using the transition probability matrix from 2000 to 2005 to predict the land-use demand from 2005 to 2020. However, predicting land-use demand from 2020 to 2065 based on the transition probability matrix from 2005 to 2020 covers 3 15-year Markov forecast periods. Figure 4 shows the predicted urban land-use demand in 2065 under the 3 scenarios SSP1, SSP2, and SSP5. The results differ from Markov's predicted value by 32.2%, 26.4%, and 19.4% respectively. The Markov model underestimates the observed value in the previous prediction and overestimates predictions based on the SSPs scenarios for 2065. This result indicates that Markov chains may be more suitable for predicting land-use demand over relatively short periods that involve fewer cycles. When the driving factors remain unchanged or the uncertainties faced in the future are relatively small, appropriately extending the forecast time or expanding the number of forecast periods can produce credible results.

This work focuses on the scaling relationship between the population size and urban area in urban systems and the future evolution of urban LULC. In this context, changes in population size are the critical factors influencing urban land demand and the spatiotemporal evolution of LULC in urban systems. Population size is affected by factors such as fertility, mortality, migration, and education level. With the continuous advancement of industrialization and urbanization, the employment pressure and living cost of urbanites are increasing. Meanwhile, newborn survival rates and people's education level continue to improve. These compounding factors make the process of urbanization lead to a significant decline in people's willingness to have children in most countries around the world [74].

The SSP scenario sets the story line of the future population and economic development based on population and human resources, economic development, lifestyle, human development, environment and natural resources, policy and system, and technological development. SSP scenarios fully describe the opportunities, potentials, and uncertainties facing the future population and economic development. Therefore, considering the development uncertainty of socioeconomic factors, the urban areas predicted by the SSP scenario and the Markov model differ significantly (Figure 4b). The implication of this finding to the environmental remote sensing science literature is that the integration of allometric scaling laws for population size and urban area is a candidate for more effective urban simulations if the uncertainties due to population size are considered.

The spatial pattern of the urban expansion rate predicted in Figure 6 provides an important reference for analyzing the consequences of urbanization, such as the heat-island effect, water shortage, and loss of cultivated land. First, under all SSP scenarios, we must focus on the urban heat-island problem in densely populated areas, such as the northeast and around Quanzhou Bay. At the same time, under the SSP5 scenario, the development path driven by fossil energy will expand industrial, residential, and commercial urban land at a super-high rate during 2020–2065. The heat-island effect of major satellite cities also must be considered. Second, the population density of Jinjiang City in 2022 was as high as 3177 people/km^2, and its spatiotemporal distribution is relatively uneven at the township scale (Figure 1c). With the continuous advancement of the urbanization process under different SSP scenarios, the population density will continue to increase and show significant spatial differentiation. Jinjiang City is a city with scarce water resources. In 2020, its per capita water resources and per capita comprehensive water consumption were 62 m^3/person and 276 m^3/person, respectively [75]. The possible water resource stress in densely populated areas under different SSP scenarios and its coping strategies should be further studied. Finally, the expansion of urban land in Jinjiang City is mainly at the expense of occupying cultivated land, which means that the city's food demand will rely more on the production of other regions. At the same time, the food demand of the region will drive the development of arable land away from Jinjiang City to other regions. We must further study this teleconnection effect of LULC changes under different SSP scenarios from the county scale [76].

The Intergovernmental Panel on Climate Change estimates that the possible rise in the global mean sea level by 2100 will range between 0.29 m and 1.10 m [77]. Jinjiang City is surrounded by the sea on three sides, the elevation of most areas in the coastal zone is 1–4 m, and the elevation of some low-lying areas is less than 1 m. In addition, Jinjiang City is densely populated in the urban center and in satellite cities around Quanzhou Bay and Weitou Bay, and large quantities of cultivated land are distributed in coastal depressions. It is thus urgent to analyze the exposure risk and loss of population and economic output that may result from the average rise in sea level and the 100-year-storm surge under different SSP scenarios.

6. Conclusions

This research proposes the MKCA$_{\text{RF-AS}}$ urban model, which combines an allometric scaling law, a Markov chain, the random forest algorithm, and cellular automata to simulate future scenarios of land use under various shared social development pathways. The FoM and urban FoM obtained by the MKCA$_{\text{RF-AS}}$ model for urban growth simulations from 2005 to 2010 are 25.5% and 35.0%, respectively, and the urban FoM is 4.06% greater than that produced by the MKCA$_{\text{ANN}}$ model. To verify the scaling relations between the population size and urban area, this research revises the Markov transition probability matrix and predicts the spatiotemporal evolution of LULC for 2020–2065. The results lead to the following conclusions:

(1) Prediction errors may occur when using the Markov model to predict long-term land-use demand spanning multiple periods. The error between the predicted urban land-use demand based on the transition probability matrix for the period 2000–2005 and the observed urban land-use demand in 2020 is 12.7%. (2) The scaling exponent relating the population growth and urban area of Jinjiang City from 2005 to 2020 is 0.933. The relative rate of urban land area expansion is less than the relative growth rate of the population, which reflects an efficient use of land in the urban system of Jinjiang. (3) The simulation accuracy of the MKCA$_{\text{RF-AS}}$ model depends strongly on land-use demand. While the Markov model produces an error of 6.28% when predicting urban land use during 2005–2010, the FoM declines by 21.42%. (4) Most townships in Jinjiang City develop at medium and high rates under scenarios SSP2 and SSP5. Based on the assumption of allometric scaling between the population size and urban area, the prediction of LULC provides an objective reference for spatial land-use planning.

Given the prevalence of allometric scaling phenomena in urban systems, this work focuses on integrating allometric scaling between the population size and urban area into the proposed model. In the future, we will explore and verify how scaling invariance affects the evolution of urban spatial forms in a larger urban area and over a longer time span.

Author Contributions: Conceptualization, J.L. and L.T.; methodology, J.L. and L.T; software, J.L.; validation, J.L.; formal analysis, L.T.; investigation, J.L.; resources, L.T. and G.S.; data curation, L.T.; writing—original draft preparation, J.L.; writing—review and editing, L.T. and G.S.; visualization, J.L.; supervision, L.T. and G.S.; project administration, L.T.; funding acquisition, L.T. All authors have read and agreed to the published version of the manuscript.

Funding: This study was supported by the National Natural Science Foundation of China (NSFC) (Grant No. 41471137).

Data Availability Statement: Not applicable.

References

1. Beillouin, D.; Cardinael, R.; Berre, D.; Boyer, A.; Corbeels, M.; Fallot, A.; Feder, F.; Demenois, J. A global overview of studies about land management, land-use change, and climate change effects on soil organic carbon. *Glob. Chang. Biol.* **2022**, *28*, 1690–1702. [CrossRef] [PubMed]
2. Ahammad, R.; Hossain, M.K.; Sobhan, I.; Hasan, R.; Biswas, S.R.; Mukul, S.A. Social-ecological and institutional factors affecting forest and landscape restoration in the Chittagong Hill Tracts of Bangladesh. *Land Use Policy* **2023**, *125*, 106478. [CrossRef]
3. De Olivera, L.C.M.; de Mendonça, G.C.; Costa, R.C.A.; de Camargo, R.A.L.; Fernandes, L.F.S.; Pacheco, F.A.L.; Pissarra, T.C.T. Impacts of urban sprawl in the Administrative Region of Ribeirão Preto (Brazil) and measures to restore improved landscapes. *Land Use Policy* **2023**, *124*, 106439. [CrossRef]
4. Sun, Y.; Liu, D.; Wang, P. Urban simulation incorporating coordination relationships of multiple ecosystem services. *Sustain. Cities Soc.* **2022**, *76*, 103432. [CrossRef]
5. Domingo, D.; Van Vliet, J.; Hersperger, A.M. Long-term changes in 3D urban form in four Spanish cities. *Landsc. Urban Plan.* **2023**, *230*, 104624. [CrossRef]
6. Jia, B.; Luo, X.; Wang, L.; Lai, X. Changes in Water Use Efficiency Caused by Climate Change, CO_2 Fertilization, and Land Use Changes on the Tibetan Plateau. *Adv. Atmos. Sci.* **2023**, *40*, 144–154. [CrossRef]
7. Black, B.; van Strien, M.J.; Adde, A.; Grêt-Regamey, A. Re-considering the status quo: Improving calibration of land use change models through validation of transition potential predictions. *Environ. Model. Softw.* **2023**, *159*, 105574. [CrossRef]
8. Chen, Y.; Weng, Q.; Tang, L.; Wang, L.; Xing, H.; Liu, Q. Developing an intelligent cloud attention network to support global urban green spaces mapping. *ISPRS J. Photogramm. Remote Sens.* **2023**, *198*, 197–209. [CrossRef]
9. Sun, S.; Parker, D.C.; Brown, D.G. From an agent-based laboratory to the real world: Effects of "neighborhood" size on urban sprawl. *Comput. Environ. Urban Syst.* **2023**, *99*, 101889. [CrossRef]
10. Feng, Y.; Gao, C.; Wang, R.; Li, P.; Xi, M.; Jin, Y.; Tong, X. A moving window-based spatial assessment method for dynamic urban growth simulations. *Geocarto Int.* **2022**, *37*, 15282–15301. [CrossRef]
11. Zhang, B.; Hu, S.; Wang, H.; Zeng, H. A size-adaptive strategy to characterize spatially heterogeneous neighborhood effects in cellular automata simulation of urban growth. *Landsc. Urban Plan.* **2023**, *229*, 104604. [CrossRef]
12. Molinero-Parejo, R.; Aguilera-Benavente, F.; Gómez-Delgado, M.; Shurupov, N. Combining a land parcel cellular automata (LP-CA) model with participatory approaches in the simulation of disruptive future scenarios of urban land use change. *Comput. Environ. Urban Syst.* **2023**, *99*, 101895. [CrossRef]
13. Halder, S.; Das, S.; Basu, S. Use of support vector machine and cellular automata methods to evaluate impact of irrigation project on LULC. *Environ. Monit. Assess.* **2023**, *195*, 3. [CrossRef] [PubMed]
14. Shojaei, H.; Nadi, S.; Shafizadeh-Moghadam, H.; Tayyebi, A.; Van Genderen, J. An efficient built-up land expansion model using a modified U-Net. *Int. J. Digit. Earth* **2022**, *15*, 148–163. [CrossRef]
15. Liu, J.; Xiao, B.; Li, Y.; Wang, X.; Bie, Q.; Jiao, J. Simulation of dynamic urban expansion under ecological constraints using a long short term memory network model and cellular automata. *Remote Sens.* **2021**, *13*, 1499. [CrossRef]
16. Wu, X.; Liu, X.; Zhang, D.; Zhang, J.; He, J.; Xu, X. Simulating mixed land-use change under multi-label concept by integrating a convolutional neural network and cellular automata: A case study of Huizhou, China. *GIScience Remote Sens.* **2022**, *59*, 609–632. [CrossRef]
17. Liang, X.; Guan, Q.; Clarke, K.C.; Liu, S.; Wang, B.; Yao, Y. Understanding the drivers of sustainable land expansion using a patch-generating land use simulation (PLUS) model: A case study in Wuhan, China. *Comput. Environ. Urban Syst.* **2021**, *85*, 101569. [CrossRef]
18. Luo, Z.; Zhang, W.; Wang, Y.; Wang, T.; Liu, G.; Huang, W. Spatial optimization of ecological ditches for non-point source pollutants under urban growth scenarios. *Environ. Monit. Assess.* **2023**, *195*, 105. [CrossRef]
19. Chasia, S.; Olang, L.O.; Sitoki, L. Modelling of land-use/cover change trajectories in a transboundary catchment of the Sio-Malaba-Malakisi Region in East Africa using the CLUE-s model. *Ecol. Model.* **2023**, *476*, 110256. [CrossRef]
20. Lin, X.; Wang, Z. Landscape ecological risk assessment and its driving factors of multi-mountainous city. *Ecol. Indic.* **2023**, *146*, 109823. [CrossRef]

21. Ou, M.; Li, J.; Fan, X.; Gong, J. Compound Optimization of Territorial Spatial Structure and Layout at the City Scale from "Production–Living–Ecological" Perspectives. *Int. J. Environ. Res. Public Health* **2023**, *20*, 495. [CrossRef]
22. Ren, Q.; He, C.; Huang, Q.; Zhang, D.; Shi, P.; Lu, W. Impacts of global urban expansion on natural habitats undermine the 2050 vision for biodiversity. *Resour. Conserv. Recycl.* **2023**, *190*, 106834. [CrossRef]
23. Isinkaralar, O.; Varol, C.; Yilmaz, D. Digital mapping and predicting the urban growth: Integrating scenarios into cellular automata—Markov chain modeling. *Appl. Geomat.* **2022**, *14*, 695–705. [CrossRef]
24. Ouyang, X.; Xu, J.; Li, J.; Wei, X.; Li, Y. Land space optimization of urban-agriculture-ecological functions in the Changsha-Zhuzhou-Xiangtan Urban Agglomeration, China. *Land Use Policy* **2022**, *117*, 106112. [CrossRef]
25. Yang, J.; Tang, W.; Gong, J.; Shi, R.; Zheng, M.; Dai, Y. Simulating urban expansion using cellular automata model with spatiotemporally explicit representation of urban demand. *Landsc. Urban Plan.* **2023**, *231*, 104640. [CrossRef]
26. Overmars, K.P.; De Koning, G.H.J.; Veldkamp, A. Spatial autocorrelation in multi-scale land use models. *Ecol. Model.* **2003**, *164*, 257–270. [CrossRef]
27. Arsanjani, J.J.; Helbich, M.; Kainz, W.; Boloorani, A.D. Integration of logistic regression, Markov chain and cellular automata models to simulate urban expansion. *Int. J. Appl. Earth Obs. Geoinf.* **2013**, *21*, 265–275. [CrossRef]
28. Isinkaralar, O.; Varol, C. A cellular automata-based approach for spatio-temporal modeling of the city center as a complex system: The case of Kastamonu, Türkiye. *Cities* **2023**, *132*, 104073. [CrossRef]
29. Van Vliet, J.; Naus, N.; Van Lammeren, R.J.; Bregt, A.K.; Hurkens, J.; Van Delden, H. Measuring the neighbourhood effect to calibrate land use models. *Comput. Environ. Urban Syst.* **2013**, *41*, 55–64. [CrossRef]
30. Liao, J.; Tang, L.; Shao, G. Multi-Scenario Simulation to Predict Ecological Risk Posed by Urban Sprawl with Spontaneous Growth: A Case Study of Quanzhou. *Int. J. Environ. Res. Public Health* **2022**, *19*, 15358. [CrossRef]
31. Zhang, S.; Zhong, Q.; Cheng, D.; Xu, C.; Chang, Y.; Lin, Y.; Li, B. Landscape ecological risk projection based on the PLUS model under the localized shared socioeconomic pathways in the Fujian Delta region. *Ecol. Indic.* **2022**, *136*, 108642. [CrossRef]
32. Cao, W.; Dong, L.; Cheng, Y.; Wu, L.; Guo, Q.; Liu, Y. Constructing multi-level urban clusters based on population distributions and interactions. *Comput. Environ. Urban Syst.* **2023**, *99*, 101897. [CrossRef]
33. Aretouyap, Z.; Abdelfattah, M.; Gaber, A. Urban sprawl analysis and shoreline extraction in Douala-Cameroon city using optical and radar sensors. *Geocarto Int.* **2022**, *37*, 14596–14608. [CrossRef]
34. Wang, H.; Guo, F. City-level socioeconomic divergence, air pollution differentials and internal migration in China: Migrants vs talent migrants. *Cities* **2023**, *133*, 104116. [CrossRef]
35. Yang, L.; Guo, J.; Cao, S. What structural factors have held back China's birth rate? *Environ. Dev. Sustain.* **2022**, 1–14. [CrossRef]
36. Longley, P.A.; Batty, M.; Shepherd, J. The size, shape and dimension of urban settlements. *Trans. Inst. Br. Geogr.* **1991**, *16*, 75. [CrossRef]
37. Kaufmann, T.; Radaelli, L.; Bettencourt, L.M.; Shmueli, E. Scaling of urban amenities: Generative statistics and implications for urban planning. *EPJ Data Sci.* **2022**, *11*, 50. [CrossRef]
38. Abdulrasheed, M.; MacKenzie, A.R.; Whyatt, J.; Chapman, L. Allometric scaling of thermal infrared emitted from UK cities and its relation to urban form. *City Environ. Interact.* **2020**, *5*, 100037. [CrossRef]
39. Lei, W.; Jiao, L.; Xu, G. Understanding the urban scaling of urban land with an internal structure view to characterize China's urbanization. *Land Use Policy* **2022**, *112*, 105781. [CrossRef]
40. Chen, Y. Multi-scaling allometric analysis for urban and regional development. *Phys. A Stat. Mech. Appl.* **2017**, *465*, 673–689. [CrossRef]
41. Woldenberg, M.J. An allometric analysis of urban land use in the United States. *Ekistics* **1973**, *36*, 282–290.
42. Dutton, G. Criteria of growth in urban systems. *Ekistics* **1973**, *36*, 298–306.
43. Wang, X.; Meng, X.; Long, Y. Projecting 1 km-grid population distributions from 2020 to 2100 globally under shared socioeconomic pathways. *Sci. Data* **2022**, *9*, 563. [CrossRef] [PubMed]
44. Li, Z.; Murshed, M.; Yan, P. Driving force analysis and prediction of ecological footprint in urban agglomeration based on extended STIRPAT model and shared socioeconomic pathways (SSPs). *J. Clean. Prod.* **2023**, *383*, 135424. [CrossRef]
45. JCSB. Bulletin of the Seventh National Population Census of Jinjiang City. 2021. Available online: http://www.jinjiang.gov.cn/xxgk/zfxxgkzl/bmzfxxgk/tjj/zfxxgkml/202105/t20210531_2565668.htm (accessed on 8 January 2023).
46. JCSB—Jinjiang City Statistics Bureau and National Bureau of Statistics of China. *Jinjiang Statistical Yearbook in 2022*; China Statistics Press: Beijing, China, 2022.
47. Jiang, T.; Su, B.; Wang, Y.; Wang, G.; Luo, Y.; Zhai, J.Q.; Huang, J.; Jing, C.; Gao, M.; Lin, Q. Gridded datasets for population and economy under Shared Socioeconomic Pathways for 2020–2100. *Adv. Clim. Chang. Res.* **2022**, *18*, 381–383.
48. Yang, J.; Huang, X. The 30 m annual land cover dataset and its dynamics in China from 1990 to 2019. *Earth Syst. Sci. Data* **2021**, *13*, 3907–3925. [CrossRef]
49. Breiman, L. Random forests. *Mach. Learn.* **2001**, *45*, 5–32. [CrossRef]
50. Ren, L.; Seklouli, A.S.; Zhang, H.; Wang, T.; Bouras, A. An adaptive Laplacian weight random forest imputation for imbalance and mixed-type data. *Inf. Syst.* **2023**, *111*, 102122. [CrossRef]
51. Pan, Y.; Kong, X.; Yuan, Y.; Sun, Y.; Han, X.; Yang, H.; Zhang, J.; Liu, X.; Gao, P.; Li, Y. Detecting the foreign matter defect in lithium-ion batteries based on battery pilot manufacturing line data analyses. *Energy* **2023**, *262*, 125502. [CrossRef]

52. Wu, J.; Zhao, R.; Sun, J. State transition of carbon emission efficiency in China: Empirical analysis based on three-stage SBM and Markov chain models. *Environ. Sci. Pollut. Res.* **2023**, 1–11. [CrossRef]
53. Girma, R.; Fürst, C.; Moges, A. Land use land cover change modeling by integrating artificial neural network with cellular Automata-Markov chain model in Gidabo river basin, main Ethiopian rift. *Environ. Chall.* **2022**, *6*, 100419. [CrossRef]
54. Bettencourt, L.M. The origins of scaling in cities. *Science* **2013**, *340*, 1438–1441. [CrossRef] [PubMed]
55. Chen, Y. Scaling, fractals and the spatial complexity of cities. In *Handbook on Cities and Complexity*; Edward Elgar Publishing: Cheltenham, UK, 2021.
56. Feng, Y.; Wang, R.; Tong, X.; Zhai, S. Comparison of change and static state as the dependent variable for modeling urban growth. *Geocarto Int.* **2022**, *37*, 6975–6998. [CrossRef]
57. Wang, H.; Guo, J.; Zhang, B.; Zeng, H. Simulating urban land growth by incorporating historical information into a cellular automata model. *Landsc. Urban Plan.* **2021**, *214*, 104168. [CrossRef]
58. Lin, J.; Li, X.; Wen, Y.; He, P. Modeling urban land-use changes using a landscape-driven patch-based cellular automaton (LP-CA). *Cities* **2023**, *132*, 103906. [CrossRef]
59. White, R.; Engelen, G. Cellular automata and fractal urban form: A cellular modelling approach to the evolution of urban land-use patterns. *Environ. Plan. A* **1993**, *25*, 1175–1199. [CrossRef]
60. Zeng, Y.; Hesketh, T. The effects of China's universal two-child policy. *Lancet* **2016**, *388*, 1930–1938. [CrossRef]
61. Pontius, R.G.; Boersma, W.; Castella, J.-C.; Clarke, K.; de Nijs, T.; Dietzel, C.; Duan, Z.; Fotsing, E.; Goldstein, N.; Kok, K. Comparing the input, output, and validation maps for several models of land change. *Ann. Reg. Sci.* **2008**, *42*, 11–37. [CrossRef]
62. Liu, X.; Liang, X.; Li, X.; Xu, X.; Ou, J.; Chen, Y.; Li, S.; Wang, S.; Pei, F. A future land use simulation model (FLUS) for simulating multiple land use scenarios by coupling human and natural effects. *Landsc. Urban Plan.* **2017**, *168*, 94–116. [CrossRef]
63. Xu, T.; Gao, J.; Coco, G. Simulation of urban expansion via integrating artificial neural network with Markov chain–cellular automata. *Int. J. Geogr. Inf. Sci.* **2019**, *33*, 1960–1983. [CrossRef]
64. Feng, Y.; Liu, Y.; Tong, X.; Liu, M.; Deng, S. Modeling dynamic urban growth using cellular automata and particle swarm optimization rules. *Landsc. Urban Plan.* **2011**, *102*, 188–196. [CrossRef]
65. Nordbeck, S. Urban allometric growth. *Geogr. Ann. Ser. B Hum. Geogr.* **1971**, *53*, 54–67. [CrossRef]
66. Lee, Y. An allometric analysis of the US urban system: 1960–80. *Environ. Plan. A* **1989**, *21*, 463–476. [CrossRef] [PubMed]
67. Chen, Y.; Feng, J. A hierarchical allometric scaling analysis of Chinese cities: 1991–2014. *Discret. Dyn. Nat. Soc.* **2017**, *2017*, 5243287. [CrossRef]
68. Bettencourt, L.M.; Lobo, J.; Helbing, D.; Kühnert, C.; West, G.B. Growth, innovation, scaling, and the pace of life in cities. *Proc. Natl. Acad. Sci. USA* **2007**, *104*, 7301–7306. [CrossRef]
69. Lynch, M.; Trickovic, B.; Kempes, C.P. Evolutionary scaling of maximum growth rate with organism size. *Sci. Rep.* **2022**, *12*, 22586. [CrossRef]
70. Jia, Y.; Tang, L.; Zhang, P.; Xu, M.; Luo, L.; Zhang, Q. Exploring the scaling relations between urban spatial form and infrastructure. *Int. J. Sustain. Dev. World Ecol.* **2022**, *29*, 665–675. [CrossRef]
71. Veregin, H.; Tobler, W.R. Allometric relationships in the structure of street-level databases. *Comput. Environ. Urban Syst.* **1997**, *21*, 277–290. [CrossRef]
72. Lv, M.; Chen, Z.; Yao, L.; Dang, X.; Li, P.; Cao, X. Potential Zoning of Construction Land Consolidation in the Loess Plateau Based on the Evolution of Human–Land Relationship. *Int. J. Environ. Res. Public Health* **2022**, *19*, 14927. [CrossRef]
73. Chen, Y. An extended patch-based cellular automaton to simulate horizontal and vertical urban growth under the shared socioeconomic pathways. *Comput. Environ. Urban Syst.* **2022**, *91*, 101727. [CrossRef]
74. Xiong, Y.; Jiao, G.; Zheng, J.; Gao, J.; Xue, Y.; Tian, B.; Cheng, J. Fertility Intention and Influencing Factors for Having a Second Child among Floating Women of Childbearing Age. *Int. J. Environ. Res. Public Health* **2022**, *19*, 16531. [CrossRef] [PubMed]
75. QWCB—Quanzhou Water Conservancy Bureau. Quanzhou Water Resources Bulletin in 2020. 2020. Available online: http://slj.quanzhou.gov.cn/zwgk/tzgg/202111/t20211119_2655072.htm (accessed on 6 April 2023).
76. Chen, Y.; Li, X.; Liu, X.; Zhang, Y.; Huang, M. Quantifying the teleconnections between local consumption and domestic land uses in China. *Landsc. Urban Plan.* **2019**, *187*, 60–69. [CrossRef]
77. Oppenheimer, M.; Glavovic, B.; Hinkel, J.; van de Wal, R.; Magnan, A.K.; Abd-Elgawad, A.; Cai, R.; Cifuentes-Jara, M.; Deconto, R.M.; Ghosh, T. Sea level rise and implications for low lying islands, coasts and communities. In IPCC Special Report on the Ocean and Cryosphere in a Changing Climate. 2019. Available online: https://www.ipcc.ch/srocc/chapter/chapter-4-sea-level-rise-and-implications-for-low-lying-islands-coasts-and-communities/ (accessed on 15 April 2023).

Article

The Assessment of Industrial Agglomeration in China Based on NPP-VIIRS Nighttime Light Imagery and POI Data

Zuoqi Chen [1,2,*], Wenxiang Xu [1,2] and Zhiyuan Zhao [1,2]

[1] Key Laboratory of Spatial Data Mining and Information Sharing of Ministry of Education, National & Local Joint Engineering Research Center of Satellite Geospatial Information Technology, Fuzhou University, Fuzhou 350108, China; 215520009@fzu.edu.cn (W.X.); zyzhao@fzu.edu.cn (Z.Z.)
[2] The Academy of Digital China, Fuzhou University, Fuzhou 350108, China
* Correspondence: zqchen@fzu.edu.cn

Abstract: Industrial agglomeration, as a typical aspect of industrial structures, significantly influences policy development, economic growth, and regional employment. Due to the collection limitations of gross domestic product (GDP) data, the traditional assessment of industrial agglomeration usually focused on a specific field or region. To better measure industrial agglomeration, we need a new proxy to estimate GDP data for different industries. Currently, nighttime light (NTL) remote sensing data are widely used to estimate GDP at diverse scales. However, since the light intensity from each industry is mixed, NTL data are being adopted less to estimate different industries' GDP. To address this, we selected an optimized model from the Gaussian process regression model and random forest model to combine Suomi National Polar-Orbiting Partnership—Visible Infrared Imaging Radiometer Suite (NPP-VIIRS) NTL data and points-of-interest (POI) data, and successfully estimated the GDP of eight major industries in China for 2018 with an accuracy (R^2) higher than 0.80. By employing the location quotient to measure industrial agglomeration, we found that a dominated industry had an obvious spatial heterogeneity. The central and eastern regions showed a developmental focus on industry and retail as local strengths. Conversely, many western cities emphasized construction and transportation. First-tier cities prioritized high-value industries like finance and estate, while cities rich in tourism resources aimed to enhance their lodging and catering industries. Generally, our proposed method can effectively measure the detailed industry agglomeration and can enhance future urban economic planning.

Keywords: industrial agglomeration; GDP; points of interest; nighttime light; Gaussian process

Citation: Chen, Z.; Xu, W.; Zhao, Z. The Assessment of Industrial Agglomeration in China Based on NPP-VIIRS Nighttime Light Imagery and POI Data. *Remote Sens.* **2024**, *16*, 417. https://doi.org/10.3390/rs16020417

Academic Editor: Magaly Koch

Received: 27 November 2023
Revised: 4 January 2024
Accepted: 17 January 2024
Published: 21 January 2024

1. Introduction

Since the reform and opening-up policy, China's industrial structure has undergone significant changes. The proportion of the secondary and tertiary industries has continuously increased, rising from 72.3% in 1978 to 90.9% in 2019 in China. The proportion of the economy dedicated to the tertiary industry has undergone significant changes, increasing from 24.6% in 1978 to 53.9% in 2019, marking a new stage in China's economic development. In the process of industrial structure upgrading, it is common to observe the emergence of industrial agglomeration. This refers to the concentration of similar or related industries within a specific geographical area, where industrial capital elements accumulate over a defined spatial range [1]. Industrial agglomeration is crucial for technological progress, economic growth, ecological sustainability, and resource optimization. It facilitates China's economic shift from the pursuit of speed to the pursuit of quality [2]. Therefore, the assessment of industrial agglomeration is of great significance in supporting the government to develop policies for promoting high-quality economic development. However, the calculation of the industrial agglomeration degree currently often relies on statistical data, suffering from problems such as inconsistent statistical standards, missing data in some regions or periods, and low update frequency. It is urgent to utilize new data sources to address these limitations.

The emergence of nighttime light (NTL) data provides a new perspective for addressing this issue. In previous studies, NTL data became a vital proxy in many socioeconomic fields. For example, Wu, et al. [3] analyzed the spatially heterogeneous relationships between NTL intensity and human activities by using NTL and points-of-interest (POI) data. Wang, et al. [4] improved population mapping by using NTL and location-based social media data and the results showed higher accuracy than ones acquired with WorldPop data. Song, et al. [5] investigated the urban extension features and drivers in the Shandong Peninsula urban agglomeration using the enhanced vegetation-adjusted nighttime light index. Meanwhile, Shi, et al. [6] analyzed the poverty situation in Chongqing City by combining lights with multi-source data to construct a comprehensive poverty index. Additionally, NTL data are often used to estimate GDP. Initially, Elvidge, et al. [7] used the NTL data from Defense Meteorological Satellite Program's Operational Linescan System (DMSP-OLS) to estimate the GDP of 21 countries. Chen and Nordhaus [8] demonstrated that DMSP-OLS NTL data could offset the scarcity of GDP data in developing countries. In addition to national scales, DMSP-OLS NTL data have been applied to estimate GDP at different scales, such as sub-national and grid scales [9–11]. With the advent of the Suomi National Polar-Orbiting Partnership—Visible Infrared Imaging Radiometer Suite (NPP-VIIRS), a new generation of NTL data with superior quality to DMSP-OLS has become widely used for GDP estimation [12–16]. For example, Shi [17] found NPP-VIIRS NTL data have higher accuracy than DMSP-OLS NTL data in GDP estimation at sub-national scales. Zhao [18] used NPP-VIIRS NTL data to estimate the GDP in South China at the grid scale, revealing high correlation coefficients (R^2 values of 0.8935 for prefecture GDP and 0.9243 for county GDP). While NTL data are widely used, no studies have been found regarding the assessment of industrial agglomeration using NTL data.

Traditional methods for assessing industrial agglomeration mainly include the Herfindahl index, spatial Gini coefficient, Ellison–Glaeser index, and location quotient [19–22]. The Herfindahl index calculates the sum of squared market shares of all companies in an industry, requiring specific enterprise output data, and it is difficult to use the existing data to meet the calculation needs of a large range [19]. The spatial Gini coefficient can be used to measure the degree of industrial agglomeration via estimation using the numerical integration of the area inside the Lorenz curve in the graph of cumulative GDP, sorted according to decreasing geographic area, for a given industry [20]. The Ellison–Glaeser index combines the Herfindahl index and the spatial Gini coefficient to measure the degree of industrial agglomeration [21]. The location quotient uses the proportion of overall economic activities in the region to the national total, which can eliminate the effects of regional scale [22]. Due to the main target of this study being the calculation of the degree of industrial agglomeration of different industries across Chinese cities, the location quotient is more effective in reducing the impact of regional differences [23]. Therefore, while considering data availability, we chose location quotient as the indicator for industrial agglomeration in this study.

To calculate the industrial agglomeration of different industries in cities across the country using location quotient, the GDP data of each industry are required. However, previous GDP estimations mainly focused on the total GDP or the GDP of the primary, secondary, and tertiary sectors [13]. As different industries can have similar light emissions for different NTL pixels or similar concentrations in the same NTL pixel, merely using the NTL data cannot estimate the GDP well with different industries [24]. Regarding overcoming this challenge, the rise of social sensing big data presents a new avenue for estimating the GDP of different industries. This data type correlates strongly with diverse human activities in urban environments [25]. POI data, as social sensing data, record the location and function of geographic entities. Recent studies utilize POI data in studies related to human activities, such as research identifying urban function areas and classifying urban land use [26–30]. POI data can reflect the spatial distribution and aggregation intensity of different types of economic activities, and so we considered the integration of the NTL and POI data for the GDP estimation in different industries.

The random forest (RF) model is among the most popular machine learning algorithms and has been widely used in estimation studies of socioeconomic indicators such as GDP, population, poverty, and electricity consumption [15,31,32]. Meanwhile, the Gaussian process regression (GPR) model, a kernel-based model, is adaptable to different data structures through the selection of appropriate covariance functions, making it particularly suitable for estimating the GDP of different industries [33]. In order to compare the estimation performance of RF and GPR model, we set them up to estimate GDP data from different industries and chose the one that performed better.

To calculate the degree of industrial agglomeration, this study considered 345 cities in China and used the NTL, POI, and related auxiliary data as inputs to the GPR and RF models to estimate the GDP in eight industries (industry, construction, retail, transportation, lodging and catering, finance, estate, and other tertiary industries) according to the Industrial Classification of National Economic Activities (GB/T4754-2017) [34]. The structure of the paper is organized as follows. The description of the study area and data processing will be described in Section 2. The retrieval algorithm GPR and RF model and the process of the model construction will be discussed in Section 3. We will present the results of the GDP estimation and industrial agglomeration for different industries in Section 4. The results will be discussed in Section 5. Finally, the conclusion will be presented in Section 6.

2. Study Area and Materials

2.1. Study Area

In total, 345 cities in China were selected as the study area, excluding Hong Kong, Macau, and Taiwan, as shown in Figure 1. In order to reduce the influence of spatial heterogeneity on GDP estimation in different zones of China, we trained the GDP models separately for the three major economic zones in China [35,36], namely, east, central, and west China (Figure 1), which have different economic development levels. East China has the highest GDP, per capital income, investments, and resident consumption, with decreasing values from east to west [37]. The number of cities for each zone is 100, 69, and 71, respectively.

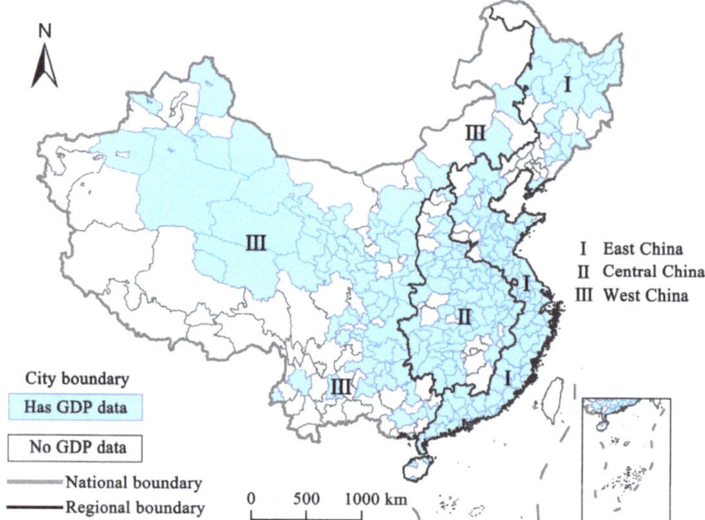

Figure 1. Study area for China in 2018. The study includes 345 cities in China, with statistical GDP data available for eight industries in 240 of these cities.

2.2. Materials

The datasets used in this study included NPP-VIIRS NTL data, POI data, road network (RD) data, digital elevation model (DEM) data, normalized difference vegetation index (NDVI) data, and statistical GDP data. NPP-VIIRS NTL data and POI data were adopted as input features. RD, DEM and NDVI data were input into the model as auxiliary variables along with NTL and POI features. The statistical GDP data were divided into two groups to train and evaluate the GDP estimation model. To maintain consistency with the latest China's Industrial Classification standard in 2017, the year 2018 was selected as our study period.

2.2.1. Nighttime Light Data

The NPP-VIIRS NTL data show significant improvements compared to the DMSP/OLS data. The onboard day/night band (DNB) was the primary band used to detect nighttime light intensity. The overpass time during the night was approximately 1:30 a.m. It operated within a wavelength range of 0.5–0.9 μm and provided data with a spatial resolution of 750 m. The captured area included regions ranging from 70°N to 65°S in latitude [38]. In this study, we used the version 1 series of China's NPP-VIIRS NTL data from 2018, provided by the Colorado School of Mines. To facilitate the calculation of various features of NTL, the NPP-VIIRS NTL data were reprojected using an Albert coordinate system with a spatial resolution of 500 m. Due to inherent noise issues in NPP-VIIRS NTL monthly composite data, such as stray light, lightning, lunar illumination, and cloud cover, we implemented a preprocessing method devised by Shi [17]. This method involved setting pixels with negative digital number (DN) values to zero and using pixels with DN values exceeding the maximum recorded in Beijing, Shanghai, and Guangzhou as outliers. Outlier pixels were assigned a new value, calculated as the maximum DN among their immediate eight neighbors. To generate the annual NPP-VIIRS NTL data of China in 2018, the average method was used to produce a composite of monthly NPP-VIIRS NTL data. Finally, four characteristics of NTL intensity were measured as the potential input features, including the lit area (NTL-Area), the mean value of NTL intensity (NTL-Mean), the standard deviation of NTL intensity (NTL-Std), and the total NTL intensity (NTL-Sum).

2.2.2. POI Data

In this study, 12,848,399 POI records of mainland China in 2018 were derived from the Baidu Map. The original POI data consisted of 23 major categories (such as finance and insurance services, enterprises, shopping, transportation service, etc.). The original POI data consisted of 23 major categories. For consistency with the classification of statistical GDP data, the POI data were reclassified into the same eight categories of GDP (Table 1).

Table 1. POI types related to the eight industries.

Category	Content
Industry	Mineral processing, mini company, factory, etc.
Construction	Decoration company, construction company, etc.
Retail	Supermarket, emporium, shopping center, convenience store, etc.
Transportation	Transit station, expressway service area, filling station, park, physical distribution, postal service, etc.
Lodging and catering	Hotel, restaurant, etc.
Finance	Bank, securities company, insurance company, etc.
Estate	Areola, real estate intermediary, estate company, etc.
Other tertiary industries	Educational institution, hospital, media, gymnasium, photography services, etc.

Then, discrete POIs point data were converted into continuous smooth density surfaces for each category using kernel density estimation (KDE). The bandwidth of KDE was set at

5000 m in this study, referring to the previous studies of Peng, et al. [39] and Ye [32]. To maintain consistency with NPP-VIIRS NTL data, the spatial resolution of the KDE results was fixed at 500 m. Subsequently, four characteristics of POI were quantified as the potential input features, including the KDE area of POI (POI-Area) representing the area where the kernel density value is greater than zero, the mean KDE value of POI (POI-Mean), the standard deviation KDE value of POI (POI-Std) and the total number of POI (POI-Num).

2.2.3. Auxiliary Data

The data of RD, DEM and NDVI, as auxiliary variables, were uniformly converted into the Albert projection coordinate system and reprojected to a spatial resolution of 500 m.

Road network data were obtained from OpenStreetMap (http://www.openstreetmap.org) accessed on 23 December 2023 and can reflect the intensity of economic activity and the level of development of infrastructure in a city [3]. This study used primary road, secondary road, tertiary road, motorway, and trunk road as auxiliary variables to calculate the road network density and length for each city.

The DEM data used in this study were Shuttle Radar Topography Mission (SRTM)–Consortium for Spatial Information (CGIAR-CSI) DEM data provided by National Aeronautics and Space Administration (NASA) and the National Imagery and Mapping Agency (NIMA) with a spatial resolution of 250 m. The DEM data can represent topographic change and influence human production and survival, thereby affecting socioeconomic development [6]. This study calculated the average elevation and slope as auxiliary variables for each city using DEM data.

The monthly NDVI data from the Geospatial Data Cloud (http://www.resdc.cn/) accessed on 23 December 2023 can provide information on vegetation coverage and ecological conditions, effectively eliminating the impact of the light emitted from nighttime decorations in urban parks [40]. This helps us to better extract illumination related to the eight major industries selected in our study. This is carried out by selecting the maximum value from monthly data as annual data for 2018 and calculating the area of NDVI values greater than 0.2 per pixel in each city to be a proxy for the vegetation coverage area used as an auxiliary variable.

2.2.4. Statistical GDP Data

Statistical GDP data from different industries were collected from the statistical yearbook of each city in 2018. In total, 240 cities had statistical GDP data for eight industries (Figure 1). Meanwhile, to reduce the heteroscedasticity of data, a logarithmic transformation was applied to the GDP data in this study.

3. Methods

The framework for calculating the degree of industrial agglomeration is shown in Figure 2. (I) Firstly, it is necessary to perform GDP Estimation: this involves selecting features from NTL and POI data as inputs for the GPR and RF models. Both models estimate the GDP of eight industries at the city level. (II) Then, it is necessary to calculate the degree of industrial agglomeration: utilizing the estimated GDP, the location quotient is calculated to assess the degree of industrial agglomeration with different industries.

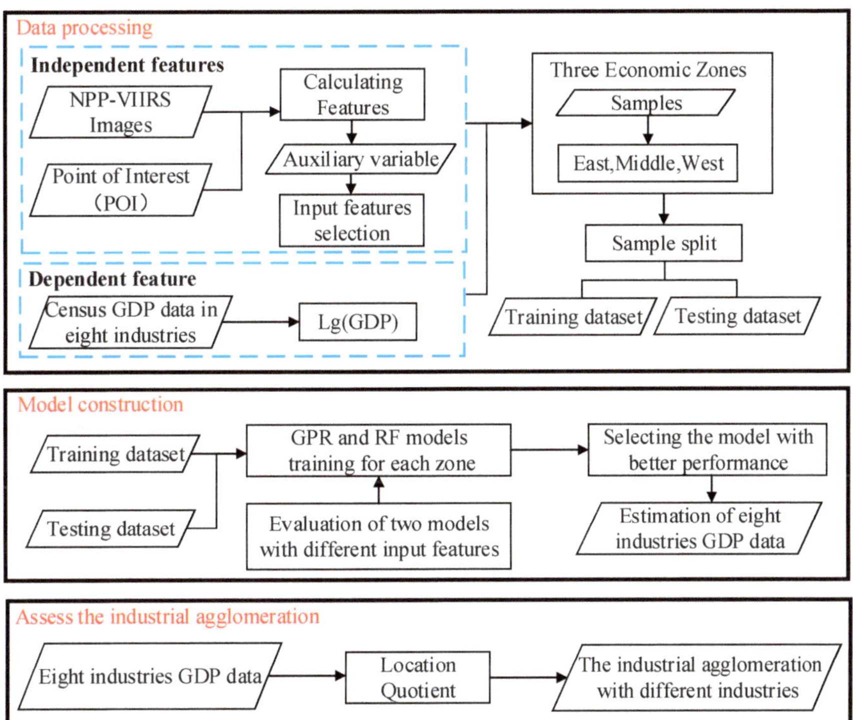

Figure 2. Flowchart of the methodology used in this study.

3.1. The Location Quotient

This study chooses the location quotient (LQ) to calculate the degree of industrial agglomeration with different industries in China. The location quotient uses the proportion of overall economic activities in the region to the national total [41]. What follows is the calculation formula for location quotient:

$$LQ_{ij} = \frac{\frac{q_{ij}}{\sum_{j=1}^{n} q_{ij}}}{\frac{Q_j}{\sum_{j=1}^{n} Q_j}}, \tag{1}$$

where LQ_{ij} represents the location quotient of industry j in region i relative to the national level, q_{ij} represents the output or GDP of industry j in region i, Q_j represents the relevant indicators of industry j at the national level, and n is the number of industries. Considering that the calculation formula requires GDP data from different industries, we must obtain GDP data from different industries.

3.2. GDP Estimation

3.2.1. Selection of NTL and POI Features

GDP estimation for different industries requires us to consider the combined effects of multiple factors. Identifying factors with a decisive impact on GDP estimation is crucial for selecting features in estimation models [42]. Accordingly, eight potential features of NTL data and POI data were chosen for correlation analysis with GDP in each industry. For the correlation analysis, we used R^2 calculated via the Formula (13). After passing the significance test, we selected the NTL and POI features, respectively, which have the highest GDP in different industries as the input features. Considering that other socioeconomic variables may also affect the model's accuracy, we not only set up a model by only using

NTL and POI features, but also added auxiliary variables for correlation analysis as a comparative group, providing a better selection of features.

3.2.2. Gaussian Process Regression

For GPR, the training dataset can be defined as $D = \{(x_i, y_i)|i = 1, 2, \ldots, n\} = \{X, Y\}$, where n is the size of training dataset, $X = [x_1, x_2, \ldots, x_n]^T$ is the input matrix which contains POI feature for a specific industry and NTL feature, and $Y = [y_1, y_2, \ldots, y_n]^T$ is the output vector with statistical GDP data for the corresponding industry. The prior distribution of D can be defined with the Gaussian Process (GP) $f(X)$, which is given by:

$$f(X) \sim GP(m(X), K(x, x')),\tag{2}$$

where x and x' denote different input variables, $m(X)$ denotes mean function, and $K(x, x')$ is covariance function.

The goal function of GPR is: $y = f(x) + \varepsilon$, where ε is the noise and can be defined as $\varepsilon \sim N(0, \sigma_n^2)$, and σ denotes the variance of the difference between x and y. Combined with the above definition of GP, it can rewritten as follows:

$$Y \sim GP\left(m(X), K(x, x') + \sigma_n^2 I\right),\tag{3}$$

where I is the identity matrix.

The testing dataset is: $D_* = \{(x_i, f(x_i))|i = n+1, n+2, \ldots, n+n_*\} = \{X_*, f_*\}$, where n_* is the number of testing dataset, and X_* and f_* are input matrix and output vector of testing dataset. In GPR, the training output Y at training points X and test output f_* at test points X_* obey the joint Gaussian distribution, which can be given as:

$$\begin{bmatrix} Y \\ f_* \end{bmatrix} \sim N\left(\begin{matrix} m(X) \\ m(X_*)' \end{matrix}\begin{bmatrix} K(X, X) + \sigma^2 I & K(X, X_*) \\ K(X_*, X) & K(X_*, X_*) \end{bmatrix}\right),\tag{4}$$

The principle of joint Gaussian distributions can estimate results using the prior joint distribution of the test output f_*, which can be given as:

$$f_*|X, Y, X_* \sim N(\overline{f_*}, cov(f_*)),\tag{5}$$

$$\overline{f_*} = m(X_*) + K(X_*, X)\left[K(X, X) + \sigma^2 I\right]^{-1} Y,\tag{6}$$

$$cov(f_*) = K(X_*, X_*) - K(X_*, X)\left[K(X, X) + \sigma^2 I\right]^{-1} K(X, X_*) + \sigma^2 I,\tag{7}$$

where $\overline{f_*}$ and $cov(f_*)$ are the mean and variance of the predicted value f_*.

For GPR, the mean function $m(X)$ was set to zero. Then, we tried and tested different covariance functions $K(x, x')$ and selected the best-performing covariance function for each industry. The covariance functions used in this study are the exponential covariance function, squared exponential covariance function, Matern 3/2 covariance function, Matern 5/2 covariance function, and rational quadratic covariance function. The covariance functions are as follows:

Exponential covariance function (*EXP*):

$$K_{EXP}(x_i, x_j) = \sigma_f^2 exp(-\frac{r}{l}),\tag{8}$$

Squared exponential covariance function (*SE*):

$$K_{SE}(x_i, x_j) = \sigma_f^2 exp(-\frac{r^2}{2l^2}),\tag{9}$$

Matern 3/2 covariance function (*M32*):

$$K_{M32}(x_i, x_j) = \sigma_f^2(1 + \frac{\sqrt{3}r}{l})exp(-\frac{\sqrt{3}r}{l}),\tag{10}$$

Matern 5/2 covariance function (*M52*):

$$K_{M52}(x_i, x_j) = \sigma_f^2(1 + \frac{\sqrt{5}r}{l} + \frac{5r^2}{3l^2})exp(-\frac{\sqrt{5}r}{l}),\tag{11}$$

Rational quadratic covariance function (*RQ*):

$$K_{RQ}(x_i, x_j) = \sigma_f^2(1 + \frac{r^2}{2\alpha l^2})^{-\alpha},\tag{12}$$

where α is the shape parameter for the rational quadratic covariance; σ_f is the standard deviation of signal; l is variance scale; and r is the absolute value of $x_i - x_j$.

3.2.3. The Random Forest

The random forest model, an ensemble learning algorithm, was proposed by Breiman [43]. It generates multiple samples through iterative self-sampling and builds corresponding decision trees based on these samples. RF regression is developed by combining these multiple decision trees, and the result is determined using the average of the estimated results from all the decision trees. The RF model has the advantage of high estimated accuracy and does not require assumptions about the prior probability distribution [44]. In this study, the same input features as the GPR model were selected to train and validate the RF model, enabling a better comparison of the accuracy between the two models. The RF model's error was minimized by optimizing parameters, such as the number of decision trees, tree depth, maximum features in a decision tree, and leaf nodes, using a grid search.

3.3. Model Construction and Evaluation

Through correlation analysis with the GDP of different industries, the optimized features for constructing subsequent GPR and RF models were selected separately for NTL and POI features with and without auxiliary variables. Note that NTL and POI features, along with auxiliary variables, were collectively input into models for training.

Subsequently, the optimized features in each industry were used as input variables and the corresponding GDP data were randomly divided into two parts: 70% for training and the remaining 30% as reference data for the evaluation. All data were grouped into three parts based on China's three economic zones. Subsequently, we trained the GPR and RF model for each industry in each zone and selected input features and model with the best performance to estimate the GDP of different industries in each city.

The root-mean-square error (*RMSE*), decision coefficient (R^2), and percent error (*pe*) were used to evaluate our results (13)–(15):

$$R^2 = \frac{\sum_i^n (x_i - \overline{x})^2(y_i - \overline{y})^2}{\sum_i^n(x_i - \overline{x})\sum_i^n(y_i - \overline{y})},\tag{13}$$

$$RMSE = \sqrt{\frac{\sum_i (x_i - y_i)^2}{n}},\tag{14}$$

$$pe = \frac{|x_i - y_i|}{x_i} * 100\%,\tag{15}$$

where x_i and y_i are the reference and estimated value, respectively, for point i; $\overline{x}$ and $\overline{y}$ are the mean of x and y, respectively; and n is the size of the datasets.

4. Results

4.1. Selection Input Features and Model Comparison

Table 2 shows that NTL-Sum and POI-Num have the best performance in each industry, with R^2 higher than 0.5 and 0.6, respectively, among the NTL and POI features without auxiliary variables. Particularly in the financial industry, their R^2 reaches 0.86 and 0.88, respectively. Thus, NTL-Sum and POI-Num are selected as input features without auxiliary variables. Table 3 shows that, after combining the auxiliary variables, the R^2 of NTL and POI features have a significant increase. This is especially true in the lodging and catering industry, where the R^2 of NTL-Sum and POI-Num increase from 0.59 and 0.62 to 0.76 and 0.77, respectively. Note that in the financial industry, the R^2 of NTL-Sum and POI-Num decrease after adding auxiliary variables, changing from 0.86 and 0.88 to 0.79 and 0.82, respectively. Except for the construction and other tertiary industries, NTL-Sum and POI-Num have the best performance, with R^2 exceeding 0.7 in each industry among NTL and POI features with auxiliary variables. Particularly in the estate industry, the highest R^2 values for these two features are 0.86 and 0.89. Therefore, for the input features with auxiliary variables, the construction industry selects NTL-Sum and POI-Std as input features. Other tertiary industries choose NTL-Area and POI-Std as input features. The remaining six industries all use NTL-Sum and POI-Num as input features.

Table 2. Decision coefficients for the relationship between each potential feature without auxiliary and GDP of different industries.

Category	NTL Features without Auxiliary Variables				POI Features without Auxiliary Variables			
	NTL-Area	NTL-Mean	NTL-Std	NTL-Sum	POI-Area	POI-Mean	POI-Std	POI-Num
Industry	0.17	0.20	0.25	**0.76 ****	0.55	0.28	0.11 *	**0.75 ****
Construction	0.40	0.23	0.33	**0.64 ****	0.44	0.19	0.08 *	**0.64 ****
Retail	0.17	0.21 *	0.36	**0.67 ****	0.56	0.24 **	0.09 **	**0.70 ****
Transportation	0.12	0.21	0.36 *	**0.70 ****	0.53	0.18	0.09	**0.73 ****
Lodging and Catering	0.18	0.23 *	0.37	**0.59 ****	0.46	0.24 **	0.08 **	**0.62 ****
Finance	0.21	0.30	0.52 **	**0.86 ****	0.58	0.31	0.13 **	**0.88 ***
Estate	0.47	0.38	0.45	**0.75 ****	0.52	0.21 **	0.04 **	**0.81 ****
Other tertiary industries	0.09	0.30	0.50	**0.76 ****	0.54	0.28	0.11 *	**0.76 ****

Note: ** model significant at the 0.01 probability level ($p < 0.01$) and * model significant at the 0.05 probability level ($p < 0.05$).

Subsequently, selected input features with and without auxiliary variables are input into the GPR model based on five covariance functions. The function yielding the highest R^2 and the lowest RMSE is deemed optimal. As shown in Tables 4 and 5, for transportation, lodging and catering, and other tertiary industries, the covariance function with the best performance is the exponential function, with input features adding auxiliary variables. The Matern 3/2 function is optimal for the estate with input features, adding auxiliary variables. And we selected Matern 5/2 as the covariance functions for the GDP estimation of the industry, with input features adding auxiliary variables. It is worth noting that, for the construction and finance industries, the model accuracy of input features without auxiliary variables is higher than that of features with auxiliary variables. Therefore, for these two industries, we selected the Matern 3/2 and exponential function without auxiliary variables as the estimation model's covariance function. Then, the R^2 values for the estimated GDP with different industries were all higher than 0.8 in the testing dataset. The GDP estimation for the finance and construction industries had the highest (0.95) and lowest (0.80) R^2, respectively. Meanwhile, the RMSEs of the estimated GDP in the industry, construction, retail, transportation, lodging and catering, finance, estate, and other tertiary industries were 234.52, 77.90, 84.96, 41.39, 24.08, 59.79, 60.89, and 262.81 (CNY 10^8), respectively.

Table 3. Decision coefficients for the relationship between each potential feature with auxiliary and GDP of different industries.

Category	NTL Features with Auxiliary Variables				POI Features with Auxiliary Variables			
	NTL-Area	NTL-Mean	NTL-Std	NTL-Sum	POI-Area	POI-Mean	POI-Std	POI-Num
Industry	0.81 **	0.78 **	0.77	**0.87 **	0.77	0.80 **	0.81 **	**0.83 **
Construction	0.62 **	0.62 **	0.62 **	**0.63 **	0.62 **	0.65 **	**0.68 **	0.62 *
Retail	0.82 **	0.79	0.79	**0.89 **	0.79	0.79	0.79	**0.81 **
Transportation	0.80 **	0.75	0.76	**0.82 **	0.76	0.76	0.78 **	**0.85 **
Lodging and Catering	0.70 **	0.69 **	0.68 **	**0.76 **	0.67	0.68 **	0.72 **	**0.77 **
Finance	0.73 **	0.72 *	0.72	**0.79 **	0.72	0.73 **	0.75 **	**0.82 **
Estate	0.82 **	0.82 **	0.81	**0.86 **	0.82 **	0.85 **	0.84 **	**0.89 **
Other tertiary industries	**0.75 **	0.74	0.74	0.74	0.76 **	0.74	**0.78 **	0.74 *

Note: ** model significant at the 0.01 probability level ($p < 0.01$) and * model significant at the 0.05 probability level ($p < 0.05$).

Table 4. Test results of the different covariance functions and model that input features without auxiliary variables.

Category	Exponential		Squared Exponential		Matern 3/2		Matern 5/2		Rational Quadratic		Random Forest	
	R^2	RMSE	R^2	RMSE	R^2	RMSE	R^2	RMSE	R^2	RMSE	R^2	RMSE
Industry	**0.88**	**257.59**	0.84	268.36	0.85	262.07	0.83	283.18	0.86	260.10	0.80	565.97
Construction	0.81	79.52	0.80	80.02	**0.82**	**77.90**	0.78	88.37	0.77	83.11	0.73	118.68
Retail	0.82	107.14	0.85	104.03	0.82	111.24	0.83	112.04	**0.86**	**98.06**	0.83	146.61
Transportation	0.82	58.23	**0.85**	**48.65**	0.81	60.05	0.82	57.02	0.83	49.03	0.84	65.05
Lodging and Catering	0.73	39.67	**0.80**	**30.04**	0.72	39.92	0.75	35.18	0.72	36.62	0.81	30.22
Finance	**0.95**	**59.79**	0.92	74.39	0.90	78.14	0.92	70.84	0.89	79.38	0.85	219.71
Estate	0.84	84.26	0.89	76.96	0.89	69.89	**0.91**	**67.85**	0.90	70.17	0.84	128.41
Other tertiary industries	**0.89**	**266.34**	0.82	317.10	0.88	283.69	0.83	305.96	0.82	309.02	0.77	520.89

Table 5. Test results of the different covariance functions and models that input features with auxiliary variables.

Category	Exponential		Squared Exponential		Matern 3/2		Matern 5/2		Rational Quadratic		Random Forest	
	R^2	RMSE	R^2	RMSE	R^2	RMSE	R^2	RMSE	R^2	RMSE	R^2	RMSE
Industry	0.85	272.68	0.84	274.64	0.88	241.86	**0.89**	**234.52**	0.89	237.31	0.84	358.87
Construction	0.80	82.03	0.77	84.26	**0.80**	**81.55**	0.76	85.69	0.76	86.14	0.77	105.68
Retail	0.83	118.18	0.87	92.44	0.81	119.22	**0.92**	**84.96**	0.80	120.79	0.90	96.32
Transportation	**0.89**	**41.39**	0.79	62.41	0.78	63.97	0.80	58.80	0.87	45.61	0.86	50.89
Lodging and Catering	**0.84**	**24.08**	0.66	50.40	0.78	31.57	0.72	40.19	0.80	29.89	0.82	28.71
Finance	**0.86**	**99.68**	0.83	108.31	0.82	112.49	0.80	124.62	0.82	111.73	0.79	128.65
Estate	0.88	77.50	0.80	119.44	**0.94**	**60.89**	0.84	83.31	0.92	68.79	0.89	75.23
Other tertiary industries	**0.89**	**262.81**	0.83	310.76	0.86	283.77	0.83	309.82	0.85	287.04	0.80	358.32

Compared to the random forest (RF) model, the GPR model can perform better, as shown in Tables 4 and 5. The GPR model exhibits a higher R^2 value and a lower RMSE value in all industries. Therefore, the following analysis of GDP distribution and industrial agglomeration is based on the GDP data estimated from the GPR model.

4.2. Estimated GDP of Eight Industries

The estimated GDP across eight industries exhibits a generally unbalanced distribution in China (Figure 3). The eastern coastal cities have a better development than the central and western cities. Nevertheless, each industry also exhibits its own spatial distribution.

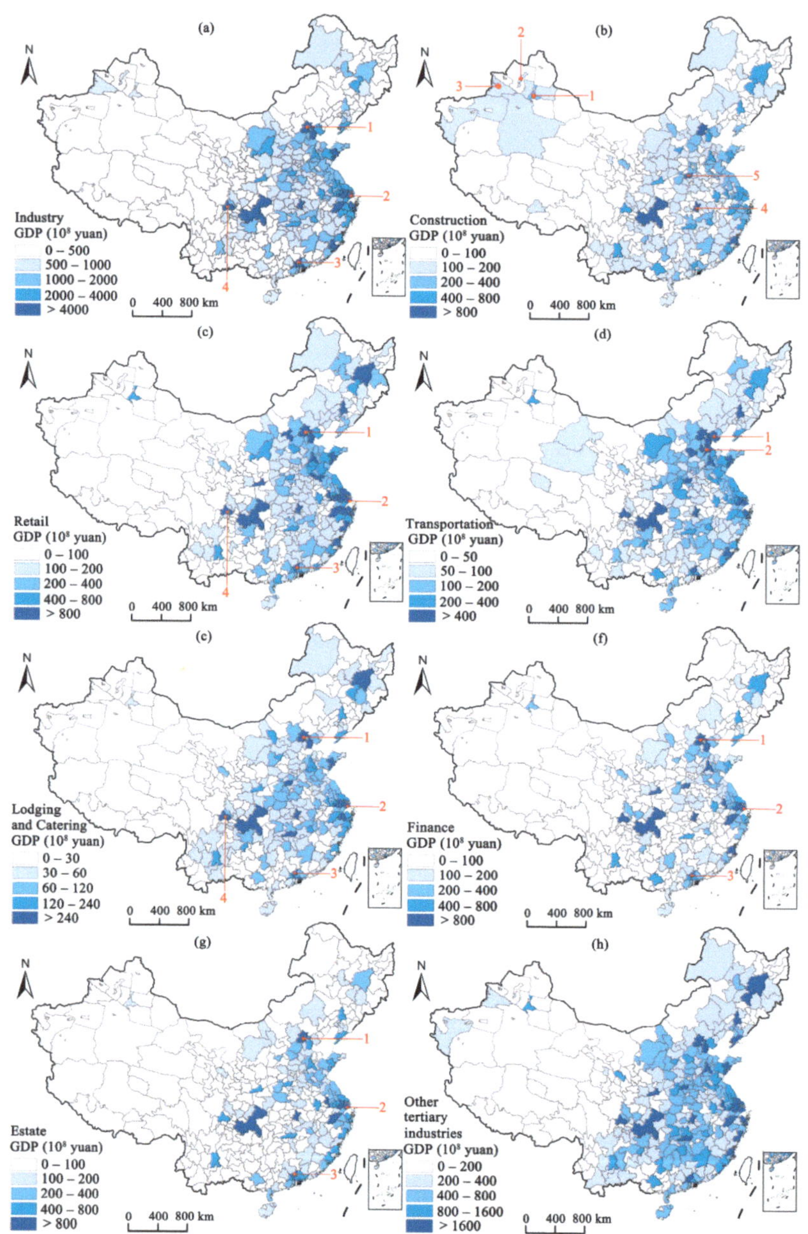

Figure 3. Estimated GDP of the (**a**) industry, (**b**) construction, (**c**) retail, (**d**) transportation, (**e**) lodging and catering, (**f**) finance, (**g**) estate, and (**h**) other tertiary industries in China. The numbers 1, 2, 3, etc. represent cities with higher GDP.

The GDP in the construction industry (Figure 3b) exhibits a pattern of concentrated development in the western region. For example, Urumqi, Karamay, and Ili in Xinjiang (red points 1, 2, 3 in Figure 3b) are not significantly lagging compared to other industries in the eastern and central regions. The better development of the construction industry in these cities can be attributed to several factors. Firstly, some of these cities are located in the economic center of Xinjiang and are rich in energy resources [45]. For example, Karamay has abundant oil resources, and its oil production reached 4.35 million tons in 2018. Different from other resource-rich provinces (e.g., Shanxi) which have been developed for many years, the infrastructure in the Xinjiang region is still not sufficient [46]. Despite occupying 1/6 of China's total land area, the public infrastructure investment in Xinjiang accounted for only 1.60% of the national total in 2018. Therefore, there is a substantial need for the construction of residential and energy infrastructure to support the ongoing development of these cities. Furthermore, the growth of the construction industry in these regions is significantly boosted by national policy initiatives. Key projects like the construction of the Sichuan–Tibet railway and the West–East Gas Pipeline Project have played crucial roles in accelerating development in these areas [47,48]. Meanwhile, it is worth noting that Wuhan and Zhengzhou (red points 4 and 5 in Figure 3b), located in the central region, serve as the capital cities of populous provinces Hubei and Henan, respectively. Compared to major metropolises like Beijing, Shanghai, and Guangzhou, Wuhan and Zhengzhou offer relatively lower housing prices, and their wage levels in their respective provinces rank first [49]. Specifically, the average wage in Wuhan is CNY 145,545, while in Zhengzhou it is CNY 50,152. This economic dynamic has made them attractive locations for employment, consequently driving up the demand for housing construction in these areas.

For transportation (Figure 3d), high GDP values are mainly concentrated in the northern coastal area, particularly within the Beijing–Tianjin–Hebei urban cluster. In this area, along with the high-GDP cities of Beijing and Tianjin, non-first-tier cities such as Tangshan and Cangzhou (red points 1, 2 in Figure 3d) also have high GDP in the transportation industry. These cities are located in the vicinity of Beijing and are influenced by its development, with well-established road transportation networks [50]. Moreover, Tangshan and Cangzhou both have their own ports, namely Tangshan Port and Huanghua Port, which ranked third and thirteenth, respectively, in cargo throughput nationwide in 2018. These ports offer advantageous water transportation conditions and are well connected to other domestic and international ports. Serving as crucial hubs for import and export trade, they offer a broad market for logistics and freight industries, driving the development of the local transportation industry.

For finance and estate (Figure 3f,g), the GDP values of most cities in the central and western regions are less than CNY 10 billion. Cities with higher GDP values in both industries are primarily concentrated in the large eastern cities. Beijing, Shanghai, and Guangzhou (red points 1, 2, 3 in Figure 3f,g) are the economic and commercial centers of the country and have all achieved a financial GDP value exceeding CNY 200 billion, hosting a great number of corporate headquarters, financial institutions, foreign investments, and a large concentration of talents that promote the development of finance in local and surrounding cities [51]. Moreover, due to the close connection between the estate and finance [52], the growth in finance has also driven the development of the estate industry.

Finally, for the industry, retail, and lodging and catering industries (Figure 3a,c,e), cities with higher GDP are primarily concentrated in major urban clusters such as the Beijing–Tianjin–Hebei urban cluster in the north, the Yangtze River Delta urban cluster in the east, the Chengdu–Chongqing urban cluster in the west, and the Pearl River Delta urban cluster in the south. These urban clusters are driven by their respective core cities, such as Beijing, Shanghai, Guangdong, and Chengdu (red points 1, 2, 3, 4 in Figure 3a,c,e). The influence of these core cities extends to other cities within their urban clusters, significantly contributing to their development in these industries.

4.3. Industrial Agglomeration Measurement of Different Industries

Based on the estimated GDP data of different industries, the degree of industrial agglomeration for each industry in China was calculated using the location quotient and The dominant industries with the highest location quotient for each city were identified, as shown in Figure 4. The degree of industrial agglomeration of eight industries was shown in Figure 5. For industry (Figure 5a), cities with a location quotient greater than 1 are mainly distributed in the central and eastern areas in China, such as Yulin in Shaanxi, Wuhu in Anhui, and Ningbo in Zhejiang (black points 1, 2, 3 in Figure 5a). The convenient transportation, well-established industrial supply chains, and abundant labor force make certain industries dominant in these cities [53].

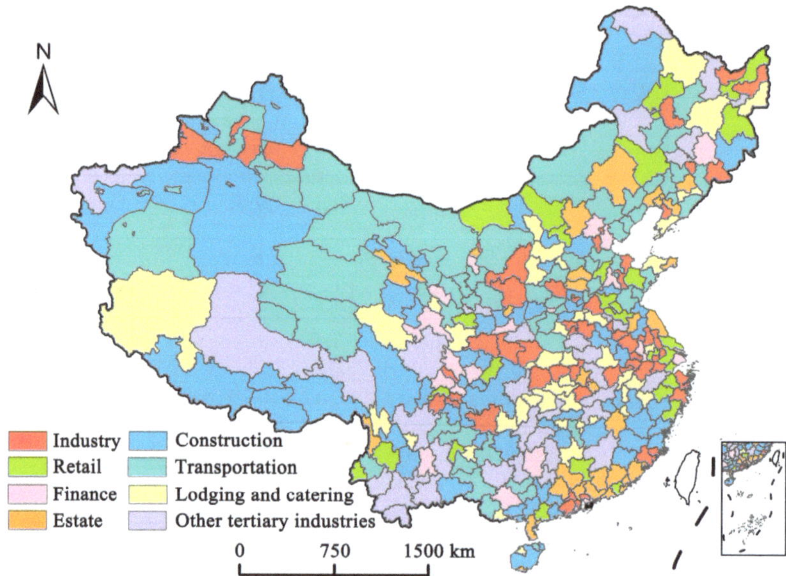

Figure 4. The industries with the highest location quotient in each China's city.

For the construction and transportation (Figure 5b,d), it can be observed that they have similar distribution patterns to the degree of industrial agglomeration. Most cities in the western region exhibit higher location quotients in both industries, while first-tier cities in the eastern region, such as Beijing and Shanghai (black points 1, 2 in Figure 5b,d), have location quotients less than 1. This discrepancy arises because many eastern cities have reached a high level of urbanization, stabilizing their construction and transportation development [54]. Eastern cities have shifted their focus to higher value-added industries, like finance, resulting in lower location quotients for construction and transportation. In contrast, western cities have relatively underdeveloped infrastructure and road networks compared to the eastern area, leaving significant room for further development. With the support of national policies, such as Sichuan–Tibet railway and West–East Gas Pipeline Project [47,48], the development of these two industries has been promoted in the western cities. For example, Figure 4 shows that in the western region, cities like Xining and Lhasa have construction as their dominant industry, while Alashan and Urumqi have transportation as their dominant industry.

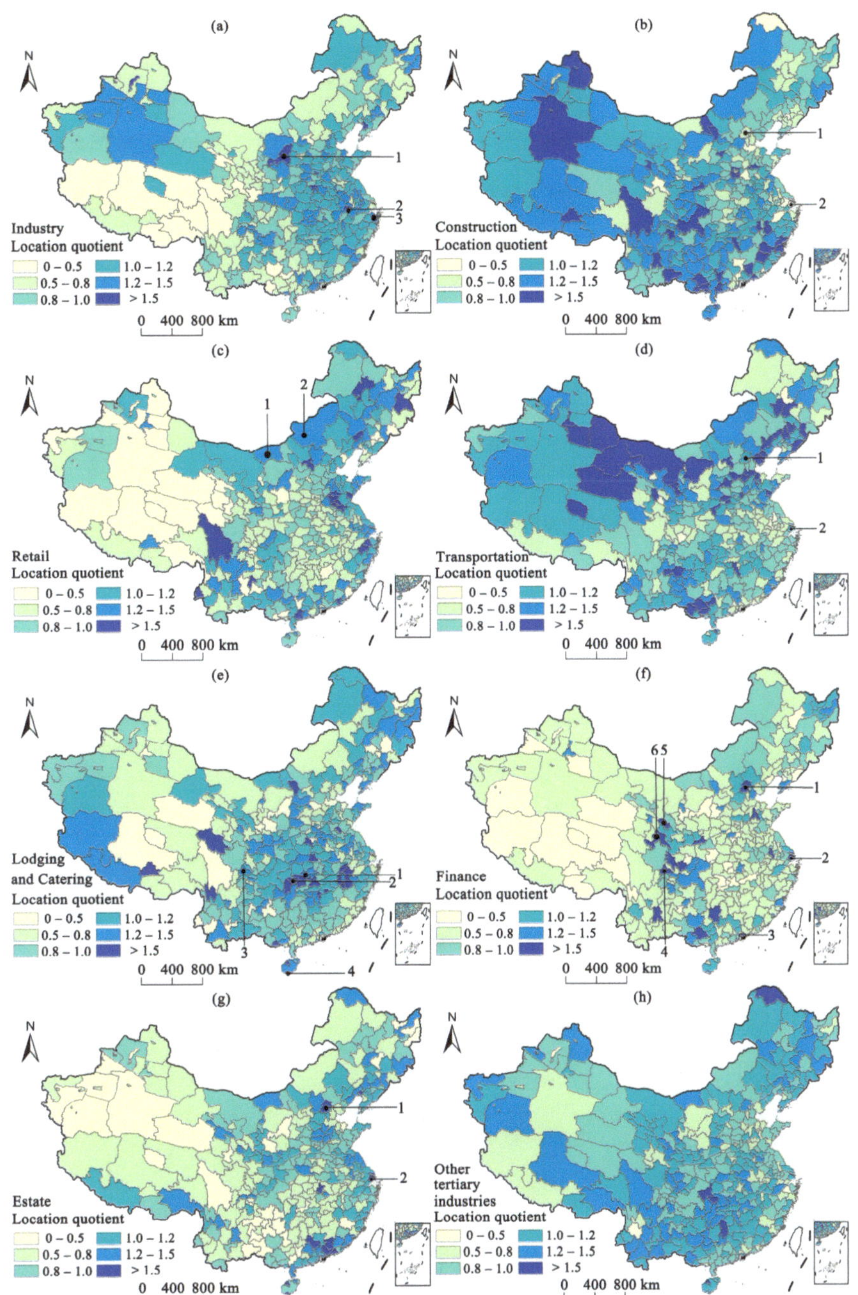

Figure 5. Location quotient of the (**a**) industry, (**b**) construction, (**c**) retail, (**d**) transportation, (**e**) lodging and catering, (**f**) finance, (**g**) estate, and (**h**) other tertiary industries in China. The numbers 1, 2, 3, etc. represent cities with higher industrial agglomeration.

The retail industry is primarily concentrated in coastal cities (Figure 5c). This concentration is due to cities' advantages in international trade and logistics, coupled with high population density and substantial consumer markets. Additionally, most cities in

Inner Mongolia, such as Bayannur and Xilingol League (black points 1, 2 in Figure 5c) exhibit a location quotient greater than 1. Furthermore, Figure 4 shows that retail is the dominant industry in Bayannur. Cities in Inner Mongolia are rich in agricultural and pastoral resources and promote the sale of these retail products nationwide by establishing e-commerce platforms, which in turn promotes the growth of the local retail industry [55]

For finance (Figure 5f), major cities like Beijing, Shanghai, Shenzhen, and Chengdu (black points 1, 2, 3, and 4 in Figure 5f), the location quotient of finance is generally greater than 1, meaning that the financial industry is the dominant industry in these cities (Figure 4). These major cities serve as economic and political centers and have well-developed financial policies that have attracted a multitude of financial institutions. This concentration has not only promoted the growth of financial GDP but also resulted in higher financial location quotients [51]. Additionally, the result indicates that several cities in the northwest, notably Lanzhou and Gannan in Gansu province (black points 5, 6 in Figure 5f), exhibit a high financial industry location entropy. Situated along the overland Silk Road, these cities are crucial for trade with Central Asia. Influenced by national Silk Road policies, they have established multiple financial pilot zones, greatly promoting the development of local finance [56]. In addition, as an important energy production area in China, the development of energy finance in western China has further promoted industrial agglomeration of finance [57].

Like the distribution of financial location quotient, the location quotients of the estate in large cities such as Beijing and Shanghai (black points 1, 2 in Figure 5g) are also greater than 1. This correlation is due to the close linkage between the estate and financial industries. Estate companies obtain financial support through loans from financial institutions, and financial institutions generate profits through these loans [58]. Therefore, cities with a high value of location quotient in the financial industry also display high values in the estate industry, reflecting the interdependence of these two industries.

Finally, looking at the lodging and catering sectors (Figure 5e), cities with high location quotients are mostly tourist cities, such as Changsha, Zhangjiajie, Chengdu, and Sanya (black points 1, 2, 3, 4 in Figure 5e). Meanwhile, lodging and catering are the dominant industries in Zhangjiajie (Figure 4). These cities which are rich in tourism resources attract many tourists which promote the development of the lodging and catering industry. Compared to the eastern regions, these tourist cities have a more vibrant nighttime economy. Lodging and catering closely related to the nighttime economy are more easily observed through NTL data [27].

Overall, we can conclude that in most cities in the central and eastern regions, industrial and retail industries are developed as locally advantageous industries. In contrast, many cities in the western region are influenced by factors such as policies prioritizing the development of the construction and transportation industries. Additionally, first-tier cities primarily focus on developing high-value industries such as finance and estate industries. Cities with abundant tourism resources tend to improve the development of local lodging and catering industry.

5. Discussion

5.1. Accuracy Assessment and Residual Analysis

The results from the comparison Tables 2 and 3 indicate that, after adding auxiliary variables, the correlation between NTL and POI features and GDP in construction and finance decreased. Additionally, in Tables 4 and 5, it is observed that the accuracy of models in these two industries also decreased after adding auxiliary variables. Analyzing the reasons for the decline in accuracy, in the case of finance, adding auxiliary variables introduced redundant information related to the financial POI features. This led to a reduction in the independent contribution of financial POI features to the interpretation of GDP, which decreased from 0.88 to 0.82. For construction, the accuracy of the model did not exhibit a significant decrease (R^2 from 0.82 to 0.80) after the addition of input features with auxiliary variables. This may be attributed to the introduction of more randomness with the addition auxiliary variables, causing additional noise in the model training process and resulting in a decline in model accuracy.

The percent error map (Figure 6) of the estimated and reference GDP values in each industry shows that most cities, especially in the eastern coastal zone, have a low error percent (0–30%). The cities with high percent errors are mainly concentrated in the west and northeast of China, which mostly have backward economic development. For example, drawing from the industries examined (Figure 6a), a high percent error is noted in northeast China, especially in the cities of Heilongjiang province, such as Jixi, Shuangyashan, and Yichun (black points 1, 2, 3 in Figure 6a), which are resource-based cities whose industrial development is highly dependent on local natural resources. With the depletion of resources and the requirements of sustainable development [59], the industry-by-industry GDP of these cities changed, resulting in their high percent errors. In addition, cities located in the northeastern and western zones suffered from population loss, which weakened consumption capacity. Thus, the GDP estimation for the tertiary industries (e.g., retail, lodging and catering, and estate) had a high percent error in these zones. Meanwhile, the GPR model had high accuracy in the eastern coastal cities with a percent error of less than 20% for each industry. Compared to the west and northeast zones, the urbanization processes and economic development of the cities in the eastern coastal zone are more balanced [60].

It is worth mentioning that some cities located in mountainous areas also have high percent errors. For example, for the construction industry (Figure 6b), some eastern coastal cities, such as Yunfu, Yangjiang, Nanping, and Longyan (black points 1, 2, 3, 4 in Figure 6b), have large percent errors because cities in mountainous areas generally fall behind cities in plains in terms of economic development [61], resulting in overestimation in these cities.

To better evaluate the accuracy of the GPR model in different industries, the percent errors were divided into three types: high accuracy (0–30%), moderate accuracy (30–50%), and inaccuracy (>50%). The detailed results of the three types are listed in Table 6. The GPR model exhibits a high capacity for estimating the GDP of the estate, finance, other tertiary industries, and retail, with a high accuracy of 80.42%, 80.00%, 75.42%, and 74.16%, respectively. Combined with Tables 4 and 5, these four industries have high R^2 in testing dataset. Moreover, the GDP estimation for the construction and lodging and catering industries exhibit high accuracy, with a GDP lower than that of the other industries (60.84% and 61.25%, respectively). Meanwhile, for the inaccuracy results, the percentages of inaccuracy are less than 20% in each industry (7.50–16.66%), which indicates the presence of a few outliers (estimated GDP away from the reference GDP). In general, the results of the decision coefficient and percent error show that the GPR model can estimate the GDP with different industries.

Table 6. Different classes of predicted accuracy for the GDP of eight industries.

Category	Percentage of Percent Error (%)		
	Inaccuracy (>50%)	Moderate Accuracy (30–50%)	High Accuracy (0–30%)
Industry	9.58	22.08	68.34
Construction	16.66	22.50	60.84
Retail	9.17	16.67	74.16
Transportation	8.33	22.50	69.17
Lodging and Catering	12.08	26.67	61.25
Finance	8.33	11.67	80.00
Estate	7.50	12.08	80.42
Other tertiary industries	8.75	15.83	75.42

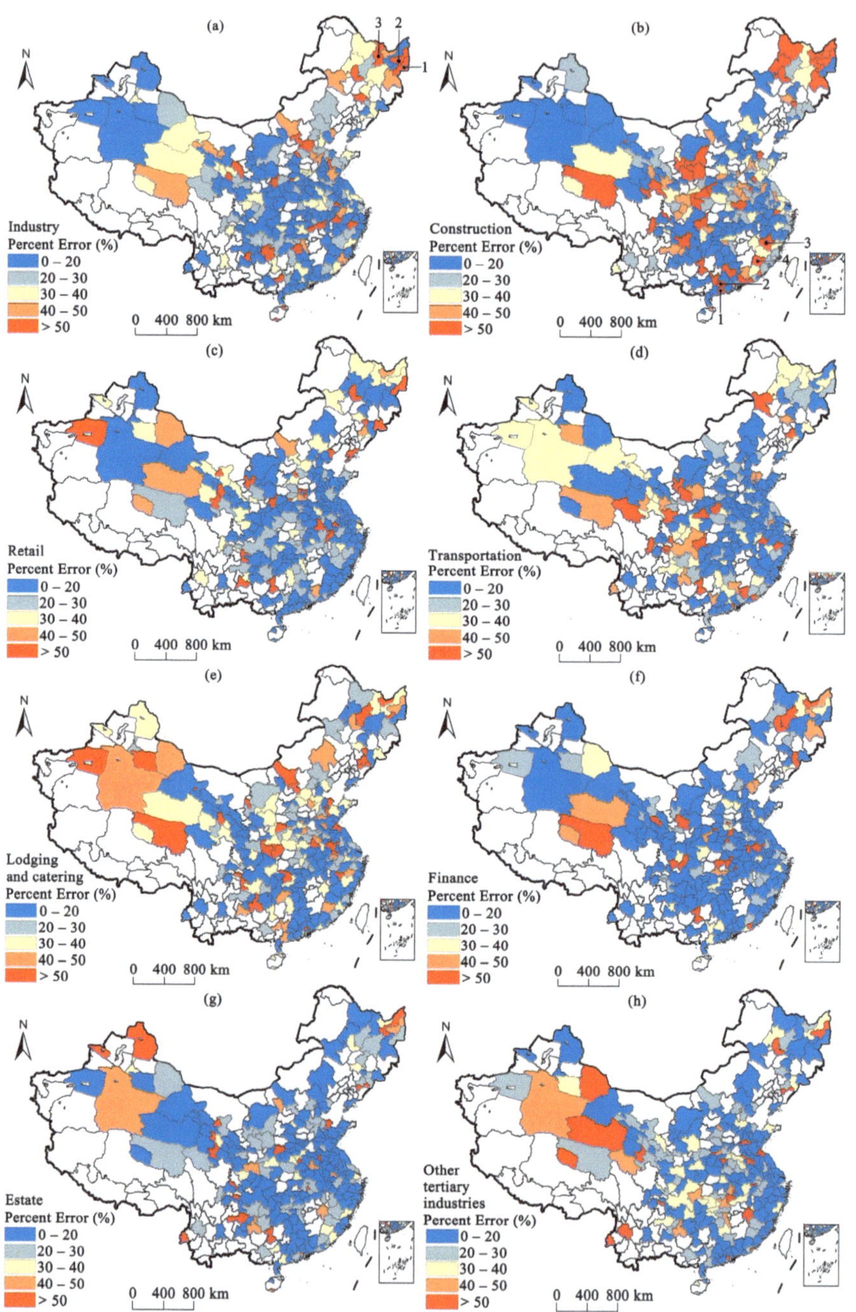

Figure 6. Percent error of the GDP with different industries for the cities with statistical GDP data in China for various industries: (**a**) industry, (**b**) construction, (**c**) retail, (**d**) transportation, (**e**) lodging and catering, (**f**) finance, (**g**) estate, and (**h**) other tertiary industries. The numbers 1, 2, 3, etc. represent cities with higher percent error.

5.2. Analysis of the Relationship between GDP and Industrial Agglomeration

To analyze the relationship between GDP and industrial agglomeration in different industries, we plotted a scatter plot (Figure 7) based on the GDP and location quotient data for each city. To mitigate the influence of outliers in the GDP data, we applied a logarithmic transformation to the GDP values. Analysis of Figure 7 reveals three distinct distribution trends. Firstly, we can observe that the industries of industry, retail, finance, and estate (Figure 7a,c,f,g) had a monotonic increasing relationship between the location quotient and GDP from west to east. Secondly, for the industries of construction and transportation (Figure 7b,d), their location quotient had no significant relationship with the GDP. Finally, it is worth noting that the scatter distribution pattern in the lodging and catering industry was quite distinctive, with cities exhibiting similar levels of location quotient but significant differences in GDP from west to east.

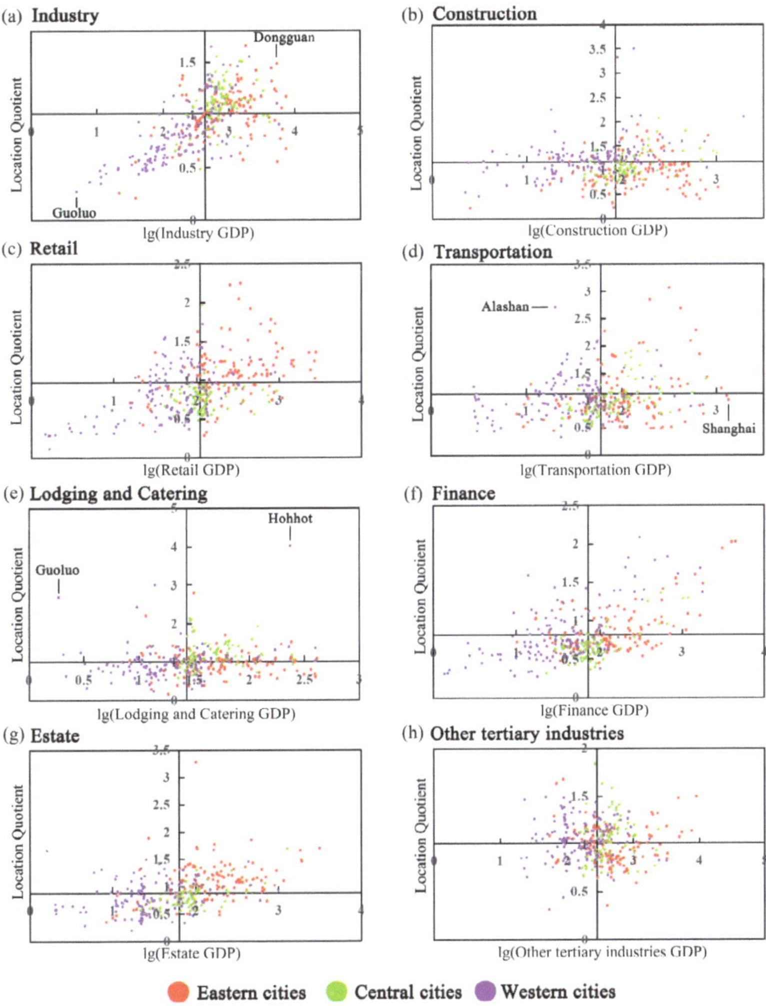

Figure 7. Scatter plots for location quotient and GDP with different industries: (**a**) industry, (**b**) construction, (**c**) retail, (**d**) transportation, (**e**) lodging and catering, (**f**) finance, (**g**) estate, and (**h**) other tertiary industries in China.

For industry (Figure 7a), the location quotient and GDP of the industry exhibit a monotonically increasing trend from west to the east in China, indicating that industrial development is progressing well in the eastern regions. The synchronous growth of GDP and location quotient suggests that industry remains a dominant sector driving the city's economic development. The main reason for this phenomenon is that the eastern regions have location advantages over the western regions. Taking Dongguan in Guangdong and Guoluo in Qinghai as examples, Dongguan not only has well-developed land transportation but also benefits from its high-quality port, which facilitates the transportation of goods. The availability of a large labor force is also an essential factor promoting industrial development. Moreover, as the pillar industry of Dongguan's economy, industry has consistently received support from the local economy and policies, resulting in a high location quotient and GDP for Dongguan [62]. In contrast, Guoluo in the western region has harsh natural conditions and a sparse population, making it unsuitable for industrial development [63]. Therefore, it has lower location quotient and GDP.

From the distribution of GDP and location quotient in the transportation in Figure 7b, it can be observed that the location quotient has no significant relationship with the GDP from west to east. However, in some specific cities in the western area, such as Alashan in Inner Mongolia, it has higher location quotients, whereas first-tier cities in the eastern region, such as Shanghai, have lower location quotients. This indicates that the development of transportation in the eastern region is gradually slowing down and the industrial focus is shifting towards the western region, resulting in higher location quotients in the western region. However, due to the overall less developed nature and lower population density of most cities in the western region, the GDP in transportation is low [64].

Regarding the lodging and catering industry data shown in Figure 7c, we can see that most cities have location quotients within the same range. This indicates that most cities do not consider lodging and catering as their primary advantageous industry for development. Cities with higher location quotients in the lodging and catering industry are mainly those with well-developed tourism resources, such as Guoluo in Qinghai and Hohhot in Inner Mongolia. Both have high location quotients in the lodging and catering industry. However, there is a substantial GDP gap between the two cities. This is because, compared to Hohhot, the market size, consumption level, and population in Guoluo are much smaller, limiting the development of GDP in the lodging and catering industry.

5.3. Policy Implications

To address the issue of imbalanced development between GDP and location quotient, we propose the following policy.

Firstly, for certain industries in cities with low location quotient and low GDP, we can enhance the infrastructure of regional road transportation networks to improve the transportation convenience and logistical efficiency of the city, thereby attracting more investments and business to settle in. In addition, the targeted development of industries based on local natural resources can be pursued. For example, in the case of Guoluo, which is surrounded by numerous rivers, the construction of hydropower stations can be considered to drive local economic development.

Secondly, to address the issue of high industry location quotient but low GDP in certain industries, GDP can be improved by strengthening communication and cooperation between the eastern and western regions, as well as among different industries. Taking the transportation industry as an example, to improve the GDP of transportation in western cities, one approach is to increase the construction of large-scale logistics centers at key transportation nodes between the eastern and western regions. This initiative can improve logistical efficiency and reduce transportation costs between the two regions, and thereby stimulate transportation growth in the west. Concurrently, encouraging the collaborative development of transportation with other industries such as industry, lodging and catering can be undertake to create industrial linkages and increase the overall benefits and economic value of transportation.

Finally, for the lodging and catering industry, the difference in location quotient among most cities is not significant, but there is a notable disparity in GDP. In such cases, on one hand, increasing the number of local distinctive tourist attractions can enhance the city's reputation and attract more tourists and investments. On the other hand, investing in supporting facilities for the lodging and catering industry and improving service quality can also attract more tourists, thereby improving GDP.

5.4. Limitations

There are still some limitations in this study. When using the POI data as the input features, our method ignores the large GDP gaps of the POI points in the same category. This can be solved by combining big data with NTL data in the future. In addition, for the GPR model, the estimated range of GDP is restricted to those areas covered by the training datasets. As such, the GPR model cannot accurately estimate GDP beyond the training data range. This problem will be addressed in the future by improving the GPR model. Meanwhile, due to the limitations of the VIIRS images in detecting low-intensity nighttime light and the impact of the satellite's overpass time at 1:30 PM, the VIIRS data are unable to detect faint and brief nighttime light emitted by human activities. Therefore, the capability of VIIRS NTL images to detect human activities in sparsely populated rural areas is limited and more suitable for extracting illumination from industrial buildings and residential areas in urban regions [65]. Consequently, this study selected the GDP of the eight major industries more relevant to urban areas for estimation, and NDVI data were used to calculate the urban green space to reduce the impact of light emitted by urban landscape lighting. In future work, the incorporation of multi-source data, such as land-use data, will be explored to estimate industries related to agriculture. Additionally, using the VIIRS imagery to extract urban NTL features could result in omissions in detecting some LED light, which leads to inaccuracies in the NTL features [66]. Therefore, we hope to address this in future work by using NTL images with a wider wavelength range.

6. Conclusions

This study utilized NTL, POI, and auxiliary data to estimate the GDP of different industries and calculated the degree of industrial agglomeration for each industry. Additionally, the study examined the spatial distribution patterns of the degree of industrial agglomeration with each industry. It was found that the central and eastern regions showed a developmental focus on industry and retail as local strengths. Conversely, many western cities emphasized construction and transportation. First-tier cities prioritized high-value industries like finance and estate, while cities rich in tourism resources aimed to enhance their lodging and catering industry. Based on the estimated results, three typical relationships between GDP and the degree of industrial agglomeration were summarized and we also analyzed the reasons for these relationships and offered policy implications. For industries which have a monotonic increasing relationship between industrial agglomeration and GDP, it is suggested to drive local economic growth through advantage industries. For industries with imbalances between GDP and industrial agglomeration, we recommend strengthening the connection between eastern and western area to improve the development of western cities. Finally, for industries such as lodging and catering, where there is little difference in the degree of industrial agglomeration but significant differences in GDP, government can develop local tourism resources and infrastructure to improve the economic level. Furthermore, the estimated GDP and industrial agglomeration of different industries can provide a scientific reference for relevant national departments and the collection of statistical data in cities.

Author Contributions: Conceptualization Z.C.; methodology, W.X. and Z.C.; formal analysis, W.X.; writing—original draft preparation, W.X. and Z.C.; writing—review and editing, Z.C., W.X. and Z.Z.; funding acquisition, Z.C. and Z.Z. All authors have read and agreed to the published version of the manuscript.

Funding: This research was funded by the National Natural Science Foundation of China (Grant No. 41801343, 42201500), Science and Technology Innovation Team of Fujian University.

Data Availability Statement: The data presented in this study are available on request from the corresponding author. The data are not publicly available due to privacy.

Acknowledgments: We appreciate the critical and constructive comments and suggestions from the reviewers that helped to improve the quality of this manuscript.

Conflicts of Interest: The authors declare no conflicts of interest.

References

1. Liu, X.; Zhang, X. Industrial agglomeration, technological innovation and carbon productivity: Evidence from China. *Resour. Conserv. Recycl.* **2021**, *166*, 105330. [CrossRef]
2. Guo, S.; Ma, H. Does industrial agglomeration promote high-quality development of the Yellow River Basin in China? Empirical test from the moderating effect of environmental regulation. *Growth Chang.* **2021**, *52*, 2040–2070. [CrossRef]
3. Wu, J.; Tu, Y.; Chen, Z.; Yu, B. Analyzing the Spatially Heterogeneous Relationships between Nighttime Light Intensity and Human Activities across Chongqing, China. *Remote Sens.* **2022**, *14*, 5695. [CrossRef]
4. Wang, L.; Fan, H.; Wang, Y. Improving population mapping using Luojia 1-01 nighttime light image and location-based social media data. *Sci. Total Environ.* **2020**, *730*, 139148. [CrossRef]
5. Song, Y.; Li, X.; Tao, G.; Liu, J. Exploring the Characteristics and Drivers of Expansion in the Shandong Peninsula Urban Agglomeration Based on Nighttime Light Data. *IEEE J. Sel. Top. Appl. Earth Obs. Remote Sens.* **2023**, *16*, 8535–8549. [CrossRef]
6. Shi, K.; Chang, Z.; Chen, Z.; Wu, J.; Yu, B. Identifying and evaluating poverty using multisource remote sensing and point of interest (POI) data: A case study of Chongqing, China. *J. Clean. Prod.* **2020**, *255*, 120245. [CrossRef]
7. Elvidge, C.D.; Baugh, K.E.; Kihn, E.A.; Kroehl, H.W.; Davis, E.R.; Davis, C.W. Relation between satellite observed visible-near infrared emissions, population, economic activity and electric power consumption. *Int. J. Remote Sens.* **1997**, *18*, 1373–1379. [CrossRef]
8. Chen, X.; Nordhaus, W.D. Using luminosity data as a proxy for economic statistics. *Proc. Natl. Acad. Sci. USA* **2011**, *108*, 8589–8594. [CrossRef] [PubMed]
9. Dai, Z.; Hu, Y.; Zhao, G. The Suitability of Different Nighttime Light Data for GDP Estimation at Different Spatial Scales and Regional Levels. *Sustainability* **2017**, *9*, 305. [CrossRef]
10. Doll, C.N.H.; Muller, J.-P.; Morley, J.G. Mapping regional economic activity from night-time light satellite imagery. *Ecol. Econ.* **2006**, *57*, 75–92. [CrossRef]
11. Forbes, D.J. Multi-scale analysis of the relationship between economic statistics and DMSP-OLS night light images. *GIScience Remote Sens.* **2013**, *50*, 483–499. [CrossRef]
12. Chen, Q. Improved GDP spatialization approach by combining land-use data and night-time light data: A case study in China's continental coastal area. *Int. J. Remote Sens.* **2016**, *37*, 4610–4622. [CrossRef]
13. Chen, Q.; Ye, T.; Zhao, N.; Ding, M.; Ouyang, Z.; Jia, P.; Yue, W.; Yang, X. Mapping China's regional economic activity by integrating points-of-interest and remote sensing data with random forest. *Environ. Plan. B Urban Anal. City Sci.* **2020**, *48*, 1876–1894. [CrossRef]
14. Shi, K.; Wu, Y.; Li, D.; Li, X. Population, GDP, and Carbon Emissions as Revealed by SNPP-VIIRS Nighttime Light Data in China With Different Scales. *IEEE Geosci. Remote Sens. Lett.* **2022**, *19*, 1–5. [CrossRef]
15. Liang, H.; Guo, Z.; Wu, J.; Chen, Z. GDP spatialization in Ningbo City based on NPP/VIIRS night-time light and auxiliary data using random forest regression. *Adv. Space Res.* **2020**, *65*, 481–493. [CrossRef]
16. Sun, J.; Di, L.; Sun, Z.; Wang, J.; Wu, Y. Estimation of GDP Using Deep Learning With NPP-VIIRS Imagery and Land Cover Data at the County Level in CONUS. *IEEE J. Sel. Top. Appl. Earth Obs. Remote Sens.* **2020**, *13*, 1400–1415. [CrossRef]
17. Shi, K. Evaluating the Ability of NPP-VIIRS Nighttime Light Data to Estimate the Gross Domestic Product and the Electric Power Consumption of China at Multiple Scales: A Comparison with DMSP-OLS Data. *Remote Sens.* **2014**, *6*, 1705–1724. [CrossRef]
18. Zhao, M. GDP Spatialization and Economic Differences in South China Based on NPP-VIIRS Nighttime Light Imagery. *Remote Sens.* **2017**, *9*, 673. [CrossRef]
19. Gašpar, A.; Seljan, S.; Kučiš, V. Measuring Terminology Consistency in Translated Corpora: Implementation of the Herfindahl-Hirshman Index. *Information* **2022**, *13*, 43. [CrossRef]
20. Xu, C.; Chen, G.; Huang, Q.; Su, M.; Rong, Q.; Yue, W.; Haase, D. Can improving the spatial equity of urban green space mitigate the effect of urban heat islands? An empirical study. *Sci. Total Environ.* **2022**, *841*, 156687. [CrossRef]
21. Arruda, V.C.L.; Marques, A.S.; Moreira, J.L.B.; Lago, T.G.S. Location and specialization indicators of animal bioenergetic potential in Paraíba (Brazil). *Energy Sustain. Dev.* **2023**, *76*, 101304. [CrossRef]
22. Dong, F.; Wang, Y.; Zheng, L.; Li, J.; Xie, S. Can industrial agglomeration promote pollution agglomeration? Evidence from China. *J. Clean. Prod.* **2020**, *246*, 118960. [CrossRef]
23. Billings, S.B.; Johnson, E.B. The location quotient as an estimator of industrial concentration. *Reg. Sci. Urban Econ.* **2012**, *42*, 642–647. [CrossRef]

24. Liu, Y.; Liu, X.; Gao, S.; Gong, L.; Kang, C.; Zhi, Y.; Chi, G.; Shi, L. Social Sensing: A New Approach to Understanding Our Socioeconomic Environments. *Ann. Assoc. Am. Geogr.* **2015**, *105*, 512–530. [CrossRef]

25. Wang, Y.; Hu, D.; Yu, C.; Di, Y.; Wang, S.; Liu, M. Appraising regional anthropogenic heat flux using high spatial resolution NTL and POI data: A case study in the Beijing-Tianjin-Hebei region, China. *Environ. Pollut.* **2022**, *292*, 118359. [CrossRef] [PubMed]

26. Andrade, R.; Alves, A.; Bento, C. POI Mining for Land Use Classification: A Case Study. *ISPRS Int. J. Geo-Inf.* **2020**, *9*, 493. [CrossRef]

27. Cui, Y.; Shi, K.; Jiang, L.; Qiu, L.; Wu, S. Identifying and Evaluating the Nighttime Economy in China Using Multisource Data. *IEEE Geosci. Remote Sens. Lett.* **2021**, *18*, 1906–1910. [CrossRef]

28. Long, Y.; Shen, Y.; Jin, X. Mapping Block-Level Urban Areas for All Chinese Cities. *Ann. Am. Assoc. Geogr.* **2015**, *106*, 96–113. [CrossRef]

29. McKenzie, G.; Janowicz, K.; Gao, S.; Yang, J.-A.; Hu, Y. POI Pulse: A Multi-granular, Semantic Signature–Based Information Observatory for the Interactive Visualization of Big Geosocial Data. *Cartogr. Int. J. Geogr. Inf. Geovis.* **2015**, *50*, 71–85. [CrossRef]

30. Wu, R.; Wang, J.; Zhang, D.; Wang, S. Identifying different types of urban land use dynamics using Point-of-interest (POI) and Random Forest algorithm: The case of Huizhou, China. *Cities* **2021**, *114*, 103202. [CrossRef]

31. Zhao, N.; Ghosh, T.; Samson, E.L. Mapping spatio-temporal changes of Chinese electric power consumption using night-time imagery. *Int. J. Remote Sens.* **2012**, *33*, 6304–6320. [CrossRef]

32. Ye, T. Improved population mapping for China using remotely sensed and points-of-interest data within a random forests model. *Sci. Total Environ.* **2019**, *658*, 936–946. [CrossRef] [PubMed]

33. Jain, A.; Nghiem, T.; Morari, M.; Mangharam, R. Learning and Control Using Gaussian Processes. In Proceedings of the 2018 ACM/IEEE 9th International Conference on Cyber-Physical Systems (ICCPS), Porto, Portugal, 11–13 April 2018; pp. 140–149.

34. *GB/T 4754—2017*; Industrial Classification for National Economic Activities. Standards Press of China: Beijing, China, 2017.

35. de Marsily, G.; Delay, F.; Gonçalvès, J.; Renard, P.; Teles, V.; Violette, S. Dealing with spatial heterogeneity. *Hydrogeol. J.* **2005**, *13*, 161–183. [CrossRef]

36. Liu, H.; Huang, B.; Gao, S.; Wang, J.; Yang, C.; Li, R. Impacts of the evolving urban development on intra-urban surface thermal environment: Evidence from 323 Chinese cities. *Sci. Total Environ.* **2021**, *771*, 144810. [CrossRef] [PubMed]

37. Dijk, M.P.v. A Different Development Model in China's Western and Eastern Provinces? *Mod. Econ.* **2011**, *02*, 757–768. [CrossRef]

38. Bennett, M.M.; Smith, L.C. Advances in using multitemporal night-time lights satellite imagery to detect, estimate, and monitor socioeconomic dynamics. *Remote Sens. Environ.* **2017**, *192*, 176–197. [CrossRef]

39. Peng, J.; Zhao, S.; Liu, Y.; Tian, L. Identifying the urban-rural fringe using wavelet transform and kernel density estimation: A case study in Beijing City, China. *Environ. Model. Softw.* **2016**, *83*, 286–302. [CrossRef]

40. Wang, N.; Du, Y.; Liang, F.; Wang, H.; Yi, J. The spatiotemporal response of China's vegetation greenness to human socio-economic activities. *J. Environ. Manag.* **2022**, *305*, 114304. [CrossRef]

41. Yin, G.; Lin, Z.; Jiang, X.; Qiu, M.; Sun, J. How do the industrial land use intensity and dominant industries guide the urban land use? Evidences from 19 industrial land categories in ten cities of China. *Sustain. Cities Soc.* **2020**, *53*, 101978. [CrossRef]

42. Alebele, Y.; Zhang, X.; Wang, W.; Yang, G.; Yao, X.; Zheng, H.; Zhu, Y.; Cao, W.; Cheng, T. Estimation of Canopy Biomass Components in Paddy Rice from Combined Optical and SAR Data Using Multi-Target Gaussian Regressor Stacking. *Remote Sens.* **2020**, *12*, 2564. [CrossRef]

43. Breiman, L. Random Forests. *Mach. Learn.* **2001**, *45*, 5–32. [CrossRef]

44. Dhibi, K.; Fezai, R.; Mansouri, M.; Trabelsi, M.; Kouadri, A.; Bouzara, K.; Nounou, H.; Nounou, M. Reduced Kernel Random Forest Technique for Fault Detection and Classification in Grid-Tied PV Systems. *IEEE J. Photovolt.* **2020**, *10*, 1864–1871. [CrossRef]

45. Chen, F.; Zhang, H.; Yang, Y.; Li, X.; He, C. Development of Different Energy Storage Systems in the Xinjiang Uygur Autonomous Region: Problems and Solutions. In Proceedings of the 2022 IEEE/IAS Industrial and Commercial Power System Asia (I&CPS Asia), Shanghai, China, 8–11 July 2022; pp. 305–310.

46. Wang, B. Social Infrastructure Development Aid for Xinjiang and Lessons for CPEC: A Case Study of Shanghai-Kashgar Paired Assistance Program. In *The Political Economy of the China-Pakistan Economic Corridor*; Springer Nature: Singapore, 2023; pp. 219–251.

47. Zhang, J.; Zheng, H.; He, W.; Huang, W. West-east gas pipeline project. *Front. Eng. Manag.* **2020**, *7*, 163–167. [CrossRef]

48. Xu, J.; Guo, J.; Sun, Y.; Tang, Q.; Zhang, J. Why China must build Sichuan-Tibet railway: From the perspective of regional comprehensive transportation network optimization. *J. Intell. Fuzzy Syst.* **2021**, *40*, 9741–9763. [CrossRef]

49. Lan, F.; Jiao, C.; Deng, G.; Da, H. Urban agglomeration, housing price, and space–time spillover effect—Empirical evidences based on data from hundreds of cities in China. *Manag. Decis. Econ.* **2021**, *42*, 898–919. [CrossRef]

50. Zhao, J.; Rong, W.; Liu, D. Urban Agglomeration High-Speed Railway Backbone Network Planning: A Case Study of Beijing-Tianjin-Hebei Region, China. *Sustainability* **2023**, *15*, 6450. [CrossRef]

51. Wang, X. The changing geographies of financial centres in China: The case of commercial banking. *Growth Chang.* **2018**, *50*, 164–183. [CrossRef]

52. Cerutti, E.; Dagher, J.; Dell'Ariccia, G. Housing finance and real-estate booms: A cross-country perspective. *J. Hous. Econ.* **2017**, *38*, 1–13. [CrossRef]

53. Kang, J.; Yang, C.; Ning, Y. Analysis of Regional Division of Labor in Value Chain Patterns and Driving Factors in the Yangtze River Delta Region Using the Electronic Information Manufacturing Industry as an Example. *Sustainability* **2023**, *15*, 4393. [CrossRef]

54. Lv, Q.; Liu, H.; Yang, D.; Liu, H. Effects of urbanization on freight transport carbon emissions in China: Common characteristics and regional disparity. *J. Clean. Prod.* **2019**, *211*, 481–489. [CrossRef]

55. Li, J.; Yan, X.; Li, Y.; Dong, X. Optimizing the Agricultural Supply Chain through E-Commerce: A Case Study of Tudouec in Inner Mongolia, China. *Int. J. Environ. Res. Public. Health* **2023**, *20*, 3775. [CrossRef] [PubMed]

56. Zhang, H. The Research on the Efficiency of Financial Support for the Development of Real Economy-A Case Study Based on the Data of the Silk Road Belt. *CONVERTER* **2021**, *2021*, 258–272.

57. Orazgaliyev, S. The Overland Silk Road: China's Energy Cooperation with Central Asia in the Context of Industry Competition. *China Int. J.* **2019**, *17*, 62–75. [CrossRef]

58. Xu, X.E.; Chen, T. The effect of monetary policy on real estate price growth in China. *Pac.—Basin Financ. J.* **2012**, *20*, 62–77. [CrossRef]

59. Hou, G.; Zou, Z.; Zhang, T.; Meng, Y. Analysis of the Effect of Industrial Transformation of Resource-Based Cities in Northeast China. *Economies* **2019**, *7*, 40. [CrossRef]

60. Fan, J.; Ma, T.; Zhou, C.; Zhou, Y.; Xu, T. Comparative Estimation of Urban Development in China's Cities Using Socioeconomic and DMSP/OLS Night Light Data. *Remote Sens.* **2014**, *6*, 7840–7856. [CrossRef]

61. Fang, Y.; Ying, B. Spatial distribution of mountainous regions and classifications of economic development in China. *J. Mt. Sci.* **2016**, *13*, 1120–1138. [CrossRef]

62. Li, X.; Hui, E.C.-m.; Lang, W.; Zheng, S.; Qin, X. Transition from factor-driven to innovation-driven urbanization in China: A study of manufacturing industry automation in Dongguan City. *China Econ. Rev.* **2020**, *59*, 101382. [CrossRef]

63. Wu, H.; Fan, W.; Lu, J. Researching on the Sustainability of Transportation Industry Based on a Coupled Emergy and System Dynamics Model: A Case Study of Qinghai. *Sustainability* **2021**, *13*, 6804. [CrossRef]

64. Chen, D.; Song, D.; Yang, Z. A review of the literature on the belt and road initiative with factors influencing the transport and logistics. *Marit. Policy Manag.* **2022**, *49*, 540–557. [CrossRef]

65. Andersson, M.; Hall, O.; Archila, M.F. How Data-Poor Countries Remain Data Poor: Underestimation of Human Settlements in Burkina Faso as Observed from Nighttime Light Data. *ISPRS Int. J. Geo-Inf.* **2019**, *8*, 498. [CrossRef]

66. Kinzey, B.R.; Perrin, T.E.; Miller, N.J.; Kocifaj, M.; Aube, M.; Lamphar, H.A. *An Investigation of LED Street Lighting's Impact on Sky Glow*; Pacific Northwest National Lab (PNNL): Richland, WA, USA, 2017.

 *remote sensing*

Article

The Impact of the Expansion and Contraction of China Cities on Carbon Emissions, 2002–2021, Evidence from Integrated Nighttime Light Data and City Attributes

Jiaqi Qian [1,2], Yanning Guan [1,*], Tao Yang [3], Aoming Ruan [2,4], Wutao Yao [1], Rui Deng [5], Zhishou Wei [1,2], Chunyan Zhang [1] and Shan Guo [1]

[1] Aerospace Information Research Institute, Chinese Academy of Sciences, Beijing 100101, China; qianjiaqi22@mails.ucas.ac.cn (J.Q.); yaowt@aircas.ac.cn (W.Y.); weizhishou21@mails.ucas.ac.cn (Z.W.); zhangcy@aircas.ac.cn (C.Z.); guoshan@irsa.ac.cn (S.G.)

[2] University of Chinese Academy of Sciences, Beijing 100049, China; ruanaoming@scbg.ac.cn

[3] The School of Architecture, Tsinghua University, Beijing 100084, China; yangtao128@tsinghua.edu.cn

[4] South China Botanical Garden, Guangzhou 510520, China

[5] School of Geographical and Earth Sciences, University of Glasgow, Glasgow G12 8QQ, UK; 2911670d@student.gla.ac.uk

* Correspondence: guanyn@radi.ac.cn

Abstract: Exploring the impact of urbanization on carbon emissions is crucial for formulating effective emission reduction policies. Using nighttime light data and attribute data from 68 Chinese cities (2002–2021), this paper develops an urban development evaluation system with the entropy method. The Lasso method is employed to select key factors affecting carbon emissions, and hierarchical regression models are utilized to analyze these factors across different city types. The results show the following: (1) The extraction of built-up areas using integrated nighttime light data yields an overall accuracy ranging from 70.90% to 98.87%, reflecting high precision. (2) Expanding cities have predominated over the past two decades, indicating a continued upward trend in urbanization in China. (3) Urban development is influenced by internal characteristics and geographic location: contracting cities are mainly inland heavy industrial centers, while expanding cities are located in economically advanced coastal regions. Additionally, it is also impacted by the growth of surrounding cities, exemplified by the imbalance between central cities and their peripheries within metropolitan areas. (4) The expansion of built-up areas is a significant factor affecting carbon emissions across all city types. For expanding cities, managing population growth and promoting tertiary sector development are recommended, while contracting cities should focus on judicious economic planning and virescence area protection.

Keywords: urbanization; contracting and expanding city; carbon emissions; nighttime light; low-carbon city; city attributes; built-up area

Citation: Qian, J.; Guan, Y.; Yang, T.; Ruan, A.; Yao, W.; Deng, R.; Wei, Z.; Zhang, C.; Guo, S. The Impact of the Expansion and Contraction of China Cities on Carbon Emissions, 2002–2021, Evidence from Integrated Nighttime Light Data and City Attributes. *Remote Sens.* **2024**, *16*, 3274. https://doi.org/10.3390/rs16173274

Academic Editor: Michael Obland

Received: 31 July 2024
Revised: 30 August 2024
Accepted: 31 August 2024
Published: 3 September 2024

1. Introduction

Urbanization refers to the process of transitioning rural populations to urban lifestyles. It is characterized by increases in urban population, the expansion of built-up urban areas, the creation of landscapes and urban environments, and changes in social structures and lifestyles [1]. Nearly 4% of the global land has been urbanized, with over 50% of the population living in urban areas [2]. The report from the C40 Cities Climate Leadership Group states that urban areas generate more than four-fifths of global greenhouse gas emissions. The total emissions from urban areas are increasing at an annual rate of 1.8% [3]. With the rapid advancement of urbanization, substantial energy consumption has led to a rise in carbon emissions. Numerous studies have confirmed that the impact of urbanization on carbon emissions is both profound and enduring [4,5].

However, the mechanisms of the impact of urbanization on carbon emissions remain unclear at the theoretical level. Three predominant conclusions have been drawn regarding the relationship between carbon emissions and urbanization: (1) The relationship between carbon emissions and urbanization is positively correlated [6–8], and this correlation may be either unidirectional or bidirectional [9,10]. (2) The relationship between carbon emissions and urbanization is negatively correlated [11]. (3) The relationship between carbon emissions and urbanization exhibits an inverted U-shaped curve [12], and this relationship changes with the progress of urbanization [13].

Existing research indicates that the impact of urbanization on carbon emissions exhibits spatial heterogeneity depending on the region. In the United Kingdom and France, there is a clear negative correlation between the urbanization rate and carbon emissions, with increasing urbanization associated with decreasing carbon emissions. In contrast, in Japan and the United States, total carbon emissions initially increase with rising urbanization rates until reaching a turning point, after which they begin to decline as urbanization continues [14,15]. Quantitative studies suggest that in developing countries, a 1% increase in the urbanization rate is associated with approximately a 0.9% increase in carbon emissions [16]. Carbon emissions are observed to peak and decline when the urbanization rate reaches about 75% [17].

The urbanization of China began late, and its urbanization rate remains lower. There is still a significant positive correlation between carbon emissions and urbanization rate [14]. Additionally, China is one of the world's largest carbon emitters. China has announced a reduction in carbon intensity by over 60% from 2005 levels by 2030. Carbon removals and emissions must be balanced by 2060 [18–21]. Studying the relationship between urban development and carbon emissions in Chinese cities can offer useful insights for global carbon reduction efforts and support goals for reaching the carbon peak and carbon neutrality. Moreover, with the large number of cities in China and the variety of urbanization types, examining this relationship can help manage differences in urban development and provide a basis for guiding cities towards high-quality, low-carbon, and sustainable growth.

Urbanization is related to population migration, changes in land use, and changes in production methods; therefore, its impact on carbon emissions is multifaceted [22,23]. Existing research suggests that changes in urban population structure and increased population density are direct factors affecting carbon emissions [23,24]. The rise in consumption levels in urban areas, coupled with an increased demand for energy, contributes to an increase in carbon emissions [25,26]. In the latter phases of industrialization, the refinement of industrial structures and technological advancements serve as instrumental factors in managing the escalation of carbon emissions [27,28]. The transformation of the economic structure has enabled the service industry to drive carbon emissions more than other industries [29]. Additionally, the expansion of urban construction land can influence carbon emissions by altering aspects such as urban transportation and green spaces [30–32]. In addition, some research indicates that the impact of urbanization on carbon emissions results from the interplay of multiple factors. For instance, economic development can improve education levels, which in turn affects public awareness and perceptions of carbon emissions [33,34]. Therefore, in order to study the relationship between carbon emissions and urbanization, it is necessary to analyze it from multiple perspectives.

Conventional data are subject to long update cycles and other limitations, making it difficult to examine the spatiotemporal characteristics of urbanization over a long time series. Compared with traditional methods, remote sensing technology has timeliness and wide spatial coverage. As a remote sensing data source that captures and records faint light radiation from the Earth's surface at night in real-time, nighttime light (NTL) data can reflect human activities accurately to some extent [35]. NTL observations have proven to be effective in tracking light sources generated by human activities. This capability aids in assessing the spatiotemporal changes in socio-economic activities and urbanization processes [36,37]. Numerous studies have used NTL data to reveal the external features of cities, such as extracting city boundaries [38] and detecting city centers and their spatial structures [39]. At the same time, NTL data have the advantages of a large scale and long time series, and are widely used to display the spatiotemporal dynamic changes of cities or urban agglomerations [40,41]. Therefore, combining nighttime light data with statistical data can not only obtain indicators of urban socio-economic changes, but also obtain long-term urban morphological structural characteristics, which is conducive to jointly analyzing the key factors affecting carbon emissions from both inside and outside the city.

This study uses multi-temporal built-up area data and nighttime light (NTL) data as sources to propose a method for integrating remote sensing data with statistical data into an urbanization development assessment system. The aim is to investigate the key factors influencing carbon emissions under different urbanization development models and to expand the understanding of the mechanisms through which urbanization impacts carbon emissions. We have three objectives:(1) to integrate two types of nighttime light data to get a long time series data set of the built-up areas of Chinese cities; (2) to establish an urban development evaluation system and classify cities according to their development status; and (3) to exam the similarities and differences in the factors affecting carbon emissions in different types of cities.

2. Materials and Methods

2.1. Study Area

In order to promote sustainable development and fulfill international environmental responsibilities, China established three batches of urban low-carbon pilot projects including 81 cities from 2010 to 2020 to undertake low-carbon initiatives (Figure 1). The data in some cities are incomplete, so this paper selects 68 of them as the focus of research. The city and its corresponding serial number can be found in Appendix A, Table A1.

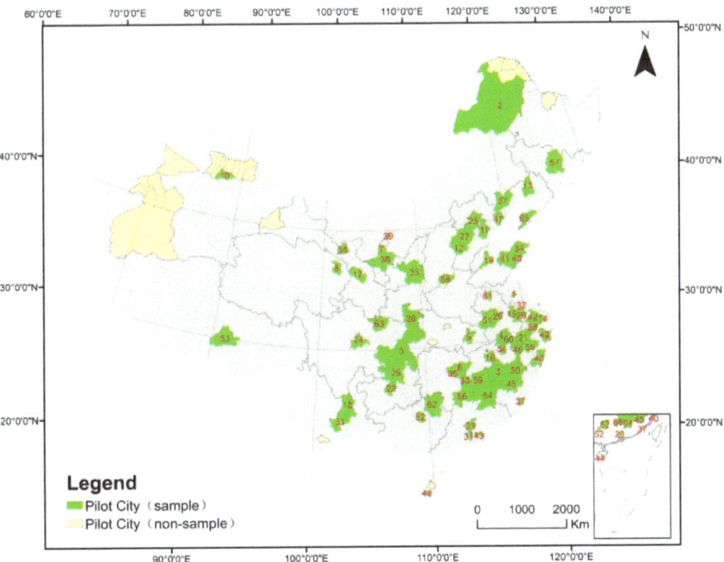

Figure 1. China's low-carbon pilot cities.

From a geographical perspective, these pilot cities span across China, ranging from the Greater Khingan Range in the north to Sanya in Hainan Province in the south. They encompass five distinct climate types: temperate continental climate, temperate monsoon climate, subtropical monsoon climate, tropical monsoon climate, and plateau mountain climate (Figure 2). At the same time, the topography of low-carbon pilot cities is quite varied. In the eastern plain regions, these cities are often grouped together, while in the central and western plateau regions, they are more spread out (Figure 3).

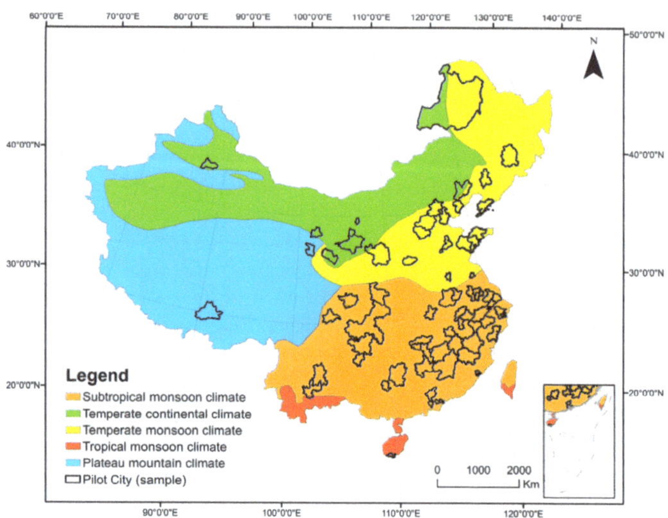

Figure 2. Climatic types of low-carbon pilot cities.

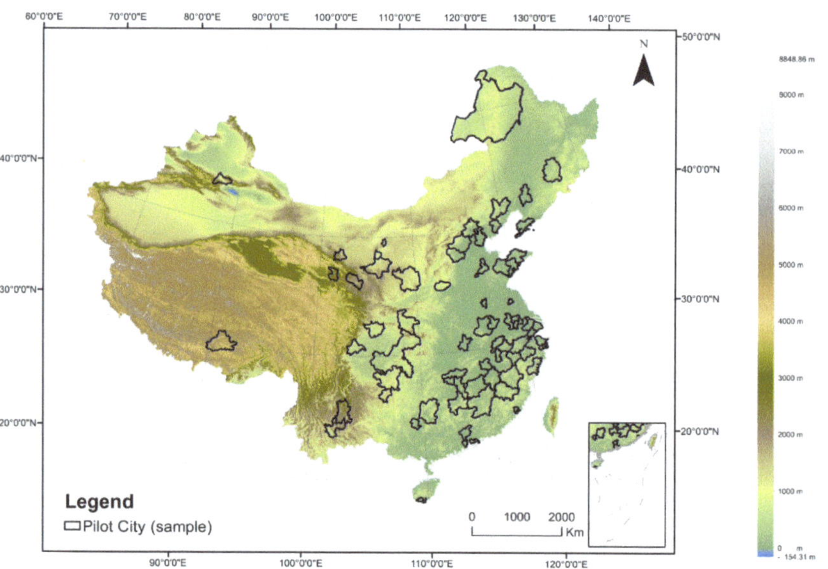

Figure 3. Topography of low-carbon pilot cities.

From a social perspective, the 81 pilot cities encompass China's four major economic zones: the Beijing–Tianjin–Hebei region, the Yangtze River Delta, the Pearl River Delta, and the Chengdu–Chongqing urban agglomeration. The varying functions and development timelines of these city clusters result in differences in their levels of development. Additionally, there are notable disparities in their scales and functions, including traditional industrial cities such as Wuhai and Zhuzhou, as well as economically open regions such as Shenzhen and Xiamen. Figures 4 and 5 reveal significant differences in economic and population scales among the cities. For example, cities like Shanghai and Beijing are densely populated and economically advanced, while cities such as Urumqi and Lhasa are sparsely populated and less economically developed.

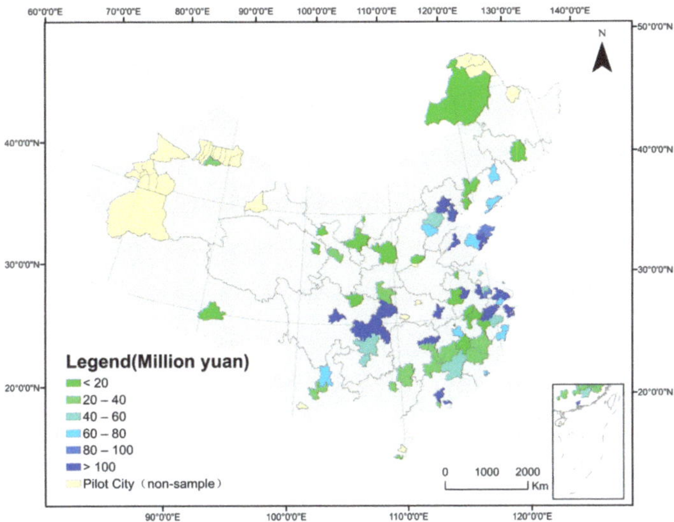

Figure 4. The GDP of low-carbon pilot cities in 2021.

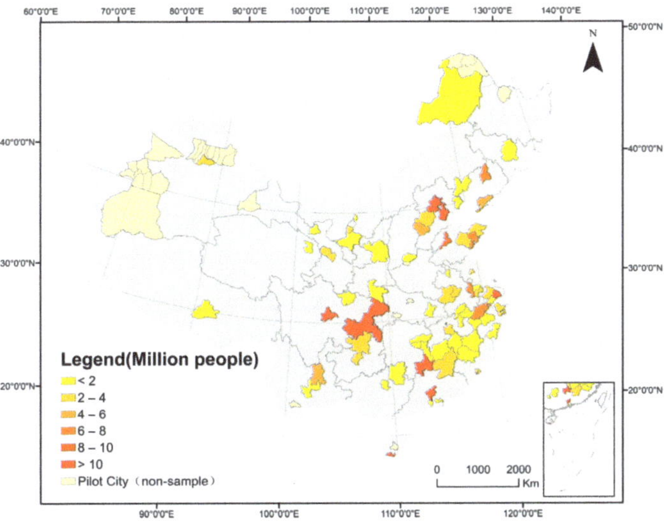

Figure 5. The population of low-carbon pilot cities in 2021.

The diverse distribution of low-carbon pilot cities facilitates the development of tailored policies, addressing the specific needs of different types of cities. The last batch of pilot cities was initiated late, resulting in limited research focused on these cities as study areas. Using these pilot cities as samples not only meets the requirement for diversity in the study area but also provides valuable insights for developing effective low-carbon policies in China.

2.2. Data Sources

There are two types of widely used nighttime light data. The DMSP/OLS nighttime light data are provided by the National Geophysical Data Center (NGDC) of the National

Oceanic and Atmospheric Administration (NOAA). The dataset, available for download, includes 34 images spanning from 1992 to 2013, with a spatial resolution of approximately 1 km and grayscale levels ranging from 0 to 63. DMSP/OLS data from 2002 to 2013 are utilized in this experiment.

The National Polar-orbiting Partnership (NPP) satellite is a new-generation Earth observation system operated by the United States, developed collaboratively by the National Aeronautics and Space Administration (NASA), the National Oceanic and Atmospheric Administration (NOAA), and the U.S. Air Force. The satellite provides monthly data with a spatial resolution of 500 m, available from 2012 onwards. This study utilizes monthly NPP-VIIRS data from 2012 to 2021, which are aggregated into annual datasets.

Specifically, DMSP/OLS data from 2012 and 2013, along with NPP-VIIRS data, are used to integrate the two types of nighttime light data.

The MODIS land cover type product MCD12Q1 is selected for the accuracy evaluation of built-up area extraction. MCD12Q1 is a MODIS land cover type product that provides an annual global distribution of land cover types with a resolution of 500 m. The data are used for the accuracy assessment of built-up area extraction. This paper extracts the built-up areas of the pilot cities from 2002 to 2021. To ensure the validity of the accuracy assessment, MODIS data from 2010 and 2020 are selected for the validation process.

The carbon emission data at the city level used in this study are from the China Emission Accounts and Datasets (CEADs). CEADs have compiled carbon accounting inventories covering China and other developing countries, with the aim of establishing a cross-validated multi-scale carbon emissions accounting methodology, and creating a regionally and locally unified, detailed carbon accounting data platform [42–44]. CEADs provide a carbon emission inventory for Chinese cities from 1997 to 2019. Because the data before 2006 are incomplete, data from 2006 to 2019 are used to analyze the factors influencing carbon emissions.

The urban statistical data and energy data are sourced from the China Urban Statistical Yearbook (2002–2021) and the China Energy Statistical Yearbook (2002–2021).

The information of the data used in this paper is shown in Table 1.

Table 1. Research data and their sources.

Dataset	Data	Source
DMSP/OLS nighttime light data	2002–2013	https://eogdata.mines.edu/products/vnl/ (accessed on 1 September 2023)
NPP-VIIRS nighttime light data	2012–2021	
The MODIS land cover type product MCD12Q1	2010, 2020	https://ladsweb.modaps.eosdis.nasa.gov
The China Emission Accounts and Datasets (CEADs)	2006–2019	https://www.ceads.net.cn/
The urban statistical data and energy data	2002–2021	http://www.stats.gov.cn/

2.3. Urban Development Evaluation System

Different dimensions of urbanization have varying influences on the carbon emissions of urban ecosystems. Currently, most scholars believe that urbanization development can be subdivided into population, economic, social, land, and ecological urbanization [45,46]. Some scholars have also described the complexity of urbanization from the perspectives of population, building structure, and greenness [47]. Based on the existing research, this paper establishes an evaluation system for urbanization development composed of a population subsystem, economic subsystem, construction subsystem, and society subsystem. The primary indicator of the evaluation system is the overall score of urban development. Population, economy, urban construction, and society subsystems constitute the secondary indicators of the evaluation system. Each subsystem contains 2–4 tertiary indicators. Figure 6 shows the composition of the urbanization development evaluation system. The following is an explanation of the selection of indicators.

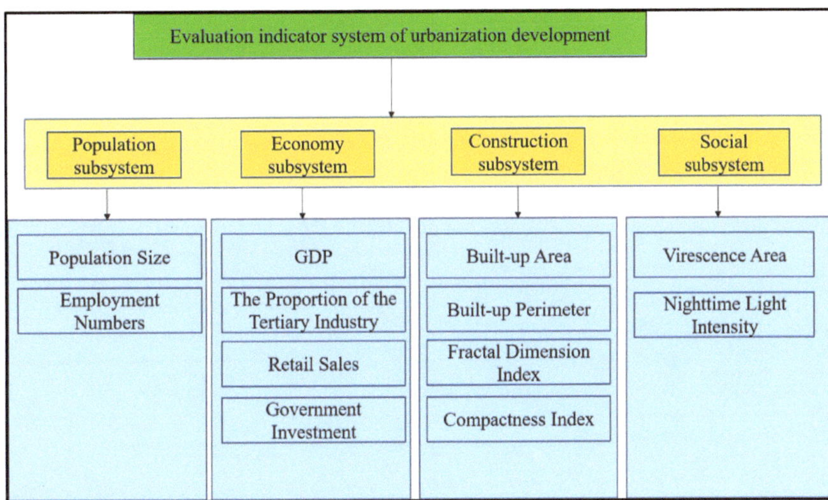

Figure 6. The evaluation system for urbanization development.

Two indicators are selected to form a population subsystem: population size (PS) and employment numbers (ENs). GDP, the proportion of the tertiary industry (PTI), retail sales (RSs) and government investment (GI) are chosen to represent the economy subsystem. For the social subsystem, virescence area (VA) represents the environmental evaluation factor and night light intensity (NLI) is selected as the human activity evaluation factor.

The built-up area (BA) and built-up perimeter (BP) regions are obtained after integrating two types of nighttime light data. The Fractal Dimension Index (FDI) and Compactness Index (CI) are also calculated. These four metrics are used to construct a development subsystem. The CI is usually used to reflect the shape characteristics of a region. The FDI can reflect the tortuosity and complexity of the shape.

The formulas for FDI and CI are as follows [48]:

$$FDI = 2\ln(P/4)/\ln A \tag{1}$$

$$CI = 2\sqrt{\pi A}/P \tag{2}$$

P represents the perimeter of the built-up area, and A represents the area of the built-up area.

The multi-index evaluation method is used to establish an urban development evaluation system, which combines various indicators by measuring the impact of each one on the city, thereby forming an evaluation system. Firstly, this paper employs the normalization of evaluation indicators to eliminate the influence of different units and dimensions among various indicators. The standardized data results lie within the range of [0, 1]. To ensure the universality of the evaluation system, this study employs the objective weighting method known as the entropy method to determine the weights of the evaluation indicators. After establishing the evaluation system, a 5-year cycle is adopted to calculate the rate of change in urban scores. Based on this rate of change, cities are classified into three categories: contracting city ($C < -0.1$), stable city ($-0.1 < C < 0.1$), and expanding city ($C > 0.1$).

The process of establishing the urbanization development evaluation system is shown in Table 2 [49].

Table 2. The process of establishing the urbanization development evaluation system.

Method	Step	Formula	Meaning
Min–Max Normalization		$X_{it} = (X - X_{t_{min}})/(X_{t_{max}} - X_{t_{min}})$	X_{it} is the standardized data for city i year t. $X_{t_{min}}$ and $X_{t_{max}}$ is the minimum and maximum values in the data of 68 cities in year t.
Information Entropy Method	The proportion of city i indicator j to all cities	$y_{ij} = x_{ij}/\sum\limits_{i,j}^{r} x_{ij}$	X_{ij} is the standardized data for city i indicator j. y_{ij} is the weight of indicator j in city i. e_{ij} is the entropy value of indicator j. r is the number of sample cities. m is the number of indicators. E_i is the score of city i.
	Entropy value of indicator j	$e_j = -(1/\ln r) \times \sum\limits_{i=1}^{r} y_{ij} \times \ln y_{ij}$	
	Information utility value of the entropy value of indicator j	$d_j = 1 - e_j$	
	Weight of indicator j	$w_j = d_j / \sum\limits_{j=1}^{m} d_j$	
	Score for the development of city i	$E_i = \sum\limits_{j=1}^{m} w_j \times x_{ij}$	
Urban development change rate		$C_{it} = (E_{it_1} - E_{it_0})/E_{it_0}$	C_{it} is the change rate of period t in city i. E_{it_0} and E_{it_1} are the scores of city i in the early and late stages of period t.

2.4. Using Nighttime Light Images to Extract Built-Up Areas

The radiometric performance and detection capabilities of different sensors on the DMSP satellites vary, leading to certain limitations in the DMSP/OLS data. Therefore, the first step involves preprocessing the DMSP/OLS data, as illustrated in Figure 7. Firstly, the imagery is extracted using Chinese vector data as a boundary. The extracted image is then converted to the Albers projection and resampled to a resolution of 1 km. As the data are sourced from multiple satellites, it is necessary to minimize differences among the data through mutual calibration. Generally, Hegang City in Heilongjiang is used as the reference area, and the F162006 image is used as the reference image. This is used to establish correspondences between the other 33 images and the reference image. The mutually corrected images still have discontinuity problems, so the images need to be further corrected for continuity. In a long time series of DMSP/OLS data, bright pixel values in the image from the previous year may become non-bright values or exhibit different DN values at the same location in the image from the subsequent year. To address this unusual fluctuation, it is necessary to conduct an interannual correction on the pixels after the continuity correction. The preprocessing process of DMSP/OLS data is shown in Table 3 [50].

The next step involves processing the NPP-VIIRS data.

The average values of data for each year from 2013 to 2021 are calculated, the annual data are synthesized, and the images are resampled to 1 km. NPP-VIIRS images contain a significant amount of faint light, and unlike DMSP/OLS data, there is no saturation effect in NPP-VIIRS data, with significant fluctuations in DN values. In order to make the NPP-VIIRS data and the DMSP/OLS data comparable, the noise and outliers of the NPP-VIIRS data need to be eliminated. Current research mostly uses 0.3 as the threshold and assigns pixels with a DN value less than 0.3 $\mu W/(cm^2 \cdot sr)$ in the image to 0, effectively eliminating noise [51]. As for the removal of outliers, it is usually assumed that the DN values for other cities will not exceed those of Beijing, Shanghai, Guangzhou, and Shenzhen [52]. Therefore, the maximum pixel values for these four cities in the corresponding years are recorded, and this maximum value is used as the threshold. Pixels exceeding this threshold are assigned a threshold value.

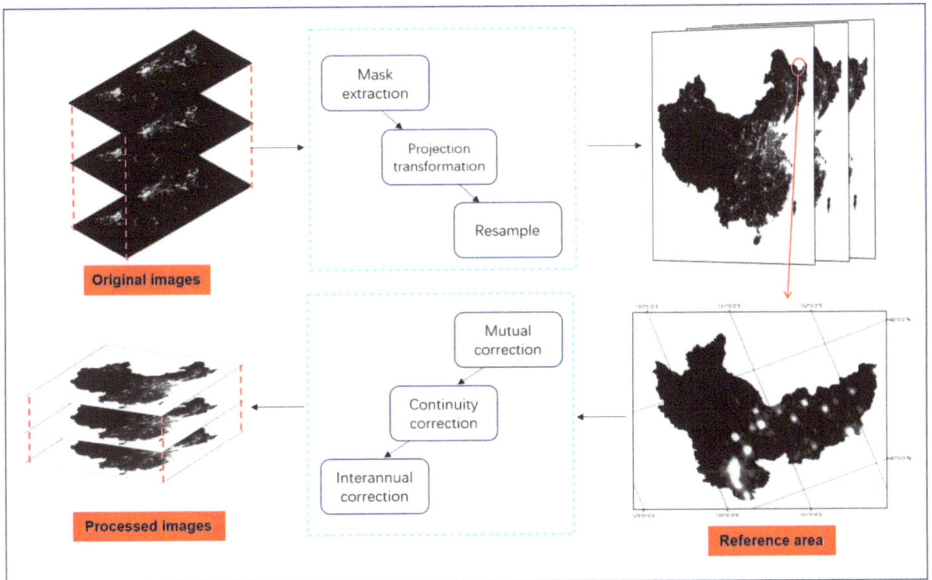

Figure 7. Preprocessing of DMSP/OLS data.

Table 3. Calculation for mutual correction, continuity correction, and interannual correction.

Step	Formula	Meaning
Mutual correction	$DN_{cal} = a \times DN^2 + b \times DN + c$	DN is the DN value of the pixel to be corrected. DN_{cal} is the DN value of the pixel after correction. a, b, c are the parameters obtained from quadratic regression.
Continuity correction	$DN_{(n,i)} = \begin{cases} 0 & DN^a_{(n,i)} = 0 \ and \ DN^b_{(n,i)} = 0 \\ (DN^a_{(n,i)} + DN^b_{(n,i)})/2 & Others \end{cases}$	$DN_{(n,i)}$ is the DN value of pixel i in the corrected image of year n. $DN^a_{(n,i)}$ and $DN^b_{(n,i)}$ are the DN value of pixel i in the original image acquired by different sensors in the year n.
Interannual correction	$DN_{(n,i)} = \begin{cases} 0 & DN_{(n+1,i)} = 0 \\ DN_{(n-1,i)} & DN_{(n+1,i)} > 0 \ and \ DN_{(n-1,i)} > DN_{(n,i)} \\ DN_{(n,i)} & Others \end{cases}$	$DN_{(n-1,i)}$, $DN_{(n,i)}$, $DN_{(n+1,i)}$ are the DN value of pixel i of the original data. from the same year's image after mutual correction and continuity correction in the $n-1$, n, $n+1$ years.

To obtain a long-term sequence of nighttime light images, NPP-VIIRS data and DMSP/OLS data need to be integrated. DMSP/OLS and NPP-VIIRS both released images for 2013. Therefore, a function relationship between the DN values of the two types of images for 2013 is established. By fitting the optimal curve, the paper explores the correlation between the two datasets, facilitating the integration of NPP-VIIRS and DMSP/OLS data.

Using the DN value range of 0–63 from DMSP/OLS data as a mask (excluding pixel values of 1 and 2), the pixel values at corresponding positions are extracted from NPP-VIIRS data. These values undergo a logarithmic transformation, and the average is calculated. Different functions are then applied for fitting (Table 4). The results show that the power function provides the best fit.

Table 4. Fitting results of DMSP/OLS data and NPP-VIIRS data.

Fit Method	Formula	Meaning	R-Squared (R2)
Exponential function	$y = 9.81e^{0.8419x}$		0.8200
Linear function	$y = 22.524x + 3.9871$	y is the DN value of DMSP/OLS images.	0.9830
Logarithmic function	$y = 15.577 \ln x + 33.829$	x is the corresponding DN value of NPP/VIIRS images	0.7510
Power function	$y = 28.078x^{0.6963}$		0.9836

This paper uses the mutation detection method to extract built-up areas. Real urban areas should maintain the integrity of their geometric shapes. The larger the light value, the greater the probability of belonging to an urban area. As the segmentation threshold increases, the polygon patches representing urban built-up areas gradually become smaller along the edges. When the segmentation threshold reaches a point, the polygon patches no longer shrink along the edges. Instead, they break and split from the inside into many smaller polygon patches, causing a sudden increase in the perimeter of the polygon representing the urban built-up area. This point is considered the threshold for extracting the urban built-up area [53].

2.5. Validation for Urban Built-Up Area Extraction Results

According to the framework proposed by Olofsson et al., the good practice recommendations for assessing accuracy comprise three components: sampling design, response design, and analysis [54]. After extracting the built-up areas, this paper evaluates the accuracy of the results following the process (Figure 8).

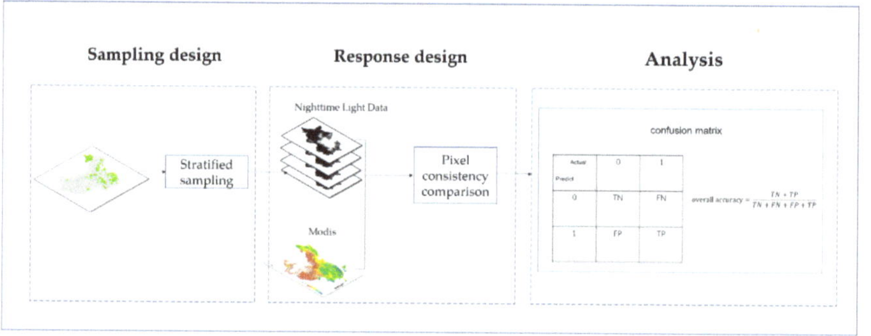

Figure 8. Process for accuracy assessment of built-up area extraction using MODIS.

For sampling design, simple random, stratified, and systematic sampling methods are commonly used in research [55]. Stratified random sampling is a probability sampling design that is relatively easy to implement and is widely used in the field of remote sensing for accuracy assessment [56]. In this study, a stratified sampling method is employed, categorizing 68 cities into three classes based on their area: large, medium, and small. Sampling is conducted within each category, resulting in the selection of nine cities for accuracy assessment. The nine cities selected are Chongqing, Beijing, and Kunming (large); Tianjin, Chengdu, and Suzhou (medium); and Guangzhou, Shanghai, and Xiamen (small).

For response design, the MCD12Q1 products of MODIS in 2010 and 2020 are used to as the reference data; these data are of a high quality and come from one of the most widely used land cover datasets globally. In this paper, the land cover class is selected and labeled as a built-up area from the MCD12Q1 product, and its consistency is compared with the built-up area pixels extracted in our study on a per-pixel basis.

For the analysis, the confusion matrix is used in this paper. The confusion matrix is a cross-tabulation method used to compare the class labels assigned by the remote sensing classification with the reference data for the sample sites, and it plays a central role in accuracy assessment [57]. The main diagonal of the confusion matrix displays correctly classified instances, while the off-diagonal elements represent misclassifications. The proportion of correctly classified pixels to the total number of pixels is referred to as the overall accuracy. A higher value indicates better classification performance [50]. In this study, the overall accuracy is used as a metric to evaluate the built-up area extraction results.

2.6. Methods for Factors Influencing Carbon Emissions

Based on the urban evaluation system established in the previous section, this article divides 68 cities into expansion and contraction types, selects socio-economic data related to carbon dioxide emissions (Table 5), and proposes a method to discuss the similarities and differences of carbon emission factors among different types of cities.

Table 5. Data on factors influencing carbon emissions in low-carbon pilot cities.

ID	Meaning	Abbreviations
Y	Carbon emission	CE
X1	Population size	PS
X2	Employment numbers	EN
X3	GDP	GDP
X4	The proportion of tertiary industry	PTI
X5	Retail sales	RS
X6	Government investment	GI
X7	Built-up area	BA
X8	Built-up perimeter	BP
X9	Fractal dimension	FDI
X10	Compactness index	CI
X11	Virescence area	VA
X12	Nighttime light intensity	NLI

2.6.1. Lasso

In order to comprehensively consider influencing factors and increase the model's interpretability, a large number of variables are usually chosen to be included in the model in research. However, as the number of variables increase, multicollinearity and variable autocorrelation will occur, which will reduce the accuracy of the model. The Lasso method is widely used to select variables. The Lasso method adds a penalty term to the model and adjusts it to affect the coefficients of the model. In this process, the coefficients of some insignificant variables will gradually approach 0. By deleting these variables, variable screening can be achieved. The specific implementation process is as follows [58]:

Y is the dependent variable, X is the independent variable, and the observed values of sample data are (X, Y). The following linear relationship can be established between them. β is the regression coefficient and ε is the random disturbance term.

$$Y = X\beta + \varepsilon \tag{3}$$

To select significant variables, it is necessary to perform Lasso estimation on the unknown parameter β in the above model. The estimated value of β must satisfy the minimum value of the following formula. $\lambda \sum_{j=1}^{p} |\beta_j|$ represents the penalty term. $\lambda \geq 0$ is an adjustment parameter, indicating the intensity of punishment. As λ becomes larger, the penalty will increase, the selection of variables will become stricter, and some insignificant variables will be eliminated.

$$\hat{\beta}_{lasso} = \mathrm{argmin}(||Y - X\beta||^2 + \lambda \sum_{j=1}^{P} |\beta_j|) = RSS + \lambda \sum_{j=1}^{P} |\beta_j| \tag{4}$$

Usually, the Lasso cross-validation (LassoCV) method can be used to adjust the values of the parameter λ. By selecting a range of λ values, calculating the corresponding cross validation errors, and choosing the parameter with the minimum CV error, the optimal tuning parameter value is determined. This article uses the K-fold cross validation method to select λ. The K-fold cross validation method divides the initial observation dataset into roughly identical K sub samples, selecting one as the validation set each time and the remaining $K - 1$ as the training set. The training is repeated K times, and the entire process will result in K test errors MSE_1, MSE_2, ..., MSE_i. The cross validation error is as follows:

$$CV_{(K)} = \frac{1}{K}\sum_{i=1}^{K} MSE_i \tag{5}$$

2.6.2. Hierarchical Regression Model

When analyzing the factors affecting carbon emissions, this article uses the hierarchical regression model. The hierarchical regression model is an effective method for analyzing the degree of influence of variables on dependent variables, as well as studying the relationship between different variables and dependent variables by analyzing the R-squared changes when variables are added at different levels [59,60].

$$\begin{cases} Y_{n,m} = \alpha_0 + \alpha_n X_{n,m} + \varepsilon_i \\ Y_{n+1,m} = \alpha_0 + \alpha_n X_{n,m} + \alpha_{n+1} X_{n+1,m} + \varepsilon_i \\ Y_{n+2,m} = \alpha_0 + \alpha_n X_{n,m} + \alpha_{n+1} X_{n+1,m} + \alpha_{n+2} X_{n+2,m} + \varepsilon_i \end{cases} \tag{6}$$

Y_i is the carbon emissions of city i. α_i is the regression coefficient for variable X_i. ε_i is the error term. n and m represent the number of regression steps and the type of city in hierarchical regression model, respectively. With the steps increasing by one, one more variable was added in the step $n + i$. The variables with significance and with a change in the R-square value were selected as independent variables to construct the specific regression model for this type of city.

3. Results

3.1. Extraction of Built-Up Areas

After extracting the boundaries of built-up areas, sample cities are selected, including Chengdu and Chongqing, and an accuracy assessment is conducted. The overall accuracy of urban area extraction ranges from a maximum of 98.87% to a minimum of 70.90% (Table 6). Therefore, it can be concluded that the extraction results using integrated nighttime light images meets the experimental requirements. Figures 9 and 10 show a comparison of the built-up areas extracted from MODIS data and nighttime light data.

Table 6. The overall accuracy calculated during the accuracy assessment.

Number	City	Time	Overall Accuracy
(a)	Chongqing	2010	90.22%
		2020	88.76%
(b)	Beijing	2010	98.87%
		2020	97.27%
(c)	Kunming	2010	97.27%
		2020	95.05%
(d)	Tianjin	2010	85.53%
		2020	85.82%

Table 6. *Cont.*

Number	City	Time	Overall Accuracy
(e)	Chengdu	2010	94.49%
		2020	86.18%
(f)	Suzhou	2010	82.73%
		2020	80.73%
(g)	Guangzhou	2010	89.12%
		2020	71.49%
(h)	Shanghai	2010	82.08%
		2020	71.49%
(i)	Xiamen	2010	79.33%
		2020	70.90%

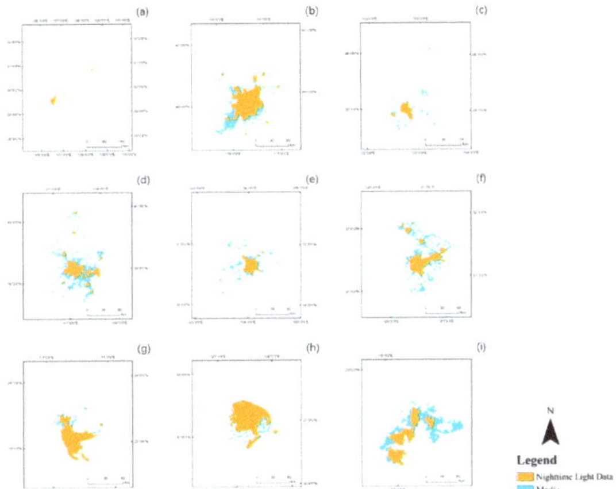

Figure 9. Comparison of built-up area extraction results using MODIS data and nighttime light data (2010).

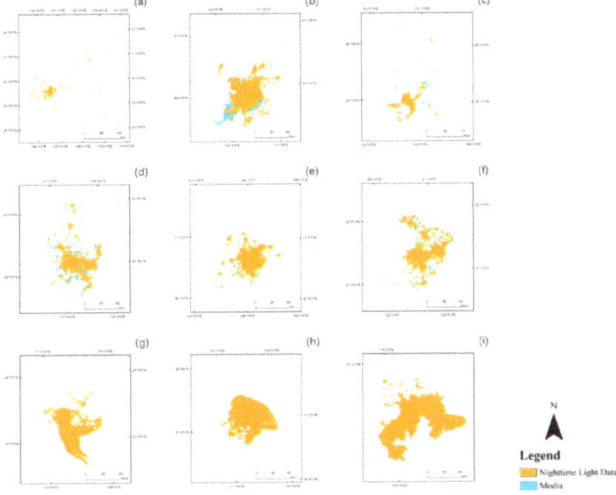

Figure 10. Comparison of built-up area extraction results using MODIS data and nighttime light data (2020).

3.2. Result of Establishment of Urban Development Evaluation System

According to the urban development evaluation system established, with a time interval of 5 years, the subsystem scores and overall development scores for 68 cities are calculated and categorized. Next, the results are discussed in sequence based on the overall development of the 68 cities, followed by the development of the four subsystems: population, economy, construction, and society.

3.2.1. Overall Development of the 68 Cities

This section presents the overall development trends of 68 cities from 2002 to 2021.

Table 7 shows the number of cities of each type in four periods. In the first period (2002–2006), the development of 20 cities was stable. And in the fourth stage (2018–2021), it increased to 38. The number of expanding cities initially increases and then decreases, while it is opposite for contracting cities. In general, the development of urbanization will undergo a period of fluctuations and gradually reach a stable state.

Table 7. The number of cities with three development types (contracting city, stable city, and expanding city).

	2002–2006	2007–2011	2012–2016	2017–2021
Contracting City	31	6	6	15
Stable City	21	30	29	38
Expanding City	16	32	33	15

Figure 11 shows the geographical distribution of different types of cities. Cities in the southeastern region, such as those in Jiangxi and Fujian provinces, exhibit synchronous development characteristics. Over the past 20 years, their development has undergone phases of contraction, followed by expansion and then stabilization. Some major cities, such as Beijing and Chongqing, have undergone more significant fluctuations in their development process. Meanwhile, cities in the northeastern and northwestern regions, such as Hulunbuir and Urumqi, have maintained a stable development pattern throughout the 20-year period. It can be seen that geographical location has a significant impact on urban development.

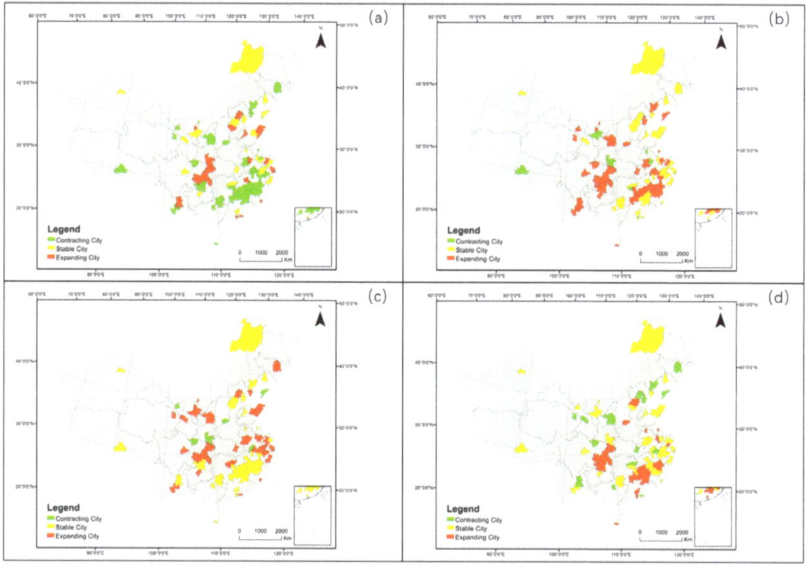

Figure 11. Urbanization development types of 68 cities in four periods. (**a**) 2002–2006; (**b**) 2007–2011; (**c**) 2012–2016; (**d**) 2017–2021.

In the meantime, we classified cities based on the scores of each subsystem and counted their numbers (Table 8).

Table 8. The number of three types of cities divided by subsystem scores.

		2002–2006	2007–2011	2012–2016	2017–2021
Population subsystem	Contracting City	13	22	4	35
	Stable City	18	35	25	17
	Expanding City	37	11	39	16
Economy subsystem	Contracting City	21	2	4	27
	Stable City	31	23	34	23
	Expanding City	16	43	30	18
Construction subsystem	Contracting City	49	9	21	7
	Stable City	14	20	25	35
	Expanding City	4	39	22	26
Society subsystem	Contracting City	9	14	7	25
	Stable City	15	27	5	15
	Expanding City	44	27	56	28

3.2.2. The Development of the Population Subsystem

This section presents the development trends of the population subsystems for 68 cities from 2002 to 2021.

Based on Figure 12, it can be observed that the population subsystem in Chinese cities exhibited an expansionary trend during the first phase (2002–2006) and the third phase (2012–2016). In contrast, a significant contraction occurred during the second phase (2007–2011) and the fourth phase (2017–2021). This indicates that the population subsystem tends to undergo a decline once it reaches a certain scale of development.

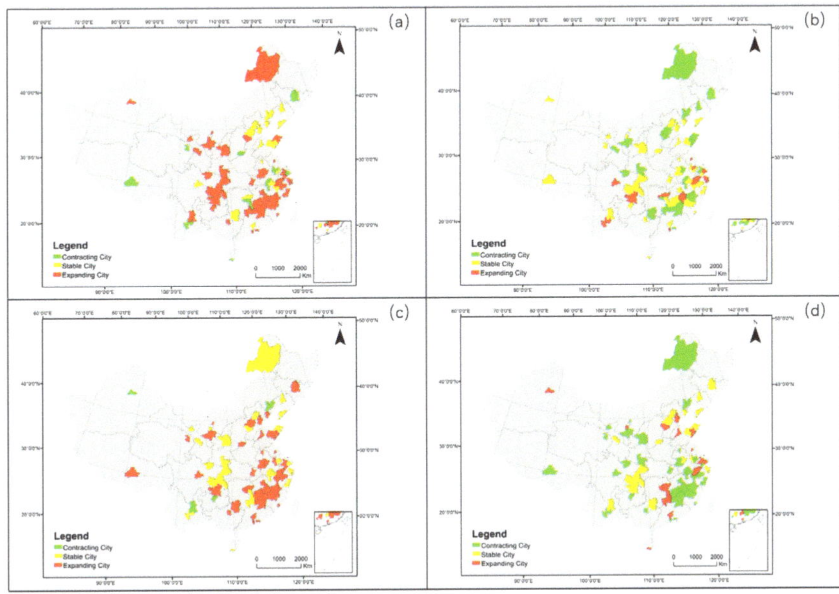

Figure 12. Types of population subsystems of 68 cities in four periods. (**a**) 2002–2006; (**b**) 2007–2011; (**c**) 2012–2016; (**d**) 2017–2021.

At the same time, the development of the population subsystem exhibits clear agglomeration characteristics. For example, cities in the southeastern regions, such as those in Jiangxi, Hunan, and Fujian Provinces, display similar patterns of change.

3.2.3. The Development of the Economy Subsystem

This section details the development of the economic subsystems for 68 cities from 2002 to 2021.

As shown in Figure 13, from 2002 to 2006, the cities with an expanding economic subsystem were mainly located in the coastal areas and the border areas. From 2007 to 2011, the economic subsystem of most cities underwent a concentrated expansion. After this period, cities with contracting economic subsystems were gradually concentrated northward, while expanding cities gradually moved southward. Overall, compared to cities in coastal regions, the economic subsystems of some resource-based cities in inland areas tend to be less stable, exhibiting more pronounced fluctuations between rapid economic growth and contraction.

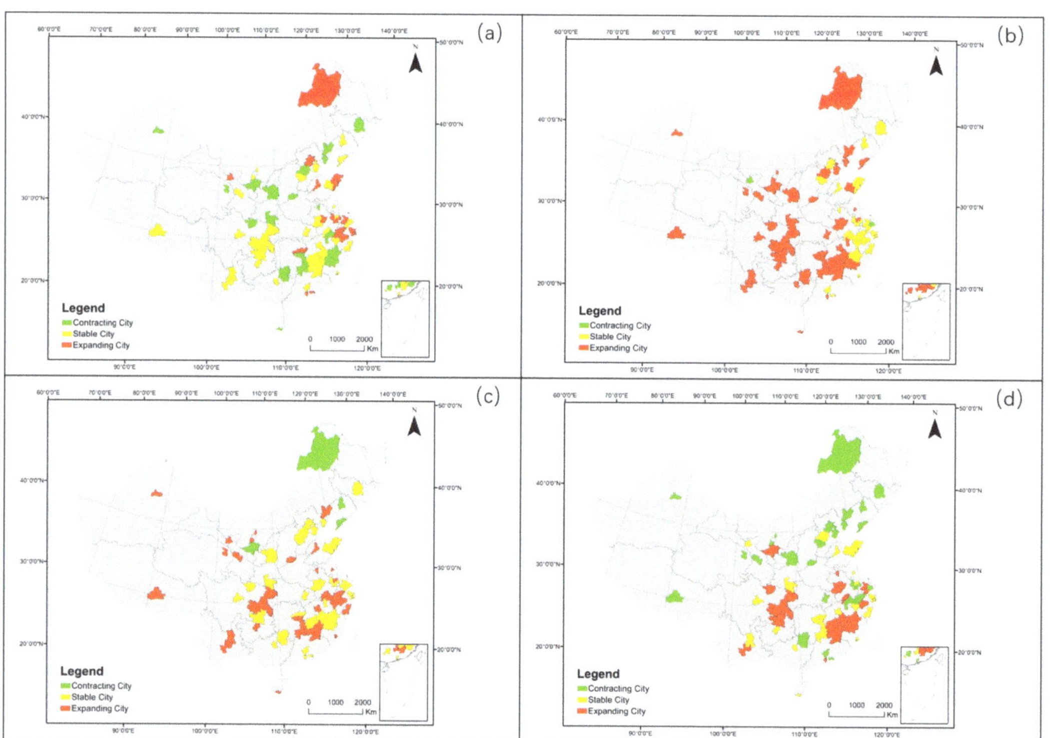

Figure 13. Types of economy subsystems of 68 cities in four periods. (**a**) 2002–2006; (**b**) 2007–2011; (**c**) 2012–2016; (**d**) 2017–2021.

3.2.4. The Development of the Construction Subsystem

This section outlines the development of the construction subsystems across 68 cities from 2002 to 2021.

As illustrated in Figure 14, the development of the construction subsystem in Chinese cities was predominantly characterized by contraction during the first phase (2002–2006). Expansion began in the second phase (2007–2011) and subsequently approached a stable state. Overall, regions with more advanced economies tend to initiate urban construction earlier. Once the demand is met, the pace of construction gradually slows down, leading to an earlier stabilization.

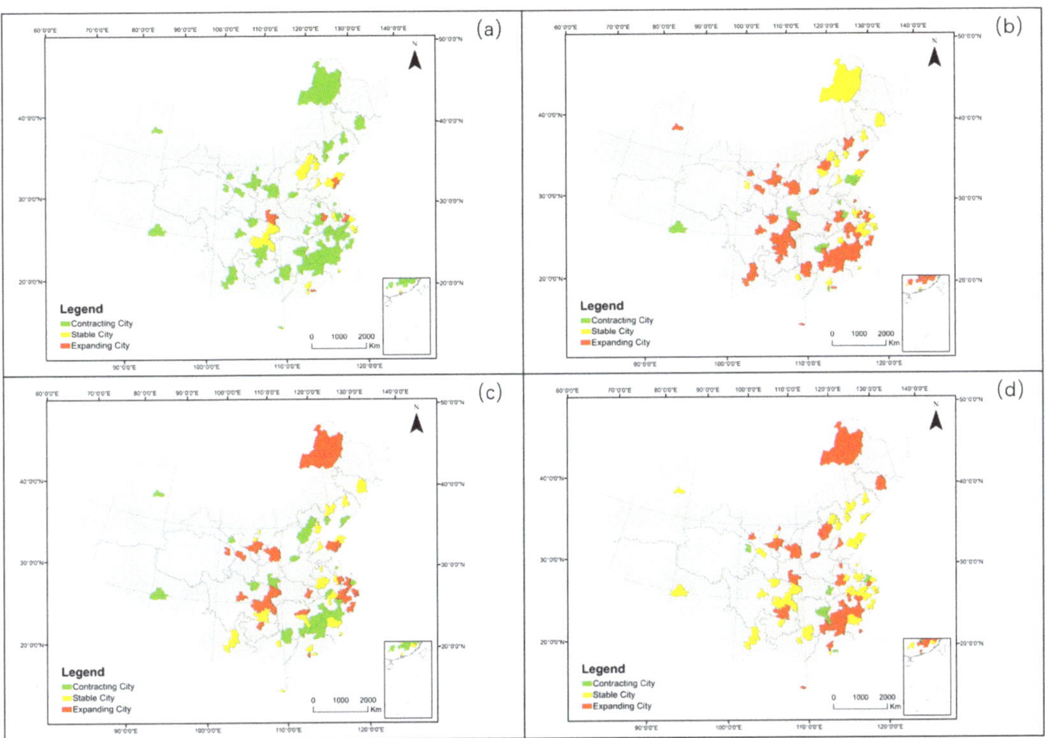

Figure 14. Types of construction subsystems of 68 cities in four periods. (**a**) 2002–2006; (**b**) 2007–2011; (**c**) 2012–2016; (**d**) 2017–2021.

3.2.5. The Development of the Society Subsystems

This section examines the development of the social subsystems for 68 cities from 2002 to 2021.

As Figure 15 shows, from 2002 to 2016, the society subsystem of the city was mainly expanding, with a number of 44, 27, and 56, respectively. In the meantime, the expanding society subsystem gradually spreads from south to north. Since 2017, the number of contracting society subsystems has begun to increase. The contracting society subsystem gradually spreads to inland areas. In general, the social subsystem of southern cities developed earlier than that of northern and western cities. The development of society subsystems in cities located in the southeast region has been in a relatively settled state of expansion.

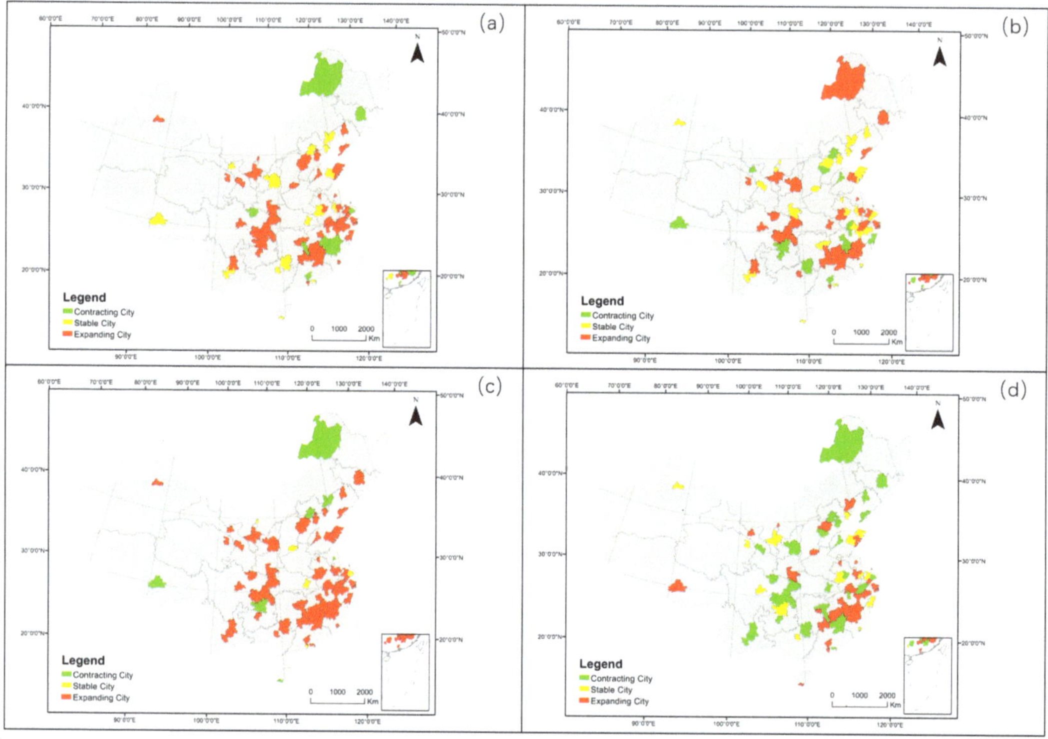

Figure 15. Types of society subsystems of 68 cities in four periods. (**a**) 2002–2006; (**b**) 2007–2011; (**c**) 2012–2016; (**d**) 2017–2021.

3.3. Influence Factor of CO_2

From 2006 to 2019, the average annual carbon emissions of 67 cities except Lhasa were 16.51 million tons. The spatial distribution of carbon emissions shows a characteristic of high in the east and low in the west (Figure 16). A total of 27 cities have average annual carbon emissions of less than 10 million tons (Mt). Cities with average annual carbon emissions of 10–30 Mt are mainly concentrated in the Liaodong Peninsula region. Cities with emissions exceeding 30 Mt are primarily concentrated in the eastern and southern regions, such as Suzhou, Hangzhou, Guangzhou, and Shenzhen. Additionally, four municipalities directly under the central government (Beijing, Shanghai, Chongqing, and Tianjin) have average annual carbon emissions exceeding 50 Mt.

According to the urban evaluation system established earlier, 68 cities from 2006 to 2019 were reclassified (Figure 17). A total of 19 cities are categorized as contracting cities, while 35 cities are classified as expanding cities. Contracting cities are primarily located in the central regions, such as Shanxi, Anhui, and Hunan province. Expanding cities are concentrated in the central and eastern regions, including the Beijing–Tianjin–Hebei urban agglomeration, the Chengdu–Chongqing metropolitan area, and the Yangtze River Delta urban cluster.

Before performing the regression analysis, Pearson correlation coefficients were used to measure the relationships between 12 influencing factors and carbon emissions. A correlation matrix was derived and visualized as a heatmap (Figure 18). To address potential issues arising from significant differences in carbon emissions and influencing factor data, all data were transformed using the natural logarithm to reduce variability.

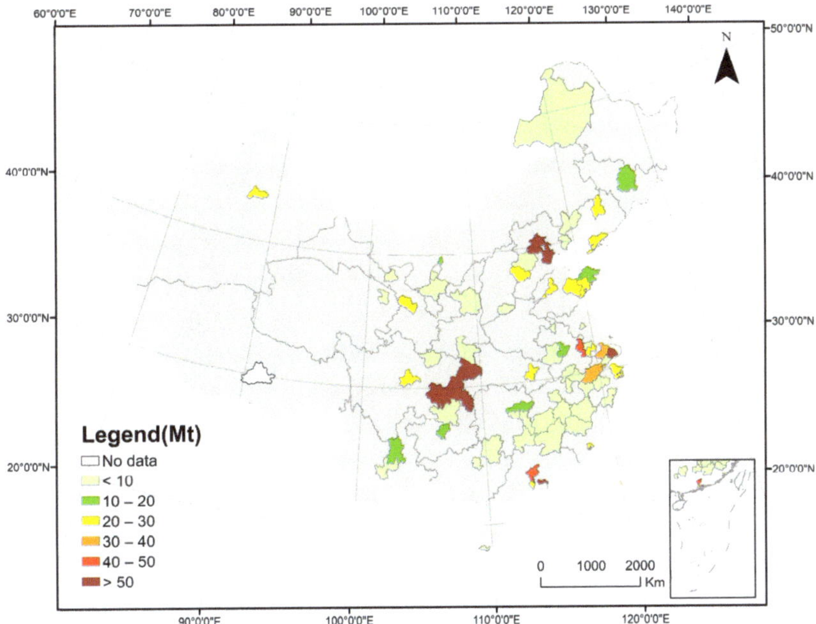

Figure 16. The average annual carbon emissions from 2006 to 2019.

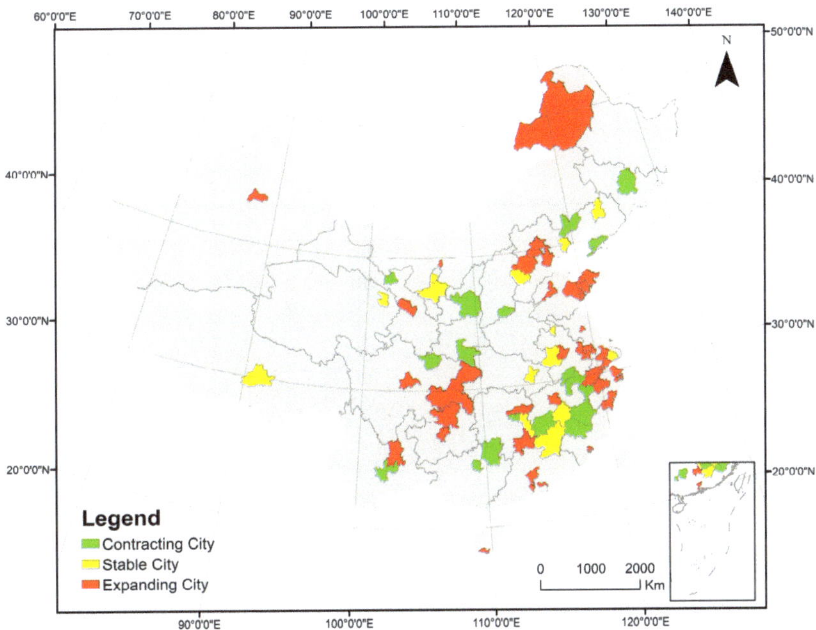

Figure 17. Urbanization development types of 68 cities from 2006 to 2019.

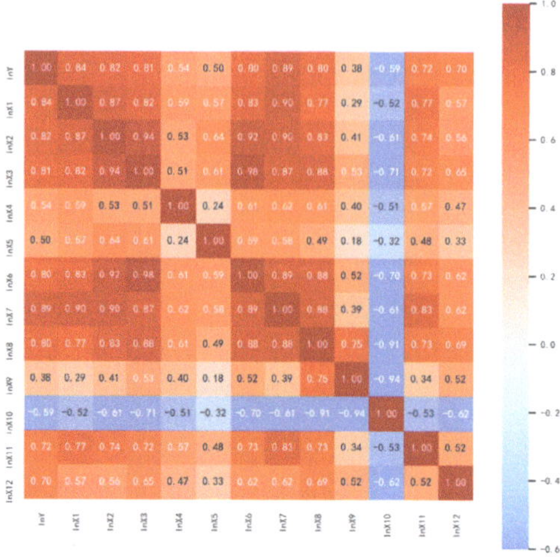

Figure 18. The heatmap of correlations between carbon emissions and the 12 influencing factors.

The heatmap reveals that the absolute values of the correlation coefficients between carbon emissions and 12 influencing factors range from 0.39 to 0.89, indicating a high level of correlation. Additionally, the absolute values of correlation coefficients among most influencing factors are also quite high. For example, the correlation coefficient between government investment and GDP is 0.98. This suggests the presence of multicollinearity among the factors.

To assess multicollinearity, the Variance Inflation Factor (VIF) was used. Typically, a VIF value greater than 10 indicates the presence of multicollinearity. The calculation results are presented in Table 9. Seven variables have VIF values exceeding 10, indicating the presence of significant multicollinearity. Therefore, a further variable selection is necessary.

Table 9. Variance Inflation Factor (VIF) of factors.

Variable	VIF
X8	83.1233
X10	58.7077
X3	54.2692
X6	45.3954
X7	34.1285
X9	19.7756
X2	13.1348
X1	6.6273
X11	3.3523
X4	3.0207
X12	2.3542
X5	1.7763

The Lasso method is used to reselect key factors influencing carbon emissions. When using the Lasso method to screen variables, one important step is to adjust the selection of parameter λ. Table 10 shows the different results corresponding to different lambda values. Df represents the degree of freedom of the model, and $\%Dev$ represents the degree of explanation of the residuals by the model. The larger the value, the better the fitting effect. It can be seen from the results in the table that the interpretability of the model continues to increase as λ decreases, but its growth rate gradually slows down. Even if the λ value keeps getting smaller and Df keeps getting larger, the residual explanation ratio does not change much. Based on this, the number of independent variables in the model can be selected as 9 or 10.

Table 10. The results of variable selection using the Lasso method.

	Df	%Dev	Lambda
1	0	0	1.062
2	1	13.49	0.9673
3	1	24.7	0.8814
...	...	...	...
53	9	83.63	0.01112
54	9	83.67	0.01013
55	9	83.69	0.00923
56	10	83.87	0.00636
57	10	83.91	0.0058
58	10	83.94	0.00528
59	11	83.97	0.00481
60	11	84.04	0.00439
61	11	84.09	0.004
62	12	84.2	0.00364
63	12	84.48	0.00276

To further select the optimal parameters, K-fold cross-validation is typically employed. Figure 19 shows the result of the K-fold cross-validation. The dashed line on the left represents the logarithmic value corresponding to the minimum root mean square error (RMSE), while the dashed line on the right indicates the logarithmic value at one standard deviation away from this minimum value. The numbers above the dashed lines indicate the corresponding number of variables. Generally, the solution corresponding to the right-hand dashed line is considered the optimal choice.

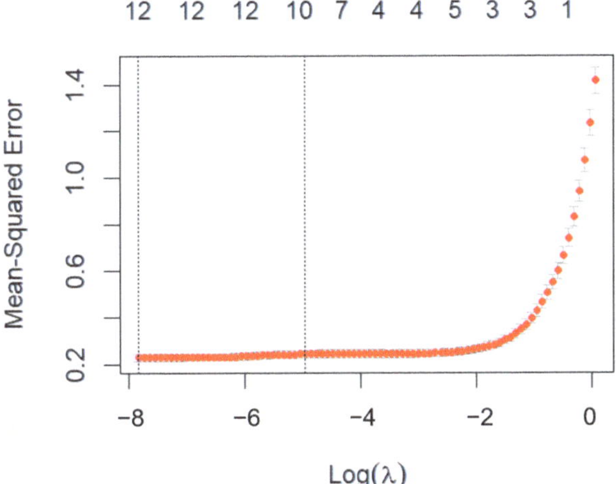

Figure 19. Cross-validation error plot for the Lasso model.

The selected λ is input into the Lasso cross-validation model, and the final results of variable selection are presented in Table 11. The coefficients for variables X6 and X8 are zero and have been discarded, while the remaining nine variables are retained. This indicates that the retained variables have a significant impact on carbon emissions.

Table 11. Table of variable selection results.

Variable	Coefficient	Result
X1	0.2060	Reserve
X2	0.0444	Reserve
X3	0.0064	Reserve
X4	−0.1708	Reserve
X5	−0.0209	Reserve
X6	0.0000	Discard
X7	0.6569	Reserve
X8	0.0000	Discard
X9	0.3582	Reserve
X10	−0.0681	Reserve
X11	−0.0181	Reserve
X12	0.3143	Reserve

Next, based on the classification results presented, the similarities and differences in the factors influencing carbon emissions between expanding and contracting cities are further discussed. Tables 12 and 13 show the hierarchical regression results of two-type cities. As shown in the table, with the inclusion of additional variables, the R-squared value of the carbon emissions influence model for contracting cities reaches 0.801, while for expanding cities, it reaches 0.837. All variables in the models pass significance tests.

According to the regression results, a multiple linear regression model is established for the proportion of carbon emissions and related factors of two-type cities as follows.

$$\ln CE_{(contracting)} = 0.543 \ln BA + 0.519 \ln NLI + 0.373 \ln EN - 0.124 \ln VA - 1.733 \ln CI + 2.672 \quad (7)$$

$$\ln CE_{(Expanding)} = 0.736 \ln BA + 0.296 \ln NLI + 0.459 \ln PS - 0.777 \ln CI - 0.531 \ln PTI + 3.613 \quad (8)$$

The fitting results indicate that for both types of cities, built-up area, nighttime light intensity, and built-up area compactness are significant factors influencing carbon emissions. Specifically, built-up area and nighttime light intensity have a positive correlation with carbon emissions, while built-up area compactness has a negative correlation.

For contracting cities, employment numbers and virescence area also significantly impact carbon emissions. Employment numbers show a positive correlation, whereas green space area exhibits a negative correlation.

In expanding cities, the population size and the proportion of the tertiary sector are more influential. Population size is positively correlated with carbon emissions, while the proportion of the tertiary industry is negatively correlated with carbon emissions.

Table 12. Results and analysis of hierarchical regressions in contracting cities.

	(1) $\ln Y$	(2) $\ln Y$	(3) $\ln Y$	(4) $\ln Y$	(5) $\ln Y$
$\ln X7$	0.939 ***	0.848 ***	0.472 ***	0.696 ***	0.736 ***
	(33.677)	(32.079)	(8.034)	(7.387)	(7.804)
$\ln X12$		0.300 ***	0.275 ***	0.290 ***	0.296 ***
		(10.015)	(8.848)	(9.413)	(9.688)
$\ln X1$			0.429 ***	0.432 ***	0.459 ***
			(7.378)	(7.563)	(8.009)
$\ln X10$				−0.777 ***	−0.777 **
				(−4.871)	(−4.781)
$\ln X4$					−0.351 **
					(−2.912)
Constant	2.063 ***	1.791 ***	1.711 ***	3.163 ***	3.613 ***
	(13.438)	(12.866)	(12.406)	(7.646)	(7.646)
Observations	377	377	377	377	377
R-squared	0.752	0.803	0.826	0.828	0.837
F	240.592	136.907	97.007	77.144	63.469

*** $p < 0.05$, ** $p < 0.1$.

Table 13. Results and analysis of hierarchical regressions in expanding cities.

	(1) ln Y	(2) ln Y	(3) ln Y	(4) ln Y	(5) ln Y
lnX7	0.949 ***	0.848 ***	0.627 ***	0.734 ***	0.543 ***
	(15.511)	(32.079)	(7.326)	(7.375)	(4.304)
lnX12		0.378 ***	0.414 ***	0.412 ***	0.519 ***
		(3.958)	(4.006)	(4.021)	(4.897)
lnX2			0.278	0.268 ***	0.373 ***
			(2.866)	(2.787)	(3.674)
lnX11				−0.104 **	−0.124 **
				(−0.258)	(−2.473)
lnX10					−1.733 ***
					(−3.006)
Constant	1.845 ***	2.048 ***	1.832 ***	1.595 ***	2.672 ***
	(7.353)	(8.270)	(7.305)	(5.821)	(6.393)
Observations	203	203	203	203	203
R-squared	0.738	0.760	0.783	0.788	0.801
F	240.592	136.907	78.558	64.720	50.016

*** $p < 0.05$, ** $p < 0.1$.

4. Discussion

4.1. Urbanization Development

Theories of urban contraction and expansion emerged in the latter half of the 20th century on a global scale. Scholars have employed changes in urban population size as the primary metric for assessing urban contraction and expansion [61,62]. With the study of urbanization processes, scholars have found that using a single indicator to evaluate urbanization processes can result in significant errors. Domestic and international research lacks a comprehensive assessment that integrates economic and social factors [63,64]. Therefore, our research offers an integrated evaluation of urban development from four dimensions—population, economy, construction, and society—extending existing theories of urban contraction and expansion.

Existing research suggests that "deindustrialization" is a primary cause of urban contraction [65]. The results of this study indicate that, over the past two decades, cities undergoing a significant contraction in China are predominantly traditional heavy industrial centers, such as Jilin, Wuhai, and Yan'an. Conversely, cities characterized by expansion are largely concentrated in regions such as the Yangtze River Delta and the Pearl River Delta, where the economy is primarily driven by the service sector and emerging industries. Additionally, the implementation of policies can have a significant impact on both urban expansion and contraction [66]. In 2008, the Chinese government proposed a policy named the Four Trillion Plan to address the global financial crisis and promote stable economic growth. Therefore, in the next five years, the economic subsystems of cities across the country will generally expand, and the direction of population subsystem expansion will also shift towards the east and south. This is also reflected in the results of this experiment. Additionally, the Beijing–Tianjin–Hebei region, the Yangtze River Delta, the Pearl River Delta, and the Chengdu–Chongqing metropolitan area, as major economic zones in China, exhibit a development pattern characterized by expansion from central cities outward. Consequently, cities situated at the center of these metropolitan areas are developed earlier and achieve stability sooner. The results of this paper also reveal that the development of urban construction subsystems follows the principle that earlier development leads to earlier stabilization. Furthermore, influenced by policies such as reform and opening-up policies, cities in the southern regions have prioritized development over those in the north, leading to an earlier expansion of the social subsystems in southern cities compared to their northern counterparts.

The expansion and contraction of cities also exhibit spatial distribution characteristics. Existing research shows that one pattern is the clustering of urban contraction and expansion; another pattern is that the expansion of central cities within metropolitan areas can induce contraction in surrounding cities [67,68]. According to the results of this paper, in southeastern China, the provinces of Jiangxi and Fujian, which share similar geographic conditions, undergo a significant interaction between their populations, leading to a flow of technology and economic activities. Within twenty years, urban development in this region has demonstrated synchronicity. This paper also indicates that, as core cities of the Chengdu–Chongqing metropolitan area, Chengdu and Chongqing attract population and resources from their surrounding areas, leading to contraction in neighboring cities such as Guangyuan, Zunyi, and Ankang.

The results of this study indicate that the expansion and contraction of cities are complex processes influenced by both the inherent attributes of cities and external factors such as policies and so on. In the meantime, these processes are closely linked to the development of surrounding cities.

4.2. Urbanization and Carbon Emissions

Current research on the relationship between urbanization and carbon emissions has explored various aspects, such as the impact of different stages of urbanization on carbon emissions and the effects of urbanization in different regions [69,70]. However, discussions on the factors influencing carbon emissions based on the expansion and contraction of urbanization are limited to smaller research scopes [71]. Therefore, the conclusions derived from this study contribute to the further development of theoretical frameworks and assist policymakers in formulating more effective carbon reduction strategies.

This study classifies 68 pilot cities into three types based on their development from 2006 to 2019 and compares the factors influencing carbon emissions in expanding and contracting cities. The results indicate that in both types of cities, an increase in built-up area and nighttime light intensity is associated with higher carbon emissions. Conversely, the compactness of the built-up area is negatively correlated with carbon emissions. Existing research suggests that urban expansion leads to an increased population, which in turn raises carbon emissions, and also results in higher energy consumption, further contributing to carbon emissions [72]. While a compact city model can reduce the residential energy consumption and the associated carbon emissions, urban expansion remains inevitable [73]. Thus, during urban development, it is crucial to avoid unchecked expansion, maintain a balance between humans and nature, and strive for optimal economic and social benefits [74,75].

In contracting cities, an increase in employment is associated with higher carbon emissions. According to previous analyses, most contracting cities in China are traditional industrial cities located in inland regions, where employment choices are predominantly industrial. In addition, the virescence area in contracting cities significantly impacts carbon emissions. Therefore, in contracting cities, it is essential to focus on environmental protection while considering industrial development, and to avoid issues such as excessive resource extraction [76].

In expanding cities, both population size and the proportion of the tertiary sector have a significant impact on carbon emissions. In China, expanding cities are largely concentrated in coastal areas, characterized by a service-oriented economy and a robust economic foundation, which attracts a substantial external population. Therefore, for expanding cities, policymakers should focus on how to balance the environmental pressures caused by the influx of the external population.

4.3. Limitations

Urbanization often leads to increased income for residents. On the one hand, higher income usually corresponds to a higher standard of living and greater consumption capacity, which can result in higher resource use and increased carbon emissions. On the other hand, higher income may also improve educational levels and enhance environmental awareness among residents [77,78]. Therefore, residents' education and individual behavior are also crucial factors influencing carbon emissions. Future studies will aim to quantify residents' psychological and behavioral characteristics to further explore the factors affecting carbon emissions from an individual perspective.

In constructing the urban development evaluation system, this paper chose the expansion of built-up areas as the primary indicator for the construction subsystem. Current research also incorporates additional metrics. For instance, the accessibility of urban transportation and the configuration of road networks can also serve as indicators for the progress of urbanization [79,80]. In future work, we will further refine the evaluation indicators to ensure that the evaluation system can more comprehensively assess urban development.

4.4. The Socio-Ecological Implications of the Results

From a socio-ecological perspective, the result indicates that while deindustrialized cities are undergoing an economic decline, service-oriented urban areas are seeing strong growth. This highlights the need for effective policies to support the economic transition of contracting cities, ensuring that they adapt sustainably and do not exacerbate social inequalities. Interactions between cities at the center of metropolitan areas and their surrounding cities illustrate the interconnected nature of urban development. This interconnectedness necessitates comprehensive regional planning that considers the wide-ranging impacts of expansion and contraction in central cities on surrounding areas. Policymakers should account for these dynamics to ensure equitable development across different regions.

Additionally, the study indicates that increases in built-up areas and energy consumption are key drivers of rising carbon emissions. Furthermore, issues related to population movement and environmental degradation also require significant attention. Therefore, governments should implement tailored low-carbon policies based on the development patterns and progress of different cities. For expanding cities, this involves integrating sustainable practices and resource efficiency. In contrast, contracting cities must prioritize environmental restoration and sustainable industrial practices.

Overall, this paper emphasizes the need for a multi-layered approach in urban policy development. On the one hand, a thorough understanding of the mechanisms driving urban growth can significantly enhance urban planning practices. By analyzing these mechanisms, planners can anticipate future challenges and opportunities and develop more forward-looking and adaptive policies. On the other hand, governments must consider both internal city factors and dynamic interactions with neighboring cities. Leveraging the advantages of developed cities to support surrounding areas is crucial for achieving regional balanced development. Policies that encourage inter-city cooperation and resource sharing can strengthen regional cohesion and promote sustainable urban growth.

5. Conclusions

This paper integrates two types of nighttime light data to create a long-term dataset of built-up areas in China from 2002 to 2021. Compared with MODIS land use data, the accuracy of the method exceeds 70%, demonstrating its effectiveness. By combining these data with social statistics, an evaluation system for urbanization development is established.

The analysis of urbanization development in 68 low-carbon pilot cities reveals that, over the past 20 years, most cities have undergone stable growth and are geographically concentrated. In contrast, cities with significant development fluctuations are more dispersed. These differences are influenced by internal city attributes, external policy factors, and the development of neighboring cities.

Regarding the factors influencing carbon emissions, the built-up area and nighttime light intensity exert a significant effect on carbon emissions across all city types. In contracting cities, employment numbers and virescence areas are significant, whereas, in expanding cities, population size and the proportion of the tertiary sector are more influential.

The study indicates that stable urban growth can lead to predictable carbon emissions and emphasizes the need for tailored policy approaches based on specific city characteristics.

Future research should focus on longitudinal studies of urbanization trends, the impact of specific policies on development and emissions, and comparative studies in other regions to evaluate the generalizability of these findings. This paper aims to enrich our understanding of urbanization and its environmental impacts, inspiring further exploration into low-carbon development strategies and contributing to the theoretical foundations of urban studies.

Author Contributions: Conceptualization, J.Q., Y.G. and T.Y.; methodology, J.Q.; validation, J.Q. and A.R.; formal analysis, J.Q.; resources, W.Y., R.D., A.R. and Z.W.; writing—original draft preparation, J.Q.; writing—review and editing, J.Q., Y.G., C.Z. and T.Y.; visualization, J.Q. and A.R.; supervision, Y.G. and S.G.; project administration, Y.G. and S.G.; funding acquisition, Y.G. and S.G. All authors have read and agreed to the published version of the manuscript.

Funding: This research was funded by Beijing Central Axis Protection Foundation, grant number 2023DYKT005.

Data Availability Statement: The data source has been declared in the article.

Acknowledgments: This research was supported by Beijing Central Axis Protection Foundation (2023DYKT005). The authors appreciate the valuable comments from the editor and reviewers.

Conflicts of Interest: The authors declare no conflicts of interest. The funders had no role in the design of the study; in the collection, analyses, or interpretation of data; in the writing of the manuscript; or in the decision to publish the results.

Appendix A

Table A1. 68 sample cities selected for this paper.

Number	City	Number	City	Number	City
1	Huai'an	24	Chengdu	47	Qinhuangdao
2	Hulun Buir	25	Beijing	48	Qingdao
3	Fuzhou	26	Changzhou	49	Ningbo
4	Chizhou	27	Baoding	50	Nanping
5	Chongqing	28	Ankang	51	Lu'an
6	Changsha	29	Zunyi	52	Liuzhou
7	Yinchuan	30	Zhuzhou	53	Lhasa
8	Xining	31	Zhongshan	54	Jingdezhen
9	Wuhan	32	Zhenjiang	55	Jinhua
10	Urumqi	33	Yuxi	56	Jinchang
11	Tianjin	34	Yantai	57	Jilin
12	Shijiazhuang	35	Yan'an	58	Jiaxing
13	Shenyang	36	Xiangtan	59	Ji'an
14	Shanghai	37	Xiamen	60	Huangshan
15	Nanjing	38	Wuzhong	61	Huaibei
16	Nanchang	39	Wuhai	62	Guilin
17	Lanzhou	40	Wenzhou	63	Guangyuan
18	Kunming	41	Weifang	64	Ganzhou
19	Jinan	42	Suzhou	65	Dalian
20	Hefei	43	Shenzhen	66	Chenzhou
21	Hangzhou	44	Sanya	67	Chaoyang
22	Guiyang	45	Sanming	68	Jincheng
23	Guangzhou	46	Quzhou		

References

1. Morikawa, H. *Urbanization and Urban System*; Damingtang Press: Hangzhou, China, 1989.
2. Liu, R.; Wang, M.; Chen, W. The influence of urbanization on organic carbon sequestration and cycling in soils of Beijing. *Landsc. Urban Plan.* **2018**, *169*, 241–249. [CrossRef]
3. Zhang, W.; Huang, B.; Luo, D. Effects of land use and transportation on carbon sources and carbon sinks: A case study in Shenzhen, China. *Landsc. Urban Plan.* **2014**, *122*, 175–185. [CrossRef]

4.   Anderson, W. Urban form, energy and the environment: A review of issues, evidence and policy. *Urban Stud.* **1996**, *33*, 7–36. [CrossRef]
5.   Shahbaz, M.; Sbia, R.; Hamdi, H.; Ozturk, I. Economic growth, electricity consumption, urbanization and environmental degradation relationship in United Arab Emirates. *Ecol. Indic.* **2014**, *45*, 622–631. [CrossRef]
6.   Hao, Y.; Zhang, Z.-Y.; Liao, H.; Wei, Y.-M.; Wang, S. Is CO_2 emission a side effect of financial development? An empirical analysis for China. *Environ. Sci. Pollut. Res.* **2016**, *23*, 21041–21057. [CrossRef]
7.   Sadorsky, P. The effect of urbanization on CO_2 emissions in emerging economies. *Energy Econ.* **2014**, *41*, 147–153. [CrossRef]
8.   Wang, S.; Fang, C.; Guan, X.; Pang, B.; Ma, H. Urbanisation, energy consumption, and carbon dioxide emissions in China: A panel data analysis of China's provinces. *Appl. Energy* **2014**, *136*, 738–749. [CrossRef]
9.   Parikh, J.; Shukla, V. Urbanization, energy use and greenhouse effects in economic development. *Glob. Environ. Chang.* **1995**, *5*, 87–103. [CrossRef]
10.   Gam, I.; Ben Rejeb, J. Electricity demand in Tunisia. *Energy Policy* **2012**, *45*, 714–720. [CrossRef]
11.   Mishra, V.; Sharma, S.; Smyth, R. Are fluctuations in energy consumption per capita transitory? Evidence from a panel of Pacific Island countries. *Energy Policy* **2009**, *37*, 2318–2326. [CrossRef]
12.   Martínez-Zarzoso, I.; Maruotti, A. The impact of urbanization on CO_2 emissions: Evidence from developing countries. *Ecol. Econ.* **2011**, *70*, 1344–1353. [CrossRef]
13.   Shafiei, S.; Salim, R.A. Non-renewable and renewable energy consumption and CO_2 emissions in OECD countries: A comparative analysis. *Energy Policy* **2014**, *66*, 547–556. [CrossRef]
14.   Xu, L.; Qu, J.; Li, H.; Zeng, J.; Zhang, H. Analysis and forecasting of carbon emissions from residential energy consumption in China. *Ecol. Econ.* **2019**, *35*, 19–23, 29.
15.   Xue, B.; Li, C.; Liu, Z.; Geng, Y.; Xi, F. Analysis of the relationship between global carbon emissions and urbanization from 1970 to 2007. *Adv. Clim. Chang. Res.* **2011**, *7*, 423–427.
16.   Cai, B.; Guo, H.; Cao, L.; Guan, D.; Bai, H. Local strategies for China's carbon mitigation: An investigation of chinese city-level CO_2 emissions. *J. Clean. Prod.* **2018**, *178*, 890–902. [CrossRef]
17.   Li, H.; Qi, Y. Comparative scenarios of carbon emissions in China by 2050. *Adv. Clim. Chang. Res.* **2011**, *7*, 271–280.
18.   Yang, M.; Guo, X.; Wu, L. Combined Land-Use and Transportation Demand Modeling Based on Equitableness. In Proceedings of the 2008 Workshop on Power Electronics and Intelligent Transportation System, Guangzhou, China, 2–3 August 2008; pp. 505–510. [CrossRef]
19.   Dhakal, S. Urban energy use and carbon emissions from cities in China and policy implications. *Energy Policy* **2009**, *37*, 4208–4219. [CrossRef]
20.   Zhang, X.-P.; Cheng, X.-M. Energy consumption, carbon emissions, and economic growth in China. *Ecol. Econ.* **2009**, *68*, 2706–2712. [CrossRef]
21.   Wang, Y.; Luo, X.; Chen, W.; Zhao, M.; Wang, B. Exploring the spatial effect of urbanization on multi-sectoral CO_2 emissions in China. *Atmos. Pollut. Res.* **2019**, *10*, 1610–1620. [CrossRef]
22.   Pan, W. What type of mixed-use and open? A critical environmental analysis of three neighborhood types in China and insights for sustainable urban planning. *Landsc. Urban Plan.* **2021**, *216*, 104221. [CrossRef]
23.   Li, H.; Peng, J.; Yanxu, L.; Yi'na, H. Urbanization impact on landscape patterns in Beijing City, China: A spatial heterogeneity perspective. *Ecol. Indic.* **2017**, *82*, 50–60. [CrossRef]
24.   Liddle, B.; Lung, S. Age-structure, urbanization, and climate change in developed countries: Revisiting STIRPAT for disaggregated population and consumption-related environmental impacts. *Popul. Environ.* **2010**, *31*, 317–343. [CrossRef]
25.   Qi, X.; Han, Y.; Kou, P. Population urbanization, trade openness and carbon emissions: An empirical analysis based on China. *Air Qual. Atmos. Health* **2020**, *13*, 519–528. [CrossRef]
26.   Gurney, K.R.; Romero-Lankao, P.; Seto, K.C.; Hutyra, L.R.; Duren, R.; Kennedy, C.; Grimm, N.B.; Ehleringer, J.R.; Marcotullio, P.; Hughes, S.; et al. Climate change: Track urban emissions on a human scale. *Nature* **2015**, *525*, 179–181. [CrossRef] [PubMed]
27.   Wang, C.-H.; Chen, N.; Chan, S.-L. A gravity model integrating high-speed rail and seismic-hazard mitigation through land-use planning: Application to California development. *Habitat Int.* **2017**, *62*, 51–61. [CrossRef]
28.   Aunan, K.; Wang, S. Internal migration and urbanization in China: Impacts on population exposure to household air pollution (2000–2010). *Sci. Total Environ.* **2014**, *481*, 186–195. [CrossRef]
29.   Dong, Y.; Yang, C.; Liu, X. The Exploration of China New Urbanization Theory. *Urban Stud.* **2017**, *24*, 26–34.
30.   Zhou, C.; Wang, S.; Wang, J. Examining the influences of urbanization on carbon dioxide emissions in the Yangtze River Delta, China: Kuznets curve relationship. *Sci. Total Environ.* **2019**, *675*, 472–482. [CrossRef]
31.   Wu, W.; Tan, W.; Wang, R.; Chen, W.Y. From quantity to quality: Effects of urban greenness on life satisfaction and social inequality. *Landsc. Urban Plan.* **2023**, *238*, 104843. [CrossRef]
32.   Wang, C.; Wang, F.; Zhang, X.; Zhang, H. Influencing mechanism of energy-related carbon emissions in Xinjiang based on the input-output and structural decomposition analysis. *J. Geogr. Sci.* **2017**, *27*, 365–384. [CrossRef]
33.   Baiocchi, G.; Minx, J.; Hubacek, K. The Impact of Social Factors and Consumer Behavior on Carbon Dioxide Emissions in the United Kingdom: A Regression Based on Input−Output and Geodemographic Consumer Segmentation Data. *J. Ind. Ecol.* **2010**, *14*, 50–72. [CrossRef]

34. Xu, Q.; Dong, Y.; Yang, R. Urbanization impact on carbon emissions in the Pearl River Delta region: Kuznets curve relationships. *J. Clean. Prod.* **2018**, *180*, 514–523. [CrossRef]
35. Elvidge, C.; Ziskin, D.; Baugh, K.; Tuttle, B.; Ghosh, T.; Pack, D.; Erwin, E.; Zhizhin, M. A Fifteen Year Record of Global Natural Gas Flaring Derived from Satellite Data. *Energies* **2009**, *2*, 595–622. [CrossRef]
36. Stewart, B.; Addison, D. *Nighttime Lights Revisited: The Use of Nighttime Lights Data as a Proxy for Economic Variables*; The World Bank Group: Washington, DC, USA, 2015. [CrossRef]
37. Yin, Z.; Li, X.; Tong, F.; Li, Z.; Jendryke, M. Mapping urban expansion using night-time light images from Luojia1-01 and International Space Station. *Int. J. Remote Sens.* **2020**, *41*, 2603–2623. [CrossRef]
38. Wang, L.; Jia, Y.; Li, X.; Gong, P. Analysing the Driving Forces and Environmental Effects of Urban Expansion by Mapping the Speed and Acceleration of Built-Up Areas in China between 1978 and 2017. *Remote Sens.* **2020**, *12*, 3929. [CrossRef]
39. Chen, Z.; Yu, B.; Song, W.; Liu, H.; Wu, Q.; Shi, K.; Wu, J. A New Approach for Detecting Urban Centers and Their Spatial Structure With Nighttime Light Remote Sensing. *IEEE Trans. Geosci. Remote Sens.* **2017**, *55*, 6305–6319. [CrossRef]
40. Ye, Y.; Yun, G.; He, Y.; Lin, R.; He, T.; Qian, Z. Spatiotemporal Characteristics of Urbanization in the Taiwan Strait Based on Nighttime Light Data from 1992 to 2020. *Remote Sens.* **2023**, *15*, 3226. [CrossRef]
41. Lu, C.; Li, L.; Lei, Y.; Ren, C.; Su, Y.; Huang, Y.; Chen, Y.; Lei, S.; Fu, W. Coupling Coordination Relationship between Urban Sprawl and Urbanization Quality in the West Taiwan Strait Urban Agglomeration, China: Observation and Analysis from DMSP/OLS Nighttime Light Imagery and Panel Data. *Remote Sens.* **2020**, *12*, 3217. [CrossRef]
42. Shan, Y.; Guan, D.; Liu, J.; Mi, Z.; Liu, Z.; Liu, J.; Schroeder, H.; Cai, B.; Chen, Y.; Shao, S.; et al. Methodology and applications of city level CO_2 emission accounts in China. *J. Clean. Prod.* **2017**, *161*, 1215–1225. [CrossRef]
43. Shan, Y.; Guan, Y.; Hang, Y.; Zheng, H.; Li, Y.; Guan, D.; Li, J.; Zhou, Y.; Li, L.; Hubacek, K. City-level emission peak and drivers in China. *Sci. Bull.* **2022**, *67*, 1910–1920. [CrossRef]
44. Shan, Y.; Liu, J.; Liu, Z.; Shao, S.; Guan, D. An emissions-socioeconomic inventory of Chinese cities. *Sci. Data* **2019**, *6*, 190027. [CrossRef] [PubMed]
45. Peng, J.; Tian, L.; Liu, Y.; Zhao, M.; Hu, Y.; Wu, J. Ecosystem services response to urbanization in metropolitan areas: Thresholds identification. *Sci. Total Environ.* **2017**, *607–608*, 706–714. [CrossRef] [PubMed]
46. Friedmann, J. Four Theses in the Study of China's Urbanization. *Int. J. Urban Reg. Res.* **2006**, *30*, 440–451. [CrossRef]
47. Tian, Y.; Tsendbazar, N.-E.; van Leeuwen, E.; Fensholt, R.; Herold, M. A global analysis of multifaceted urbanization patterns using Earth Observation data from 1975 to 2015. *Landsc. Urban Plan.* **2022**, *219*, 104316. [CrossRef]
48. Zhang, S.; Li, X.; Shi, Y. Spatial expansion of urban built-up areas in Zhengzhou based on GIS and its influencing factors. *Henan Sci.* **2017**, *35*, 1883–1888. [CrossRef]
49. Yang, X.; Liu, X.; Wang, R.; Liu, S. Evaluation of sustainable urbanization status in the Yangtze River Economic Belt. *Resour. Dev. Mark.* **2023**, *39*, 188–198.
50. Yu, M.; Guo, S.; Guan, Y.; Cai, D.; Zhang, C.; Fraedrich, K.; Liao, Z.; Zhang, X.; Tian, Z. Spatiotemporal heterogeneity analysis of yangtze river delta urban agglomeration: Evidence from nighttime light data (2001–2019). *Remote Sens.* **2021**, *13*, 1235. [CrossRef]
51. Zhao, M.; Zhou, Y.; Li, X.; Zhou, C.; Cheng, W.; Li, M.; Huang, K. Building a Series of Consistent Night-Time Light Data (1992–2018) in Southeast Asia by Integrating DMSP-OLS and NPP-VIIRS. *IEEE Trans. Geosci. Remote Sens.* **2020**, *58*, 1843–1856. [CrossRef]
52. Li, X.; Xu, H.; Chen, X.; Li, C. Potential of NPP-VIIRS Nighttime Light Imagery for Modeling the Regional Economy of China. *Remote Sens.* **2013**, *5*, 3057–3081. [CrossRef]
53. Imhoff, M. Using nighttime DMSP/OLS images of city lights to estimate the impact of urban land use on soil resources in the United States. *Remote Sens. Environ.* **1997**, *59*, 105–117. [CrossRef]
54. Olofsson, P.; Foody, G.M.; Herold, M.; Woodcock, S.V.S.C.E.; Wulder, M.A. Good practices for estimating area and assessing accuracy of land change. *Remote Sens. Environ. Interdiscip. J.* **2014**, *148*, 42–57. [CrossRef]
55. Stehman, S.V.; Selkowitz, D.J. A spatially stratified, multi-stage cluster sampling design for assessing accuracy of the alaska (USA) national land cover database (NLCD). *Int. J. Remote Sens.* **2010**, *31*, 1877–1896. [CrossRef]
56. Cakir, H.I.; Khorram, S.; Sac, N. Correspondence analysis for detecting land cover change. *Remote Sens. Environ. Interdiscip. J.* **2006**, *102*, 306–317. [CrossRef]
57. Stehman, S.V. Estimating area from an accuracy assessment error matrix. *Remote Sens. Environ.* **2013**, *132*, 202–211. [CrossRef]
58. Zhao, J.; Li, J.; Wang, P.; Hou, G. Research on the carbon peak path of Henan Province based on the Lasso-BP neural network model. *Environ. Eng.* **2022**, *40*, 151–156.
59. Chikán, A.; Czakó, E.; Kiss-Dobronyi, B.; Losonci, D. Firm competitiveness: A general model and a manufacturing application. *Int. J. Prod. Econ.* **2022**, *243*, 108316. [CrossRef]
60. Zheng, W.; Xu, X.; Wang, H. Regional logistics efficiency and performance in China along the Belt and Road Initiative: The analysis of integrated DEA and hierarchical regression with carbon constraint. *J. Clean. Prod.* **2020**, *276*, 123649. [CrossRef]
61. Joseph, S.; Jonathan, L. Greening the rust belt: A green infrastructure model for right sizing america's shrinking cities. *J. Am. Plan. Assoc.* **2008**, *74*, 451–466.
62. Turok, I.; Mykhnenko, V. The trajectories of european cities, 1960–2005. *Cities* **2007**, *24*, 165–182. [CrossRef]
63. Wiechmann, T.; Pallagst, K.M. Urban shrinkage in Germany and the USA: A comparison of transformation patterns and local strategies. *Int. J. Urban Reg. Res.* **2012**, *36*, 261–280. [CrossRef]

64. Hollander, J.B.; Németh, J. The bounds of smart decline: A foundational theory for planning shrinking cities. *Hous. Policy Debate* **2011**, *21*, 349–367. [CrossRef]

65. Haase, A.; Rink, D.; Grossmann, K.; Bernt, M.; Mykhnenko, V. Conceptualizing urban shrinkage. *Environ. Plan. A* **2014**, *46*, 1519–1534. [CrossRef]

66. Martinez-Fernandez, C.; Audirac, I.; Sylvie, F.; Emmanuèle Cunningham-Sabot. Shrinking Cities: Urban Challenges of Globalization. *Int. J. Urban Reg. Res.* **2012**, *36*, 213–225. [CrossRef]

67. Du, Z.W.; Li, X. New phenomena of growth and contraction in the rapid urbanization of the Pearl River Delta. *Acta Geogr. Sin.* **2017**, *72*, 1800–1811.

68. Zhang, S.; Wang, C.; Wang, J.; Yao, S.; Zhang, F.; Yin, G.; Xu, X. Comprehensive measurement of urban shrinkage in China and its temporal and spatial heterogeneity. *Chin. J. Popul. Resour. Environ.* **2020**, *30*, 72–82.

69. Anser, M.K.; Alharthi, M.; Aziz, B.; Wasim, S. Impact of urbanization, economic growth, and population size on residential carbon emissions in the saarc countries. *Clean Technol. Environ. Policy* **2020**, *22*, 923–936. [CrossRef]

70. Shen, Y.; Wang, C.; Gao, C.; Ding, L. Analysis of the spatiotemporal distribution characteristics and influencing factors of carbon emissions in the Zhejiang Bay Area Economic Belt based on urbanization. *J. Nat. Resour.* **2020**, *35*, 329–342.

71. Xue, J.; Zhang, X. Research on the Evolution and Influencing Factors of Urban Carbon Emissions under the "Dual Carbon" Goal: A Case Study of the Yellow River "Several Bend" Urban Agglomeration. *Frontier* **2023**, 125–136.

72. Liddle, B. Demographic dynamics and per capita environmental impact: Using panel regressions and household decompositions to examine population and transport. *MPIDR Work. Pap.* **2003**, *26*, 23–39. [CrossRef]

73. Chen, H.; Jia, B.; Lau, S. Sustainable urban form for Chinese compact cities: Challenges of a rapid urbanized economy. *Habitat Int.* **2008**, *32*, 28–40. [CrossRef]

74. Di Vittorio, A.V.; Simmonds, M.B.; Nico, P. Quantifying the effects of multiple land management practices, land cover change, and wildfire on the California landscape carbon budget with an empirical model. *PLoS ONE* **2021**, *16*, e0251346. [CrossRef] [PubMed]

75. Lin, Q.; Zhang, L.; Qiu, B.; Zhao, Y.; Wei, C. Spatiotemporal Analysis of Land Use Patterns on Carbon Emissions in China. *Land* **2021**, *10*, 141. [CrossRef]

76. Tao, Y.; Li, F.; Wang, R.; Zhao, D. Effects of land use and cover change on terrestrial carbon stocks in urbanized areas: A study from Changzhou, China. *J. Clean. Prod.* **2015**, *103*, 651–657. [CrossRef]

77. Wang, J.; He, A. Psychological Attribution and Policy Intervention Paths of Consumers' Low-Carbon Consumption Behavior: An Exploratory Study Based on Grounded Theory. *Nankai Bus. Rev.* **2011**, *14*, 80–89, 99.

78. Ma, Q.; Men, Y. The Impact of Consumption Values and Social Consumption Culture on Public Low-Carbon Consumption Behavior under a Low-Carbon Background. *Commer. Econ. Res.* **2022**, 69–73.

79. Gim, T.H.T. Analyzing the city-level effects of land use on travel time and co 2 emissions: A global mediation study of travel time. *Int. J. Sustain. Transp.* **2021**, *16*, 496–513. [CrossRef]

80. Lu, J.; Li, B.; Li, H.; Al-Barakani, A. Expansion of city scale, traffic modes, traffic congestion, and air pollution. *Cities* **2020**, *108*, 102974. [CrossRef]

remote sensing

Article

Exploring the Relationship between Temporal Fluctuations in Satellite Nightlight Imagery and Human Mobility across Africa

Grant Rogers [1,*], Patrycja Koper [1], Cori Ruktanonchai [1,2], Nick Ruktanonchai [1,2], Edson Utazi [1], Dorothea Woods [1], Alexander Cunningham [1], Andrew J. Tatem [1], Jessica Steele [1], Shengjie Lai [1] and Alessandro Sorichetta [3]

[1] WorldPop, School of Geography and Environmental Science, University of Southampton, Southampton SO17 1BJ, UK; patrycjak363@gmail.com (P.K.); cewarren6@vt.edu (C.R.); nrukt00@vt.edu (N.R.); a.d.cunningham@soton.ac.uk (A.C.); js1m14@soton.ac.uk (J.S.); shengjie.lai@soton.ac.uk (S.L.)
[2] Department of Population Health Sciences, Virginia Polytechnic Institute, State University, Blacksburg, VA 24061, USA
[3] Dipartimento di Scienze della Terra "Ardito Desio", Università degli Studi di Milano, Via Mangiagalli 34, 20133 Milan, Italy; alessandro.sorichetta@unimi.it
* Correspondence: grant.e.rogers@gmail.com

Abstract: Mobile phone data have been increasingly used over the past decade or more as a pretty reliable indicator of human mobility to measure population movements and the associated changes in terms of population presence and density at multiple spatial and temporal scales. However, given the fact mobile phone data are not available everywhere and are generally difficult to access and share, mostly because of commercial restrictions and privacy concerns, more readily available data with global coverage, such as night-time light (NTL) imagery, have been alternatively used as a proxy for population density changes due to population movements. This study further explores the potential to use NTL brightness as a short-term mobility metric by analysing the relationship between NTL and smartphone-based Google Aggregated Mobility Research Dataset (GAMRD) data across twelve African countries over two periods: 2018–2019 and 2020. The data were stratified by a measure of the degree of urbanisation, whereby the administrative units of each country were assigned to one of eight classes ranging from low-density rural to high-density urban. Results from the correlation analysis, between the NTL Sum of Lights (SoL) radiance values and three different GAMRD-based flow metrics calculated at the administrative unit level, showed significant differences in NTL-GAMRD correlation values across the eight rural/urban classes. The highest correlations were typically found in predominantly rural areas, suggesting that the use of NTL data as a mobility metric may be less reliable in predominantly urban settings. This is likely due to the brightness saturation and higher brightness stability within the latter, showing less of an effect than in rural or peri-urban areas of changes in brightness due to people leaving or arriving. Human mobility in 2020 (during COVID-19-related restrictions) was observed to be significantly different than in 2018–2019, resulting in a reduced NTL-GAMRD correlation strength, especially in urban settings, most probably because of the monthly NTL SoL radiance values remaining relatively similar in 2018–2019 and 2020 and the human mobility, especially in urban settings, significantly decreasing in 2020 with respect to the previous considered period. The use of NTL data on its own to assess monthly mobility and the associated fluctuations in population density was therefore shown to be promising in rural and peri-urban areas but problematic in urban settings.

Keywords: night-time lights; Google Aggregated Mobility Research Dataset; human mobility; Africa; rural and urban classification

Citation: Rogers, G.; Koper, P.; Ruktanonchai, C.; Ruktanonchai, N.; Utazi, E.; Woods, D.; Cunningham, A.; Tatem, A.J.; Steele, J.; Lai, S.; et al. Exploring the Relationship between Temporal Fluctuations in Satellite Nightlight Imagery and Human Mobility across Africa. *Remote Sens.* **2023**, *15*, 4252. https://doi.org/10.3390/rs15174252

Academic Editor: Christiane Weber

Received: 27 April 2023
Revised: 26 July 2023
Accepted: 29 July 2023
Published: 30 August 2023

1. Introduction

The acquisition of data pertaining to human mobility and presence is of critical importance within numerous fields of research for producing socioeconomic and development indicators, estimating greenhouse gas emissions, mapping urban extents, and assessing the spread, prevalence, and incidence of various human diseases, among others, driving a demand to refine the processes by which these mobility metrics are measured. With rates of human mobility increasing in their volumes and reach at both global and local scales, methods and datasets for quantifying them, particularly in data-sparse middle- and low-income settings, are becoming an important need. Moreover, with seasonal changes in human mobility that drive disease dynamics (Grenfell et al., 2001; Wesolowski et al., 2012; Wesolowski, Metcalf et al., 2015; Wesolowski, Qureshi et al., 2015) [1–4] the demand for resources (Steele et al., 2021) [5] and impact infrastructure planning needs (Strano et al., 2018) [6] can be particularly challenging to quantify (Lai et al., 2022; Mao et al., 2015; Song et al., 2021; Woods et al., 2022) [7–10]. Since its public distribution in recent decades, satellite-derived night-time light (NTL) imagery has proved itself a reliable proxy of human presence, where large bright areas correspond to higher populations compared to dimly lit areas (Bharti and Tatem, 2018; Bharti et al., 2011; Bustos, 2015) [11–13]. Furthermore, NTL imagery has been used as a global indicator of anthropogenic activity and development (Elvidge et al., 2012) [14] and, due to its historical availability and regular acquisition, enables comparative studies to be made over both short and long time periods (Doll et al., 2000; Ebener et al., 2005) [15,16]. As the technology has matured, so has the quality and availability of the NTL data for the scientific and operational communities with the new Visible Infrared Imaging Radiometer Suite (VIIRS) instrument, aboard the joint National Aeronautics and Space Administration (NASA) and National Oceanic and Atmospheric Administration (NOAA) Suomi National Polar-orbiting Partnership (Suomi NPP) and NOAA-20 satellites, offering several refinements compared to the older Defense Meteorological Satellite Program-Operational Linescan System (DMSP-OLS), such as increased spatial resolution of both the Ground Instantaneous Field of View (i.e., 0.55 versus 25 km^2 at Nadir) and the corresponding generated global grids (i.e., 15 versus 30 arc-second grid cell corresponding to ~500 m versus ~1 km at the equator) and temporal resolution (i.e., monthly versus annual) of cloud-free composites, as well as the full filtering of data impacted by stray light (Elvidge et al., 2013) [17].

Previous research has highlighted the potential of multi-temporal NTL imagery for measuring changes in population presence and density over time as a result of mobility. This has included seasonal labour migration into towns and cities in the Sahel region of Africa (Lai, Farnham et al., 2019) [18] and its impact on infectious disease dynamics (Bharti et al., 2011) [12], net migration at NUTS III level in Europe (Chen 2020) [19], seasonal flows of tourists (Stathakis and Baltas, 2018; Tselios and Stathakis, 2020) [20,21], COVID-19 lockdown in global megacities (Xu, et al., 2021) [22], and induced displacement (Lu et al., 2016) [23].

In each case, while the evidence is clear on NTL data capturing aspects of population presence and density changes induced by mobility, there are often other factors, such as disaster- or conflict-induced power outages (Montoya-Rincon et al., 2022) [24], that can be hard to disentangle, and thus, translation into quantitative direct measures of mobility can be challenging. Moreover, the saturation of brightness values in highly urbanised settings can also affect the relationship between changes in brightness, or lack thereof, and mobility. To improve our understanding of the value of NTL data for assessing human mobility and the associated changes in population presence and density, comparisons with alternative datasets are required.

Data on the aggregated movements of mobile phones over time have often been shown to be a reliable and accurate source of quantitative estimates of human movement patterns from subnational to global scales (Lai, Farnham et al., 2019; Lai, zu Erbach-Schoenberg et al., 2019; Ruktanonchai et al., 2018) [18,25,26]. Such data are typically obtained and derived either from Call Detail Records (CDRs), whereby anonymized and aggregated billing records of communications routed through cell towers are measured (Bengtsson et al., 2011; Buckee et al., 2013; Ruktanonchai et al., 2016) [27–29], or from aggregations of smartphone-derived GPS location data (Lai, zu Erbach-Schoenberg et al. 2019; Ruktanonchai et al., 2018) [25,26]. Each has their own set of biases and uncertainties which impact the accuracy and reliability of the assessed human mobility patterns (Lai, zu Erbach-Schoenberg et al., 2019) [25]. The Google Aggregated Mobility Research Dataset (GAMRD) data, providing a measurement of human movements as quantized flow metrics are principally derived from smartphones and represent the result of anonymous and aggregated phone locations for users who have opted into Google's Location History feature, which is off by default (Ruktanonchai et al., 2018) [26]. Previous research has indicated that there is a strong nonlinear relationship between GAMRD and NTL data (Dickinson et al., 2020) [30]. However, these studies only obtained and analysed mobility data for a short time period (e.g., 6–12 months) in a single country. The degree to which this relationship varies across locations and degrees of urbanisation has not been explored, particularly in low- and middle-income settings and at the monthly timescale. Based on multiple-year (2018–2020) and large-scale mobility data and NTL data at fine spatial resolution across 12 African countries, the current study seeks to address this through (i) examining the NTL-GAMRD relationship across Africa for two time periods (i.e., 2018/19 and 2020) and (ii) determining how the degree of urbanisation affects the correlations.

2. Materials and Methods

The GAMRD data contain anonymized mobility flows aggregated over users who have turned on their Location History setting that is off by default. The dataset aggregates flows between S2 cells which are here further aggregated by the level 2 administrative unit of origin and destination within and between 12 African countries (Figure 1).

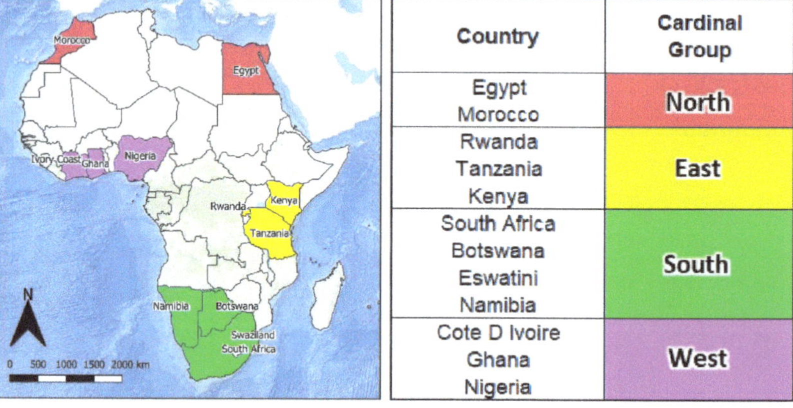

Country	Cardinal Group
Egypt	North
Morocco	
Rwanda	East
Tanzania	
Kenya	
South Africa	South
Botswana	
Eswatini	
Namibia	
Cote D Ivoire	West
Ghana	
Nigeria	

Figure 1. The 12 African countries selected for the current study and grouped according to the United Nations Geoscheme for Africa.

To produce this dataset, machine learning is applied to log data to automatically segment it into semantic "trips" (Bassolas et al., 2019) [31]. To provide strong privacy guarantees, all trips are anonymized and aggregated using a differentially private mechanism (Wilson et al., 2020) [32] to aggregate flows over time (Google, n.d.) [33]. This research was carried out on the resulting heavily aggregated and differentially private data. No individual user data was ever manually inspected; only heavily aggregated flows of large populations were handled.

All anonymized trips are processed in aggregate to extract their origin and destination location and time. For example, if users travelled from location A to location B within time interval t, the corresponding cell (A, B, t) in the tensor would be n∓err, where err is Laplacian noise. The automated Laplace mechanism adds random noise drawn from a zero-mean Laplace distribution and yields a (ϵ, δ)-differential privacy guarantee of $\epsilon = 0.66$ and $\delta = 2.1 \times 10^{-29}$ per metric. Specifically, for each week W and each location pair

(A, B), the number of unique users who took a trip from location A to location B during week W is calculated. To each of these metrics, Laplace noise from a zero-mean distribution of scale 1/0.66 is added. All metrics for which the noisy number of users is lower than 100 are removed, following the process described in (Wilson et al., 2020) [32], and the rest are published. This yields that each published metric satisfies (ϵ, δ)-differential privacy with values defined above. The parameter ϵ controls the noise intensity in terms of its variance, while δ represents the deviation from pure ϵ-privacy. The closer they are to zero, the stronger the privacy guarantees.

The GAMRD dataset used in this study covered the years 2018, 2019, and 2020 and initially contained weekly data representing relative population flows which were subsequently aggregated to a monthly timescale to allow direct comparison with the monthly VIIRS NTL data. The GAMRD data for 2020 were initially supplied in S2 Geometry (S2 Geometry, 2018) [34] and were then converted to the GCS-WGS84 coordinate system. Based on the origin and destination coordinates of the trips and using shapefiles representing level 2 administrative units, the relative flows were aggregated into 3 unique GAMRD-based flow metrics (i.e., internal flow, inward flow, and outward flow) as described in Table 1. Therefore, for every level 2 administrative unit of each country of interest, three distinct GAMRD-based flow metrics were available for each month over the study period (i.e., 2018–2020).

Table 1. The three Google Location History Flow Metrics created for the current study based on the direction and nature of movement within and between level 2 administrative units.

GAMRD Flow Metric	Description
Internal Flow	The internal flow within the same administrative unit (as this value increases, population total is unchanged but is more mobile)
Inward Flow	The external flow to the administrative units from others either within the same country or abroad (as this value increases, population within this admin unit increases)
Outward Flow	The external flow increases, population within this admin unit decreases)

Although it would have been preferable to combine GAMRD data from all years (i.e., 2018, 2019, and 2020), this was not possible due to the different spatial aggregation methods used to produce them, and so the study results were necessarily split into two time-periods: 2018/19 and 2020. The 2018/19 data represented flows originally calculated between S2 cells whilst the 2020 data were originally provided based on 1 km cells in WGS84. Although both groups were later reformatted to represent flows between level 2 administrative units in GCS-WGS84, the machine learning-based algorithm used to calculate the raw flows produced two unique datasets that can be justifiably compared to each other and to other data (namely, the VIIRS-NTL data in this study) but cannot be directly combined. Finally, the data referring to December 2019 were removed due to quality issues.

A Python script (Py v3.6) was created to download and extract VIIRS-NTL imagery for each country of interest and thereafter apply postprocessing stages in preparation for zonal statistics. The NTL data were provided by the Colorado School of Mines as monthly composites in geotiff format with the globe divided into 6 tiles (Elvidge et al., 2017) [35]. Monthly composites were filtered to exclude data impacted by stray light, lightning, lunar illumination, and cloud-cover where the monthly series is run globally using two different configurations. The first excludes any data impacted by stray light. The second includes these data if the radiance values have undergone the stray-light correction procedure. These two configurations, one of which includes the stray-light corrected data, will have more data coverage toward the poles, but will be of reduced quality with the decision of which configuration to use being dependent on the context. For each of the months from 2012–2020, for the monthly non-tiled versions, the annual masks for each year were applied to all the months for that year. For example, the 2020 lit mask was applied on all the months of 2020 (Mills et al., 2013) [36]. According to Elvidge et al. (2013) [17], in contrast to the DMSP overpass time which is near 7.30 pm, the SNPP overpass time is near 1.30 am and peak lighting is prior to 10 pm (after which there is some decline in the quantity of outdoor lighting, but we also agree with Eldvige et al. (2013) [17] that VIIRS data strongly indicate that there is still plenty of lighting being detected after midnight which may or may not only link to public infrastructure lights). After using the annual composites for removing ephemeral lights (unrelated to electric lighting) and background (non-lights) from monthly composites which were already processed for removing persistent gas flares, as well as the impact of sunlit, moonlit, stray lights, lightening, high energy particle, overglow, and cloud-cover, the monthly composites should only include electric lights, which may or may not be related to population presence and thus be affected by human mobility in various ways in different contexts (i.e., urban, peri-urban, vs. rural). At the time of the study design and data analysis (mid-2019), VIIRS annual composite data were not available for all three years of 2018–2020, and only the data that were available up until 2016 had been postprocessed to remove ephemeral lights, such as volcanic activity, fires, and atmospheric noise (Elvidge et al., 2017) [35]. However, the version 1 series of the monthly composites were not filtered to screen out lights from aurora, fires, boats, and other non-residential lights, thus requiring additional postprocessing (Li et al., 2013; Wang et al., 2017) [37,38]. The downloaded files were composed of a primary radiance raster (*rade9.tif) containing floating-point radiance values with units in nanoWatts/cm^2/sr and a corresponding coverage raster (*cvg.tif) of integer values representing the number of observations made on each pixel in each month to be used for quality control.

After the NTL rasters were downloaded, the postprocessing steps illustrated in Figure 2 were implemented following the recommendations set out in previous studies using VIIRS-NTL data (Li et al., 2013; Wang et al., 2017) [37,38]: firstly, the respective radiance and coverage rasters were clipped to the extent of each country of interest then buffered to 100 km to allow the preservation of pixels when projecting from WGS84 to UTM. Radiance pixels were then converted to zero if their values were either negative or zero in the 2016 annual composite raster. In cases where the coverage raster indicated that no observations were made in a particular pixel, the corresponding radiance pixel was converted to no data.

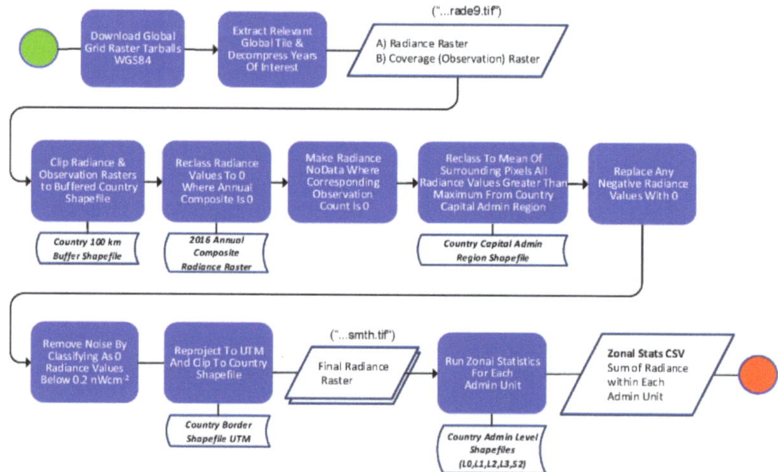

Figure 2. Workflow for extracting and postprocessing VIIRS night-time lights imagery.

To remove any signals created by non-residential lights (such as gas flares), the maximum pixel value in the capital city region of each country for each month was determined. Working under the assumption that no residential lights would be brighter than these radiance values outside of the capital city, any pixels outside of the capital region greater than these values were converted to the mean of the surrounding pixels. The final step was to remove background noise from the data by removing all values lower than 0.2 nWcm^{-2} (between 50 degrees north and south) (Elvidge et al., 2017) [35]. All rasters were then projected to UTM Albers and clipped using country shapefiles. The postprocessed and smoothed radiance rasters were labelled with an appropriate suffix (*smth.tif) ready for zonal statistics. Shapefiles representing level 2 administrative units, as provided by the GADM v3.6 (Warmerdam, 2008) [39], were used in conjunction with the postprocessed radiance rasters for zonal statistics. The "Sum" metric was used to determine the total radiance (Sum of Lights or SoL) per month within each level 2 administrative unit, with results eventually exported to a CSV file for further analysis.

Whilst the current study would ideally include all African countries, the geographical and temporal range was limited by the availability of the GAMRD data. The criteria by which countries were determined to have sufficient GAMRD data were that the data should have an average spatial coverage per country of more than 85% and that less than 10% of all administrative regions of each country have no data. By following these criteria, twelve countries for 2018, 2019, and 2020 were selected for a correlative analysis between the three GAMRD flow metrics and the corresponding NTL SoL values calculated for each level 2 administrative unit. In addition, a previous global study (Dickinson et al., 2020) [30] found that in different parts of the world, the relationship between mobility and light production differed considerably, and analyses should account for such regional variations. The twelve selected countries spanned a broad geographic range across the African continent and provided a convenient means for grouping for subsequent analysis according to the United Nations Geoscheme for Africa (UN Statistics Division, 2022) [40] which separates African countries according to cardinal direction as shown in Figure 1.

To allow collective analysis over multiple countries, each level 2 administrative unit within each country was categorised according to its degree of urbanisation. By grouping administrative units in this manner, it was possible to demonstrate how NTL and GAMRD correlations vary according to the degree of urbanisation, such as highly populated urban areas vs. sparsely populated rural areas. The GHS Settlement Model (GHS-SMOD) raster provides a classification raster of global coverage that gives for every 1 km^2 raster pixel a value corresponding to one of eight possible rural/urban classifications (Florczyk et al., 2019) [41] as illustrated in Figure 3.

Code	RGB	Grid level term	Spatial entity (polygon) Technical term	Other cells Technical term	Municipal level term Technical term
30	255 0 0	URBAN CENTRE GRID CELL	URBAN CENTRE *DENSE, LARGE CLUSTER*		CITY *LARGE SETTLEMENT*
23	115 38 0	DENSE URBAN CLUSTER GRID CELL	DENSE URBAN CLUSTER *DENSE, MEDUM CLUSTER*		DENSE TOWN *DENSE, MEDIUM SETTLEMENT*
22	168 112 0	SEMI-DENSE URBAN CLUSTER GRID CELL	SEMI-DENSE URBAN CLUSTER *SEMI-DENSE, MEDIUM CLUSTER*		SEMI-DENSE TOWN *SEMI-DENSE, MEDIUM SETTLEMENT*
21	255 255 0	SUBURBAN OR PERI-URBAN GRID CELL		SUBURBAN OR PERI-URBAN GRID CELLS *SEMI-DEMSE GRID CELLS*	SUBURBS OR PERI-URBAN AREA *SEMI-DENSE AREA*
13	55 86 35	RURAL CLUSTER GRID CELL	RURAL CLUSTER *SEMI-DENSE, SMALL CLUSTER*		VILLAGE *SMALL SETTLEMENT*
12	171 205 102	LOW DENSITY RURAL GRID CELL		LOW DENSITY RURAL GRID CELLS *LOW DENSITY GRID CELLS*	RURAL DISPERSED AREA *LOW DENSITY AREA*
11	205 245 122	VERY LOW DENSITY RURAL GRID CELL		VERY LOW DENSITY RURAL GRID CELLS *VERY LOW DENSITY GRID CELLS*	MOSTLY UNINHABITED AREA *VERY LOW DENSITY AREA*
10	122 182 245	WATER GRID CELL	-	-	-

Figure 3. The GHS Settlement Model (GHS-SMOD) rural/urban classification definitions.

As the GHS-SMOD raster provides rural/urban classifications at the 1 km^2 pixel level, an aggregation procedure is required to determine the "overall" degree of urbanisation of each level 2 administrative unit. An R-Script (R v4.0.3) was created for this purpose and took as input: the GHS-SMOD raster as provided by the GHSL-SMOD Project (Florczyk et al., 2019) [41], the 2020 Population Raster as provided by WorldPop (WorldPop School of Geography and Environmental Science, University of Southampton; Department of Geography and Geosciences, University of Louisville; Departement de Geographie, Universite de Namur; and the Center for International Earth Science Information, n.d.) [42] and the level 2 administrative unit shapefiles for the country of interest. The GHS-SMOD raster was extracted into eight separate rasters for each rural/urban classification and converted to binary format (i.e., 0.1). The WorldPop population raster was then multiplied for each of the eight binary rural/urban classification rasters. The resultant product rasters therefore contained the population count in each rural/urban classification, with the corresponding final level 2 administrative unit values obtained through summation via Zonal Statistics. The eight rural/urban classifications were combined into 3 groups: Group 1 (Classes 10, 11, 12, 13), Group 2 (Classes 21, 22, 23), and Group 3 (Class 30). The final classification was then determined via a nested hierarchy and majority approach whereby each unit was assigned to: Group 1 if Group 1 > 50% total country population, Group 2 if Group 1 < 50% and Group 3 < 50% total country population, or Group 3 if Group 3 > 50% total country population. Within the highest group, the individual highest classification value provided the final rural/urban classification for each administrative unit. A simplified illustration of the procedure using Kenya as an example is shown in Figure 4.

Calculating the Urban Classification of Administrative Units (Kenya as an Example)

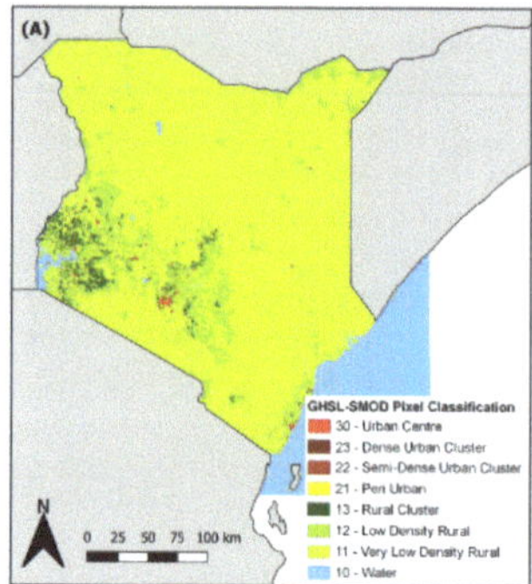

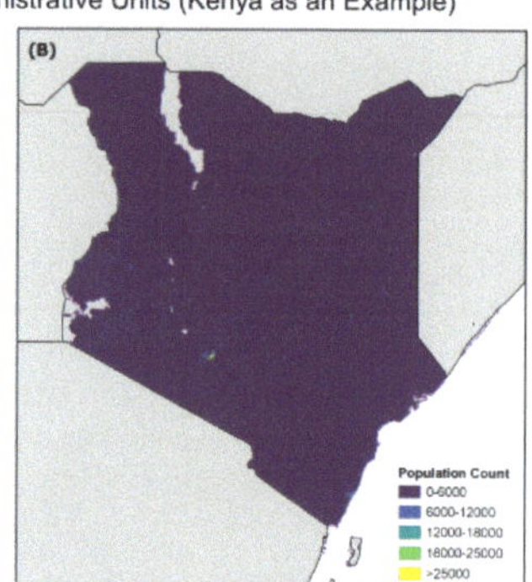

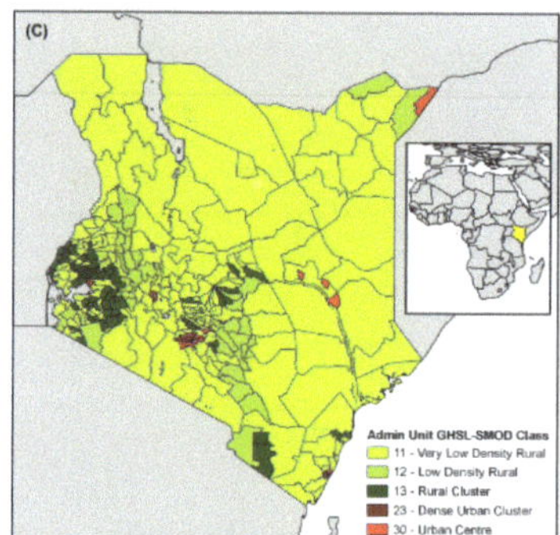

Figure 4. Visual overview of the process to determine urban classification at the administrative unit level: The GHSL-SMOD raster (**A**) is multiplied by the corresponding population raster (**B**) which is ran through zonal statistics to obtain the majority urban classification for the admin units of each country's level 2 shapefile (**C**).

For each administrative unit during the study period 2018–2019 and 2020, NTL data were used to calculate the SoL value per month, while the GAMRD data provided the anonymized and aggregated flows per month. To determine the relationship between GAMRD and NTL data, their monthly values were used as input for a correlation test using the Spearman's product-moment correlation coefficient (ρ) and its corresponding p-value. By grouping administrative units together according to the rural/urban classification, a high sample size was available for the correlation tests.

Furthermore, we examined the relationship between GAMRD and NTL data in a Gaussian Regression model-based framework to understand the amount of variation in the GAMRD data that could be explained by the NTL data. Both variables were log-transformed to improve the relationship between them and to improve normality. Two models were fitted: (i) a full model that includes NTL and degree of urbanisation as covariates and also accounts for temporal correlation and random variation between countries, and (ii) a reduced model that includes NTL as the only covariate. The reduced model was fitted in a frequentist framework while the full model was fitted in a Bayesian framework using the INLA package in R (Lindgren and Rue, 2015) [43]. The predictive ability of both models was evaluated using a hold-out cross-validation exercise in which we used 80% of the data for model fitting and 20% for validation. The Pearson's correlation coefficient and the R-squared statistic were then computed using the observed and predicted values. Both the full and reduced models were fitted for each GAMRD-based flow metric (i.e., internal flow, inward flow, and outward flow) separately.

3. Results

3.1. Correlation over Combined Countries

To gain a broad overview of how the NTL SoL values correlate with the three GAMRD flow metrics, all twelve selected African countries were included for the two time periods (i.e., 2018–2019 and 2020). Correlation strength definitions based on the Spearman's correlation coefficient (ρ) were taken from Akoglu (Akoglu, 2018) [44]. These are ρ (0.1–0.3) = weak correlation, ρ (0.4–0.6) = moderate correlation, and ρ (0.7–0.9) = strong correlation. An example of a typical correlation plot of NTL SoL radiance values ~ GAMRD Inward Flows is shown in Figure 5.

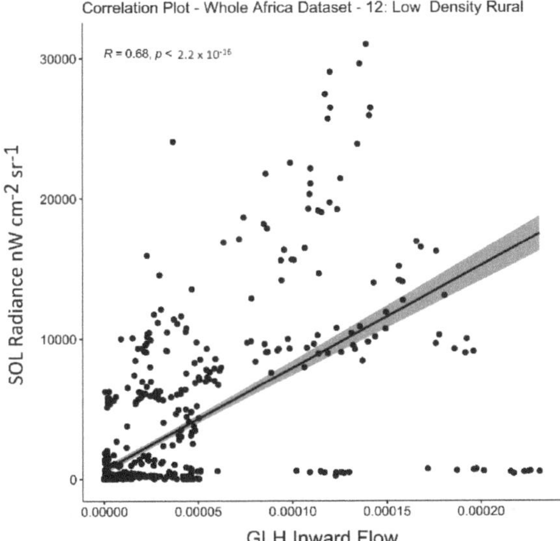

Figure 5. A sample of a correlation plot between the NTL SoL radiance values and GAMRD Inward Flows for the low-density rural class (12 in Figure 3) across all twelve selected African countries during the 2018–2019 period.

The GAMRD Internal Flow metric for 2018–2019 showed moderate to strong positive correlations in the rural group (11, 12, 13 in Figure 3) and dense-urban class (23 in Figure 3), with weak to moderate correlations in the urban (30 in Figure 3) and peri-urban (21 in Figure 3) classes. For 2020, the correlations were marginally higher in the rural group (11, 12, 13 in Figure 3) and marginally lower in the urban group (21, 23, 30 in Figure 3).

The GAMRD Inward Flow metric for 2018–2019 showed moderate positive correlations in the rural group (11, 12, 13 in Figure 3) and dense-urban class (23 in Figure 3) with weak to moderate correlations in the urban (30 in Figure 3) and peri-urban (21 in Figure 3) classes. For 2020, the correlations were considerably lower in the very-low-density rural (11 in Figure 3) class, marginally less in the rural (12, 13 in Figure 3) group and considerably less in the urban group (21, 23, 30 in Figure 3).

The GAMRD Outward Flow metric for 2018–2019 showed moderate negative correlations in the rural group (11, 12, 13 in Figure 3) and dense-urban (23 in Figure 3) class with weak to moderate correlations in the urban (30 in Figure 3) and peri-urban (21 in Figure 3) classes. For 2020, the correlations were considerably less in the very-low-density rural (11 in Figure 3) class, marginally less in the rural group (12, 13 in Figure 3) and considerably less in the urban group (21, 23, 30 in Figure 3). The Spearman's correlation coefficients for each rural/urban classification were placed on a bar chart with the relative proportions of each rural/urban classification included for reference as shown in Figure 6.

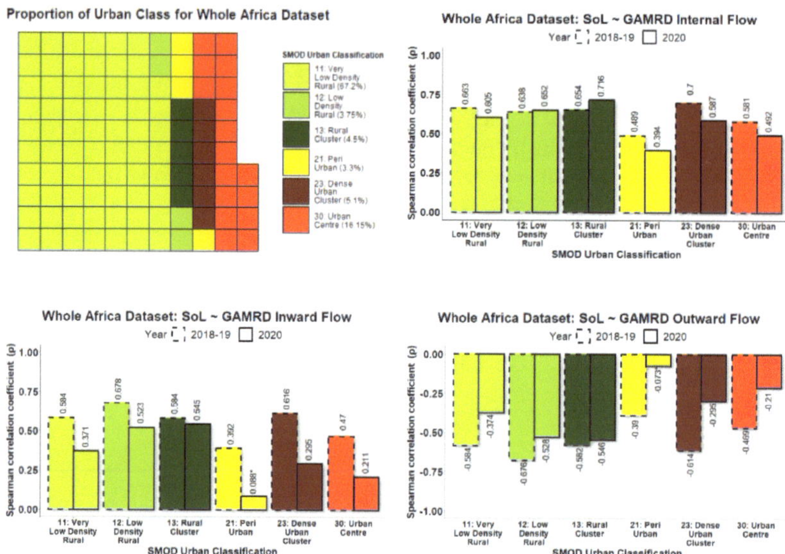

Figure 6. The proportion of rural/urban classifications for all level 2 administrative units within the 12 selected Africa countries (top left). The Spearman's correlation coefficients for each rural/urban classification for the NTL SoL radiance values and GLH flow metrics. All values $p < 0.0001$ except those indicated by an asterisk *.

The results indicated that the rural groups (11, 12, 13 in Figure 3) and dense-urban class (23 in Figure 3) have the highest correlation between the NTL SoL radiance values and GAMRD flow metrics, with the urban group (21, 30 in Figure 3) having the lowest correlations. The differences in correlations between the two time periods were considerable, with 2020 having far smaller correlations coefficients across almost all rural/urban classifications compared to those of 2018–2019.

Using the Gaussian regression model for each time period (i.e., 2018–2019 and 2020), significant positive relationships were observed between the GAMRD flow metrics and NTL SoL radiance values for both internal and inward flows, and significant negative relationships for outward flows. Moreover, significant differences were found between the "urban centre" class and all other rural/urban classifications. The fitted full models showed good predictive power based on the cross-validation exercise results which are shown in Table 2. For 2018–2019, the out-of-sample R^2 values of the fitted models were 0.29, 0.28, and 0.27 for the internal, inward, and outward flows, while the corresponding correlations were 0.54, 0.53, and 0.52, respectively. This means that models were able to explain at least 27% of the total variation in the GAMRD flows using NTL SoL as a covariate, whilst adjusting for the other sources of variation in the data. However, with the reduced model, this predictive power reduces to at most 9%, highlighting the importance of accounting for the degree of urbanisation and other sources of variation in the data in the analysis. For 2020, similar results were obtained, although the predictive powers of both the full models ($\leq$26%) and the reduced models (4%) were lower. This can be explained by the fact that while NTL SoL radiance values remained relatively stable throughout all the years (2018, 2019, and 2020), the GAMRD flows significantly decreased.

Table 2. Out-of-sample validation NTL-GAMRD statistics based on hold-out cross-validation exercise.

Year	GAMRD Metric	Correlation		R^2	
		Full Model	Reduced Model	Full Model	Reduced Model
2018–2019	Internal	0.54	0.31	0.29	0.09
	Inward	0.53	0.28	0.28	0.08
	Outward	0.52	0.28	0.27	0.08
2020	Internal	0.48	0.21	0.23	0.04
	Inward	0.51	0.20	0.26	0.04
	Outward	0.47	0.19	0.23	0.04

3.2. Annual Correlation Variation

To gain a deeper insight into why the correlation coefficients for NTL-GAMRD were so different between the two time periods of 2018/19 and 2020, the data were split monthly to analyse any change in correlation during the year, as illustrated in Figure 7. During 2018, the Spearman's correlation coefficient (ρ) remained relatively stable throughout most of the year for all three GAMRD flow metrics except during April, when the values decreased substantially for the rural group (11, 12, 13 in Figure 3) and "urban centre" class (30 in Figure 3) and increased slightly for the "peri-urban" and "dense urban cluster" classes. During 2019, the Spearman's correlation coefficient (ρ) remained relatively stable throughout the year across all the rural/urban classifications with the only noticeable perturbation in values during October. During 2020, two large perturbations in the Spearman's correlation coefficient (ρ) were observed centred around the months of April and September.

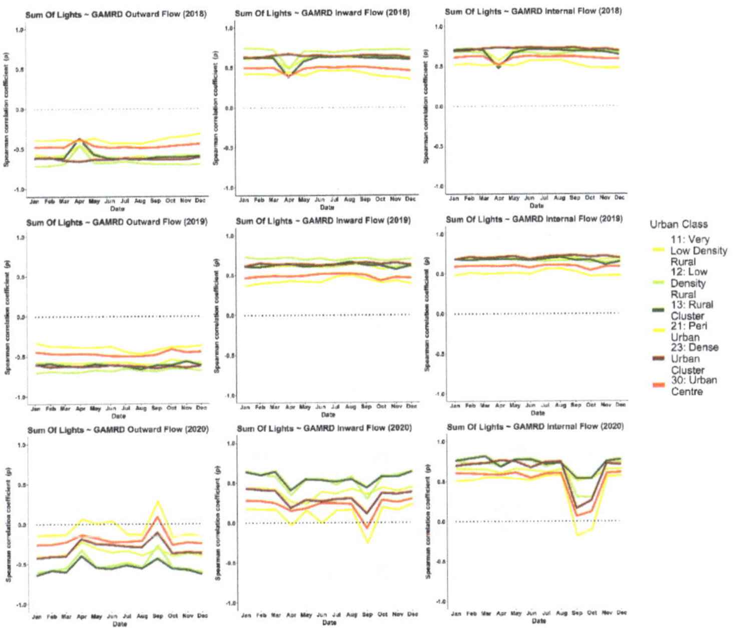

Figure 7. Spearman's correlation coefficient variation per month between NTL SoL radiance values and GLH flow metrics for the 12 selected Africa countries over the years 2018, 2019, and 2020.

3.3. Sum of Lights and GAMRD Annual Value Variation

To determine the possible source of the inter-annual variation of the correlation, the NTL SoL radiance and GAMRD flow totals were analysed separately across each year of 2018, 2019, and 2020. For each year, a combined NTL SoL metric was calculated by grouping all administrative units according to their degree of urbanisation across all countries and then summing the corresponding NTL SoL radiance values for each month, as shown in Figure 8. Whilst the individual values themselves were not analysed, the degree to which this metric varied through the year may highlight months of particularly high variance of the NTL SoL radiance. The variation of the combined NTL SoL metric was most noticeable within the administrative units classified as "urban centres" (30 in Figure 3), whilst for all other rural/urban classifications, the metric was relatively stable throughout all the years 2018, 2019, and 2020.

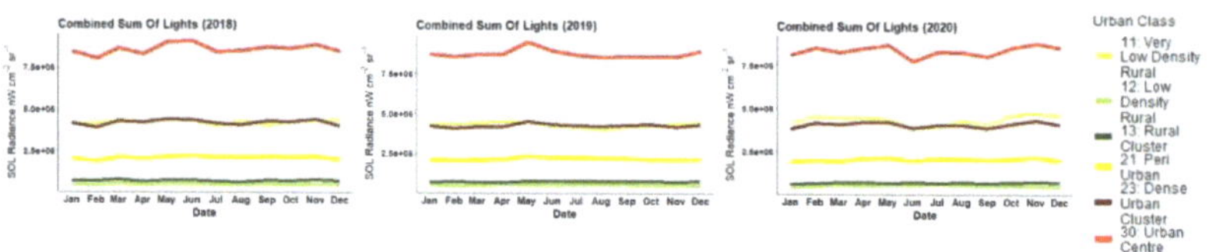

Figure 8. The combined NTL SoL metric for all administrative units in each month in 2018, 2019, and 2020 across all twelve selected African countries.

Similarly, for each of the three GAMRD flow metrics (i.e., internal flow, inward flow, and outward flow), administrative units were grouped according to their degree of urbanisation across all countries, and the corresponding flow values for each month were summed together as shown in Figure 9. As with the combined NTL SoL metric, the total flow values were not directly analysed, but rather their variation throughout the year was used as an indicator of periods of potentially unusual human mobility. During 2018 and 2019, the GAMRD flow totals remained relatively stable for all the rural/urban classifications except the "urban centre" class (30 in Figure 3). During 2020, the "urban centre" class (30 in Figure 3) showed drastic variations across a large range for most months, with two particularly large shifts in April and September, whilst the remaining rural/urban classes remained relatively stable throughout the year.

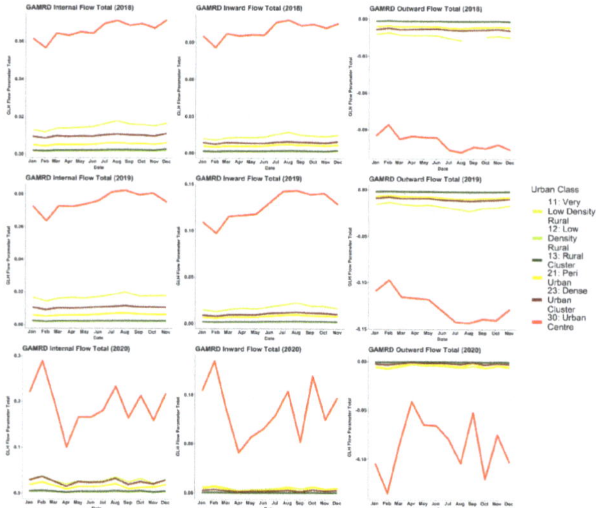

Figure 9. The total GLH flow parameters summed over all administrative units and grouped according to urban classification.

3.4. Correlation over Country Groups

According to the United Nations Geoscheme for Africa, the twelve study countries were separated to four cardinal groups (north, south, east, and west) and the NTL-GAMRD correlation analysis was repeated as illustrated in Figures A1–A4 in Appendix A. The Northern Africa group had a high degree of urbanisation with 37.43% of administrative units classified as "urban centres" (30 in Figure 3). Correlation results were generally higher within the rural group (11, 12, 13 in Figure 3) for both the 2018–2019 and 2020 time periods, with correlations for most classes noticeably lower for 2020 than for 2018–2019. The Eastern Africa group had a high degree of rural presence with 66.47% of administrative units classified as "very low density rural" (11 in Figure 3). Correlation results were highest within the rural group (11, 12, 13 in Figure 3) and "urban centres" (30 in Figure 3) for both the 2018–2019 and 2020 time periods, with correlations for most classes noticeably lower for 2020 than for 2018–2019. Correlations for 2020 were comparable with correlations for 2018–2019 across most classes with both positive and negative variance. The Southern Africa group had a balanced rural/urban proportion with neither urban nor rural classifications dominating the administrative units. Correlation results were highest within the rural group (12, 13 in Figure 3) and urban group (23, 30 in Figure 3) for both the 2018–2019 and 2020 time periods, with correlations for 2020 considerably lower than correlations for 2018–2019 for all classes. The Western Africa group had a high degree of rural presence with 82.58% of administrative units classified as "very low density rural" (11 in Figure 3). Correlation results were highest within the urban group (21, 23, 30 in Figure 3) for both the 2018–2019 and 2020 time periods, with correlations for 2020 noticeably lower than correlations for 2018–2019 for most classes. Correlations for 2020 were comparable with correlations for 2018–2019 across most classes with both positive and negative variance.

4. Discussion

NTL has been widely used for population spatial distribution mapping, and a strong correlation between NTL and GAMRD data has been demonstrated in previous studies (Dickinson et al., 2020) [30]; however, the variation of this relationship according to the degree of urbanisation was previously unexplored. The longitudinal Google Aggregated Mobility Research Dataset (GAMRD) and the VIIR NTL data for Africa in 2018–2020, according to the degree of urbanisation, provide a good opportunity to improve our understanding of the value of NTL data for assessing human mobility and the associated changes in population presence in low- and middle-income countries. The diversity of the study's countries in different regions does ensure that this study contains wide variance in socioeconomic, geographic, and demographic contexts. Our study, conducted from 2019–2021, has demonstrated the high variability in correlations between administrative-unit-level NTL radiance values and GAMRD flow metrics across a broad geographic range and within different rural/urban classifications. Administrative units classified as rural and semi-rural were shown to have on average the highest NTL-GAMRD correlation whilst administrative units classified as "urban centres" had the lowest (not including the peri-urban class, which had several low p-values). This is likely due to the saturation and greater stability of lighting and NTL brightness values within urban centres/areas. Indeed, large urban centres/areas in Sub-Saharan Africa and elsewhere tend to be more consistently lit throughout the year and are often bright enough in their core to saturate NTL brightness values (Zhao et al., 2019) [45]. This means that changes in brightness due to human mobility and the associated population presence and density changes are less likely to occur in urban centres/areas than in small towns/rural and peri-urban areas, where population arrivals may lead to an increase in brightness due to electric lighting or fires in residential areas (Bharti et al., 2011) [12].

What was most noticeable in the study was the significant difference in correlation strength between the two time periods of 2018–2019 and 2020. Correlations across most rural/urban classifications and particularly "urban centres" were considerably lower in 2020 than in 2018–2019. Whilst the variation of the NTL SoL radiance values across all urban classes remained relatively stable throughout the year, the GHL flow metrics referring to 2020 showed that the corresponding flow values were far more erratic than those in 2018–2019, with the months of April and September being most prominent in their deviation, creating a consequential effect for the NTL-GAMRD correlations during these months. These changes might be attributed to the implementation of lockdown measures during the COVID-19 pandemic, with the first wave in March–April (Haider et al., 2020) [46] and the second wave in September–October (Kuehn, 2021) [47]. This has implications for the reliability of NTL data as a proxy for human mobility during periods of unusual human activity, such as lockdown periods. In addition, NTL data have several drawbacks such as delayed access to real-time data, low light detection thresholds, and the requirement for additional postprocessing; however, these are expected to continue to be addressed as the technology further develops and novel data sources become available (Zhao et al., 2019) [45].

In addition, we only found one similar study conducted by Dickinson et al., 2020 [30]. Based on linear regression and random forest models, they used Google's human mobility data in 2016 to predict VIIRS satellite imagery and then assessed how accurately this simulated global NTL imagery could be used to predict GDP across regions in 2015–2016. They demonstrated that the relationship between human mobility and VIIRS NTL was nonlinear and varied considerably around the globe. The differences across regions were made clear by the improvement in the model performance when modelling each region independently rather than constructing a single global model. Our study further measured the degree to which this relationship varied across locations with different levels of urbanisation and development in 2018–2020. However, we found that compared with urban settings, there was a higher association between NTL data and mobility changes in rural and peri-urban areas. In addition, a reduced NTL-GAMRD correlation strength in 2020 was observed, especially in urban settings, most probably because of the monthly NTL SoL radiance values remaining relatively similar in 2018–2019 and 2020 but the human mobility significantly decreasing in 2020 with respect to the previous considered period. Our study provides new insights about changes in mobility and NTL as well as their association across settings during a global crisis such as the pandemic.

Furthermore, it is important to highlight that GAMRD data present several limitations and potential biases as well. Indeed, such data are limited to mobile internet coverage and smartphone users who have opted into Google's Location History feature, which is off by default, and thus, they may not be representative of the population as a whole. Similarly, their representativeness may vary by location and be particularly low in rural areas characterised by low population densities. Additionally, GAMRD data are still likely to be biased towards educated males living in urban areas (Lai, zu Erbach-Schoenberg et al., 2019) [25]. Moreover, comparisons across rather than within locations are only descriptive, since these regions can differ in substantial ways. Another primary drawback of GAMRD data is the difficulty of obtaining them due to the restrictive data sharing policies implemented to protect individual privacy and GAMRD data being subject to differential privacy algorithms designed to protect user's anonymity, which obscure fine details. However, considering potential biases of representativeness among populations across regions, it is important to include subnational and up-to-date statistics on smart-device ownership and internet penetration in future research where possible. Mobile phone subscribers and smartphone adoption are expected to continue growing in low- and middle-income countries, and surveys for measuring mobile phone/smartphone penetration and social media coverage may be necessary to obtain more precise metrics for each country and subnational region (e.g., administrative unit level 1 or 2). In addition, considering the potential biases of representativeness among populations, rather than grouping countries together, it may be of interest to analyse each country separately to avoid generalisations in the future.

5. Conclusions

Following the global COVID-19 pandemic and the consequent restrictions on human mobility, importance has risen dramatically for datasets that can successfully explore short-term and intra-annual human mobility and assess the associated population presence and density changes. With several proxies available, it is useful to understand the limitations and accuracy of each dataset that can be used for mobility research, which motivates the current study. Results have indicated that VIIRS NTL data may be best-suited for the analysis of human mobility within more rural areas and that during periods of unusual human activity (such as the lockdown periods in 2020), VIIRS NTL data may not provide the necessary spatial resolution for detailed study. In addition, as the NTL-GAMRD correlation was found to be potentially weaker in "urban centres" areas, this highlights the importance of integrating additional geospatial datasets that are able to capture different scales of variation into a larger multivariate model, instead of our current simple modelling framework.

As refinements in NTL technology become available and new datasets are released with higher spatial resolution and enhanced postprocessing, it is hoped that these limitations may be overcome. Despite the demonstrated efficacy of the GAMRD in "urban centres" areas and during lockdown periods, with NTL data continuing to be publicly available with wide geographic coverage, its use can remain important as a proxy of human mobility and the associated population presence and density changes until alternative datasets, such as mobile phone locations, can be more easily accessed by the scientific and operational communities.

Author Contributions: Conceptualization, A.J.T., J.S. and A.S.; methodology, G.R., P.K., C.R., N.R., E.U., A.C., J.S. and A.S.; software, G.R. and E.U.; formal analysis, G.R. and E.U.; investigation, G.R.; data curation, G.R., C.R. and N.R.; writing—original draft preparation, G.R.; writing—review and editing, G.R., E.U., S.L., D.W., A.C., A.J.T., J.S. and A.S.; visualization, G.R.; supervision, A.S.; project administration, A.J.T., J.S. and A.S.; funding acquisition, A.J.T., J.S. and A.S.; A.S. only worked with the GAMRD data while being employed at the University of Southampton and has not touched the data since he left. All authors have read and agreed to the published version of the manuscript.

Funding: This work was funded by the Bill & Melinda Gates Foundation (OPP1134076, INV-024911). The funders had no role in study design, data collection and analysis, decision to publish, and preparation of the manuscript.

Data Availability Statement: The code used for the analysis described in this study is available at the following GitHub repository: The data on night light data and urbanisation classification are available from https://ghsl.jrc.ec.europa.eu/ghs_smod2019.php (accessed on 28 July 2023). The Google Aggregated Mobility Research Dataset used for this study is available with permission from Google LLC. Ethical clearance for collecting and using secondary data in this study was granted by the institutional review board of the University of Southampton (48002). All data were supplied and analysed in an anonymous format, without access to personal identifying information.

Acknowledgments: The authors wish to thank Google LLC for sharing the mobility dataset and the Colorado School of Mines for providing early access to the 2020 VIIRS-NTL data. Thanks also to the Joint Research Centre of the European Commission for their assistance in developing the necessary scripts for the rural/urban classification of the administrative units.

Conflicts of Interest: The authors declare no conflict of interest.

Appendix A

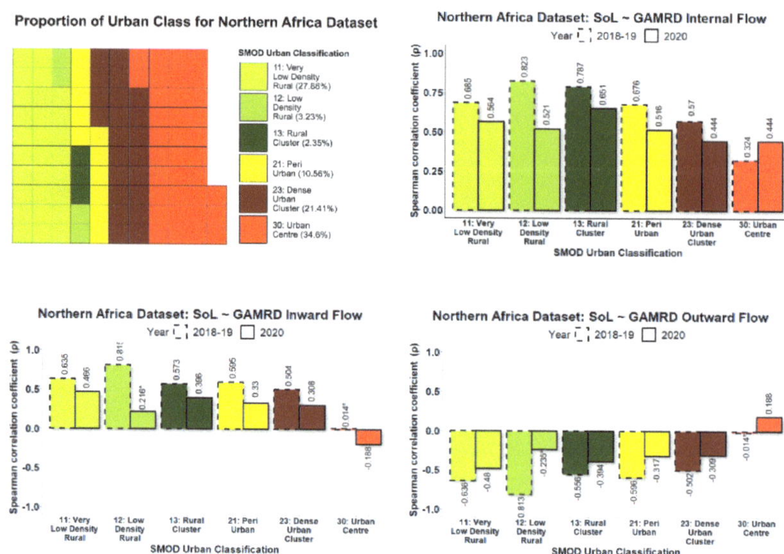

Figure A1. The proportion of rural/urban classifications for all level 2 administrative units within the Northern African countries listed in Figure 1 (top left). The Spearman's correlation coefficients for each rural/urban classification according to the NTL SoL radiance values and GLH flow metrics. All values $p < 0.001$ except those indicated by an asterisk *.

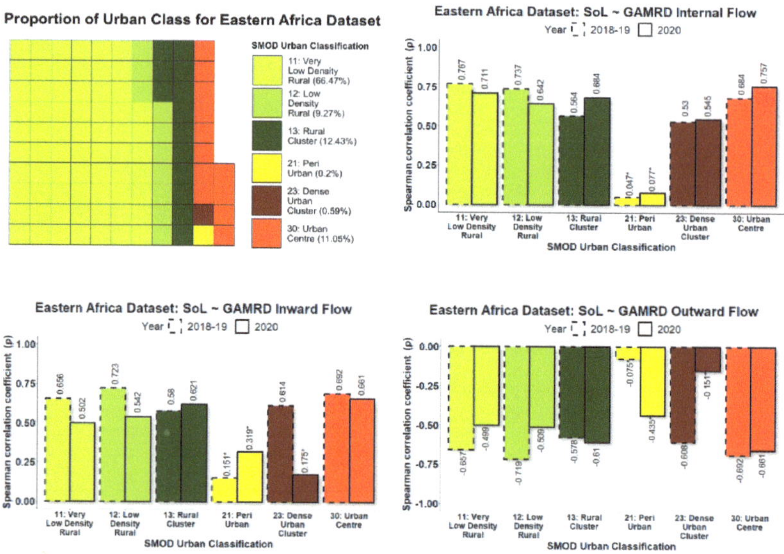

Figure A2. The proportion of rural/urban classifications for all level 2 administrative units within the Eastern African countries listed in Figure 1 (top left). The Spearman's correlation coefficients for each rural/urban classification according to the NTL SoL radiance values and GLH flow metrics. All values $p < 0.001$ except those indicated by an asterisk *.

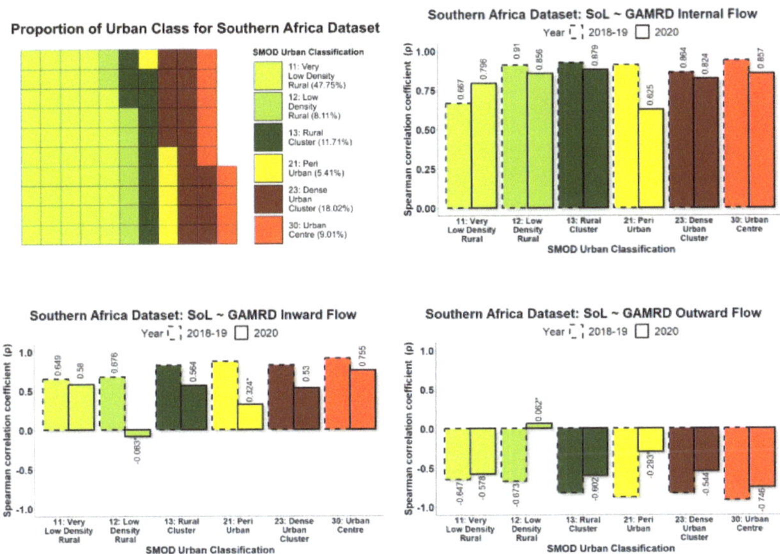

Figure A3. The proportion of rural/urban classifications for all level 2 administrative units within the Southern Africa countries listed in Figure 1 (top left). The Spearman's correlation coefficients for each rural/urban classification according to the NTL SoL radiance values and GLH flow metrics. All values $p < 0.001$ except those indicated by an asterisk *.

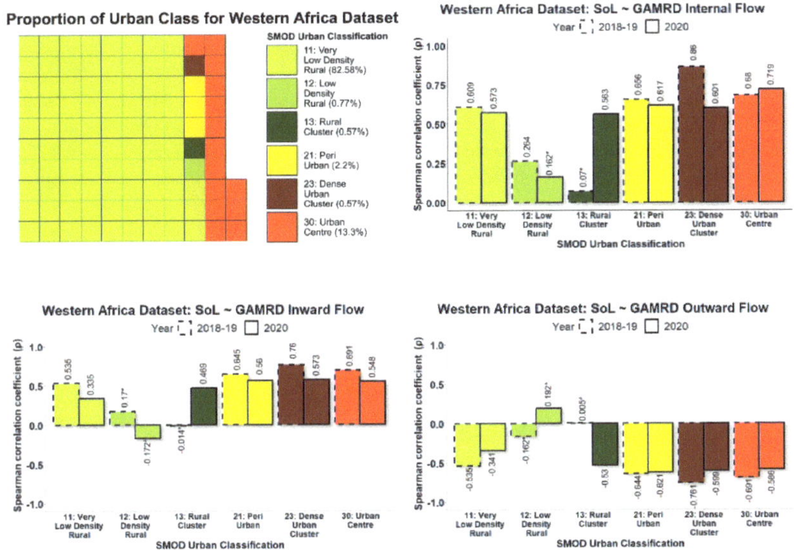

Figure A4. The proportion of rural/urban classifications for all level 2 administrative units within the Western Africa countries listed in Figure 1 (top left). The Spearman's correlation coefficients for each rural/urban classification according to the NTL SoL radiance values and GLH flow metrics. All values $p < 0.001$ except those indicated by an asterisk *.

References

1. Grenfell, B.T.; Bjørnstad, O.N.; Kappey, J. Travelling waves and spatial hierarchies in measles epidemics. *Nature* **2001**, *414*, 716–723. [CrossRef]
2. Wesolowski, A.; Eagle, N.; Tatem, A.J.; Smith, D.L.; Noor, A.M.; Snow, R.W.; Buckee, C.O. Quantifying the Impact of Human Mobility on Malaria. *Science* **2012**, *338*, 267–270. [CrossRef]
3. Wesolowski, A.; Metcalf, C.J.E.; Eagle, N.; Kombich, J.; Grenfell, B.T.; Bjørnstad, O.N.; Lessler, J.; Tatem, A.J.; Buckee, C.O. Quantifying seasonal population fluxes driving rubella transmission dynamics using mobile phone data. *Proc. Natl. Acad. Sci USA* **2015**, *112*, 11114–11119. [CrossRef]
4. Wesolowski, A.; Qureshi, T.; Boni, M.F.; Sundsøy, P.R.; Johansson, M.A.; Rasheed, S.B.; Engø-Monsen, K.; Buckee, C.O. Impact of human mobility on the emergence of dengue epidemics in Pakistan. *Proc. Natl. Acad. Sci. USA* **2015**, *112*, 11887–11892. [CrossRef] [PubMed]
5. Steele, J.E.; Pezzulo, C.; Albert, M.; Brooks, C.J.; zu Erbach-Schoenberg, E.; O'Connor, S.B.; Sundsøy, P.R.; Engø-Monsen, K. Nilsen, K.; Graupe, B.; et al. Mobility and phone call behavior explain patterns in poverty at high-resolution across multiple settings. *Humanit. Soc. Sci. Commun.* **2021**, *8*, 288. [CrossRef]
6. Strano, E.; Viana, M.P.; Sorichetta, A.; Tatem, A.J. Mapping road network communities for guiding disease surveillance and control strategies. *Sci. Rep.* **2018**, *8*, 4744. [CrossRef]
7. Lai, S.; Sorichetta, A.; Steele, J.; Ruktanonchai, C.W.; Cunningham, A.D.; Rogers, G.; Koper, P.; Woods, D.; Bondarenko, M.; Ruktanonchai, N.W.; et al. Global holiday datasets for understanding seasonal human mobility and population dynamics. *Sci. Data* **2022**, *9*, 17. [CrossRef]
8. Mao, L.; Wu, X.; Huang, Z.; Tatem, A.J. Modeling monthly flows of global air travel passengers: An open-access data resource. *J. Transp. Geogr.* **2015**, *48*, 52–60. [CrossRef]
9. Song, B.; Yan, X.-Y.; Tan, S.; Sai, B.; Lai, S.; Yu, H.; Ou, C.; Lu, X. Human mobility models reveal the underlying mechanism of seasonal movements across China. *Int. J. Mod. Phys. C* **2021**, *33*, 2250054. [CrossRef]
10. Woods, D.; Cunningham, A.; Utazi, C.E.; Bondarenko, M.; Shengjie, L.; Rogers, G.E.; Koper, P.; Ruktanonchai, C.W.; zu Erbach-Schoenberg, E.; Tatem, A.J.; et al. Exploring methods for mapping seasonal population changes using mobile phone data. *Humanit. Soc. Sci. Commun.* **2022**, *9*, 247. [CrossRef]
11. Bharti, N.; Tatem, A.J. Fluctuations in anthropogenic nighttime lights from satellite imagery for five cities in Niger and Nigeria. *Sci. Data* **2018**, *5*, 180256. [CrossRef] [PubMed]
12. Bharti, N.; Tatem, A.J.; Ferrari, M.J.; Grais, R.F.; Djibo, A.; Grenfell, B.T. Explaining seasonal fluctuations of measles in Niger using nighttime lights imagery. *Science* **2011**, *334*, 1424–1427. [CrossRef] [PubMed]
13. Bustos, M.F.A. Population, Demography and Nighttime Lights an Examination of the Effects of Population Decline on Settlement Patterns in Europe. 2015. Available online: http://www.cfe.lu.se (accessed on 28 July 2023).
14. Elvidge, C.D.; Baugh, K.E.; Anderson, S.J.; Sutton, P.C.; Ghosh, T. The Night Light Development Index (NLDI): A spatially explicit measure of human development from satellite data. *Soc. Geogr.* **2012**, *7*, 23–35. [CrossRef]
15. Doll, C.N.H.; Muller, J.-P.; Elvidge, C.D. Night-time imagery as a tool for global mapping of socioeconomic parameters and greenhouse gas emissions. *AMBIO A J. Hum. Environ.* **2000**, *29*, 157–162. [CrossRef]
16. Ebener, S.; Murray, C.; Tandon, A.; Elvidge, C.C. From wealth to health: Modelling the distribution of income per capita at the sub-national level using night-time light imagery. *Int. J. Health Geogr.* **2005**, *4*, 5. [CrossRef]
17. Elvidge, C.D.; Baugh, K.E.; Zhizhin, M.; Hsu, F.-C. Why VIIRS data are superior to DMSP for mapping nighttime lights. *Proc. Asia-Pac. Adv. Netw.* **2013**, *35*, 62. [CrossRef]
18. Lai, S.; Farnham, A.; Ruktanonchai, N.W.; Tatem, A.J. Measuring mobility, disease connectivity and individual risk: A review of using mobile phone data and mHealth for travel medicine. *J. Travel Med.* **2019**, *26*, taz019. [CrossRef]
19. Chen, X. Nighttime Lights and Population Migration: Revisiting Classic Demographic Perspectives with an Analysis of Recent European Data. *Remote Sens.* **2020**, *12*, 169. [CrossRef]
20. Stathakis, D.; Baltas, P. Seasonal population estimates based on night-time lights. *Comput. Environ. Urban Syst.* **2018**, *68*, 133–141. [CrossRef]
21. Tselios, V.; Stathakis, D. Exploring regional and urban clusters and patterns in Europe using satellite observed lighting. *Environ. Plan. B Urban Anal. City Sci.* **2020**, *47*, 553–568. [CrossRef]
22. Xu, G.; Xiu, T.; Li, X.; Liang, X.; Jiao, L. Lockdown induced night-time light dynamics during the COVID-19 epidemic in global megacities. *Int. J. Appl. Earth Obs. Geoinf.* **2021**, *102*, 102421. [CrossRef] [PubMed]
23. Lu, X.; Wrathall, D.J.; Sundsøy, P.R.; Nadiruzzaman; Wetter, E.; Iqbal, A.; Qureshi, T.; Tatem, A.J.; Canright, G.S.; Engø-Monsen, K.; et al. Detecting climate adaptation with mobile network data in Bangladesh: Anomalies in communication, mobility and consumption patterns during cyclone Mahasen. *Clim. Change* **2016**, *138*, 505–519. [CrossRef]
24. Montoya-Rincon, J.P.; Azad, S.; Pokhrel, R.; Ghandehari, M.; Jensen, M.P.; Gonzalez, J.E. On the Use of Satellite Nightlights for Power Outages Prediction. *IEEE Access* **2022**, *10*, 16729–16739. [CrossRef]
25. Lai, S.; zu Erbach-Schoenberg, E.; Pezzulo, C.; Ruktanonchai, N.W.; Sorichetta, A.; Steele, J.; Li, T.; Dooley, C.A.; Tatem, A.J. Exploring the use of mobile phone data for national migration statistics. *Palgrave Commun.* **2019**, *5*, 34. [CrossRef]
26. Ruktanonchai, N.W.; Ruktanonchai, C.W.; Floyd, J.; Tatem, A.J. Using Google Location History data to quantify fine-scale human mobility. *Int. J. Health Geogr.* **2018**, *17*, 28. [CrossRef] [PubMed]

27. Bengtsson, L.; Lu, X.; Thorson, A.; Garfield, R.; von Schreeb, J. Improved response to disasters and outbreaks by tracking population movements with mobile phone network data: A post-earthquake geospatial study in haiti. *PLoS Med.* **2011**, *8*, e1001083. [CrossRef]

28. Buckee, C.O.; Wesolowski, A.; Eagle, N.N.; Hansen, E.; Snow, R.W. Mobile phones and malaria: Modeling human and parasite travel. *Travel Med. Infect. Dis.* **2013**, *11*, 15–22. [CrossRef]

29. Ruktanonchai, N.W.; DeLeenheer, P.; Tatem, A.J.; Alegana, V.A.; Caughlin, T.T.; Zu Erbach-Schoenberg, E.; Lourenço, C.; Ruktanonchai, C.W.; Smith, D.L. Identifying Malaria Transmission Foci for Elimination Using Human Mobility Data. *PLoS Comput. Biol.* **2016**, *12*, e1004846. [CrossRef]

30. Dickinson, B.; Ghoshal, G.; Dotiwalla, X.; Sadilek, A.; Kautz, H. Inferring Nighttime Satellite Imagery from Human Mobility. In Proceedings of the AAAI Conference on Artificial Intelligence, New York, NY, USA, 7–12 February 2020; Volume 34, pp. 394–402. [CrossRef]

31. Bassolas, A.; Barbosa-Filho, H.; Dickinson, B.; Dotiwalla, X.; Eastham, P.; Gallotti, R.; Ghoshal, G.; Gipson, B.; Hazarie, S.A.; Kautz, H.; et al. Hierarchical organization of urban mobility and its connection with city livability. *Nat. Commun.* **2019**, *10*, 4817. [CrossRef]

32. Wilson, R.J.; Zhang, C.Y.; Lam, W.; Desfontaines, D.; Simmons-Marengo, D.; Gipson, B. Differentially Private SQL with Bounded User Contribution. *Proc. Priv. Enhancing Technol.* **2020**, *2020*, 230–250. [CrossRef]

33. Google. How Google Anonymises Data. *Retrieved from Google Privacy&Terms.* Available online: https://policies.google.com/technologies/anonymization (accessed on 28 July 2023).

34. S2 Geometry. 2018. Available online: http://s2geometry.io/ (accessed on 28 July 2023).

35. Elvidge, C.D.; Baugh, K.; Zhizhin, M.; Hsu, F.C.; Ghosh, T. VIIRS night-time lights. *Int. J. Remote Sens.* **2017**, *38*, 5860–5879. [CrossRef]

36. Mills, S.; Weiss, S.; Liang, C. VIIRS day/night band (DNB) stray light characterization and correction. In *Earth Observing Systems XVIII*; SPIE: Bellingham, WA, USA, 2013; Volume 8866, p. 88661P.

37. Li, X.; Xu, H.; Chen, X.; Li, C. Potential of NPP-VIIRS Nighttime Light Imagery for Modeling the Regional Economy of China. *Remote Sens.* **2013**, *5*, 3057–3081. [CrossRef]

38. Wang, R.; Wan, B.; Guo, Q.; Hu, M.; Zhou, S. Mapping Regional Urban Extent Using NPP-VIIRS DNB and MODIS NDVI Data. *Remote Sens.* **2017**, *9*, 862. [CrossRef]

39. Warmerdam, F. The geospatial data abstraction library. In *Open Source Approaches in Spatial Data Handling*; Springer: Berlin, Germany, 2008; pp. 87–104.

40. UN Statistics Division. *Methodology: Standard Country or Area Codes for Statistical Use (M49)*; Questions & Answers; UN Statistics Division: New York, NY, USA, 2022.

41. Florczyk, A.J.; Corbane, C.; Ehrlich, D.; Freire, S.; Kemper, T.; Maffenini, L.; Melchiorri, M.; Pesaresi, M.; Politis, P.; Schiavina, M.; et al. *GHSL Data Package 2019*; JRC Technical Report; European Commission, Publications Office of the European Union: Luxembourg, 2019.

42. WorldPop—School of Geography and Environmental Science, University of Southampton; Department of Geography and Geosciences, University of Louisville; Departement de Geographie, Universite de Namur and Center for International Earth Science Information, Global High Resolution Population Denominators Project. 2018. Available online: https://hub.worldpop.org/geodata/summary?id=24767 (accessed on 28 July 2023). [CrossRef]

43. Lindgren, F.; Rue, H. Bayesian Spatial Modelling with R-INLA. *J. Stat. Softw.* **2015**, *63*, 1–25. [CrossRef]

44. Akoglu, H. User's guide to correlation coefficients. *Turk. J. Emerg. Med.* **2018**, *18*, 91–93. [CrossRef] [PubMed]

45. Zhao, N.; Samson, E.L.; Liu, Y. Population bias in nighttime lights imagery. *Remote Sens. Lett.* **2019**, *10*, 913–921. [CrossRef]

46. Haider, N.; Osman, A.Y.; Gadzekpo, A.; O Akipede, G.; Asogun, D.; Ansumana, R.; Lessells, R.J.; Khan, P.; Hamid, M.M.A.; Yeboah-Manu, D.; et al. Lockdown measures in response to COVID-19 in nine sub-Saharan African countries. *BMJ Glob. Health* **2020**, *5*, e003319. [CrossRef]

47. Kuehn, B.M. Africa Succeeded Against COVID-19's First Wave, but the Second Wave Brings New Challenges. *Jama* **2021**, *325*, 327–328. [CrossRef]

remote sensing

MDPI

Article

Highway Ecological Environmental Assessment Based on Modified Remote Sensing Index—Taking the Lhasa–Nyingchi Motorway as an Example

Xinghan Wang [1,2,†], Qi Liu [2,†], Pengfei Jia [3,*], Xifeng Huang [4], Jianhua Yang [5], Zhengjun Mao [6] and Shengyu Shen [7,8]

[1] College of Civil Engineering, Tianjin University, Tianjin 300350, China; wxhan@tju.edu.cn
[2] Research Center on Flood & Drought Disaster Reduction of the Ministry of Water Resources, China Institute of Water Resources and Hydropower Research, Beijing 100038, China; liuqi@iwhr.com
[3] Citic Construction Co., Ltd., Beijing 100027, China
[4] Shaanxi Provincial Water Resources Department, Xi'an 710004, China; hxf_sxsl@163.com
[5] Academy of Eco-Civilization Development for Jing-Jin-Ji Megalopolis, Tianjin Normal University, Tianjin 300387, China; yangjh15@mail.bnu.edu.cn
[6] College of Geology and Environment, Xi'an University of Science and Technology, Xi'an 710054, China; mzj@xust.edu.cn
[7] Department of Soil and Water Conservation, Changjiang River Scientific Research Institute (CRSRI), Wuhan 430010, China; shenshengyu@mail.crsri.cn
[8] Research Center on Mountain Torrent & Geologic Disaster Prevention of the Ministry of Water Resources, Wuhan 430010, China
[*] Correspondence: jiapf@citic.com
[†] Co-first author: Qi Liu.

Citation: Wang, X.; Liu, Q.; Jia, P.; Huang, X.; Yang, J.; Mao, Z.; Shen, S. Highway Ecological Environmental Assessment Based on Modified Remote Sensing Index—Taking the Lhasa–Nyingchi Motorway as an Example. *Remote Sens.* **2024**, *16*, 265. https://doi.org/10.3390/rs16020265

Academic Editor: Peng Fu

Received: 5 November 2023
Revised: 31 December 2023
Accepted: 1 January 2024
Published: 10 January 2024

Abstract: The Lhasa to Nyingchi Expressway in Xizang made efforts to protect the ecological environment during its construction, but it still caused varying degrees of damage to the fragile ecosystems along the route. Accurately assessing the process of change in the ecological environment quality in this region holds significant research value. This study selected the Linzhi-to-Gongbo'gyamda section of the Lhasa-to-Nyingchi Expressway as the research area. Firstly, based on the remote sensing ecological index (RSEI), this study constructed an ecological environmental quality evaluation system for the Xizang region. Subsequently, using the Google Earth Engine (GEE) platform, sub-indicators were extracted, and the combination weighting method of game theory was employed to determine indicator weights. This process resulted in the calculation of the MRSEI for the study area from 2012 to 2020. Finally, by utilizing the spatial distribution of the MRSEI, monitoring the level of MRSEI changes, and employing the transition matrix, this study analyzed the changing trend of the ecological environmental quality from 2012 to 2020. The results indicate that the MRSEI are 0.5885, 0.5951, 0.5296, 0.6202, 0.59, 0.5777, 0.5898, 0.5703, and 0.5987, showing a gradual increasing trend with an initial decrease followed by an ascent. This trend is mainly attributed to concentrated road construction and subsequent ecological restoration, leading to an improvement in the restoration effect. Simultaneously, the ecological environmental quality remains relatively stable, with 69.5% of the region showing no change, and the remaining 30.5% experiencing improvement exceeding degradation. Specifically, there were significant improvements in the land with ecological quality levels categorized as poor, fair, moderate, and good. The types of degradation primarily involved lands originally classified as excellent and good degrading to good and moderate levels, respectively. The above results serve as a theoretical reference for the ecological restoration project of the Lhasa-to-Nyingchi Expressway.

Keywords: GEE; RSEI; ecological environmental quality; Lhasa–Nyingchi Motorway

1. Introduction

The Xizang region is an important ecological security zone in China, with natural protected areas accounting for one-third of the total area. The Lhasa–Nyingchi Motorway in the Xizang region not only benefits local economic development and the development and utilization of natural resources but also contributes to border stability and the common prosperity of all ethnic groups. However, on one hand, transportation construction will destroy the permafrost environment that has existed for many years. For example, the heat absorption of asphalt on the road and the exhaust emissions of vehicles along the corridor will increase the surface temperature of the area, thereby accelerating the melting of permafrost [1–3]. On the other hand, it will damage the surface vegetation and topsoil, accelerating soil erosion [4]. Due to the limited carrying capacity of resources and the environment in the Xizang region, it is difficult to restore the ecological environment once it is damaged [5]. Therefore, coordinating the construction of transportation infrastructure with the quality protection of the local ecological environment is a necessary path for the development of the Xizang region's economy. Timely evaluation of changes in the quality of the ecological environment in the area where linear engineering is located can help improve the negative impact of human activities on the environment [6].

In 2006, the Chinese Ministry of Environmental Protection proposed the ecological environment index (EI) for ecological environment assessment, but this index also has many issues in its application process, such as subjective weighting and lack of visualizable evaluation results [7]. Remote sensing technology has become an indispensable means of ecological monitoring and assessment [8]. Taking advantage of the spatial and temporal coverage of satellite data, Xu H Q [7] proposed an improved method called the remote sensing ecological index (RSEI) [7] based on the EI index. The evaluation indicators in this method mainly come from remote sensing indices. This method has been widely used in ecological environmental quality assessment research and has a high level of credibility [9–11]. For example, RSEI [7] has been applied in the evaluation of the surface water ecological environment in Freetown [11], the assessment of ecological environmental quality in the Samara region of Russia [12], and the evaluation of ecological conditions during the rainy and dry seasons in Kota Semarang [13]. In addition, researchers have also improved the evaluation algorithms and indicators based on the RSEI [7] from different perspectives. In terms of algorithm models, the RSEI initially mainly used the first principal component (PC1) of the principal component analysis (PCA) to construct the ecological index. However, the variance contribution rate of PC1 extracted in different studies varies greatly, which cannot guarantee a high contribution rate. In other words, the interpretation of the evaluation indicators obtained after dimensionality reduction is more unstable compared to the original indicators. Therefore, some scholars have made improvements to the model, such as using a modified remote sensing ecological index (MRSEI) to evaluate the ecological environment of the Xilingol League Grassland in China [14]. Machine learning has also been introduced, and the RSEI [7] time series data for Beijing, China, has been calculated using PCA combined with the random forest algorithm [15]. In terms of indicator systems, Karimi [16] used the land surface ecological status composition index (LESCI) based on the vegetation-impervious-soil triangle model to assess the surface ecological conditions in Iran, as well as some cities in Europe and North America [16]. Xing et al. used net primary productivity, vegetation index, and light index to construct an enhanced remote sensing ecological index for an ecological environment assessment of Hainan Island, China, based on local characteristics [17].

Determining appropriate indicator weights is essential in ecological environmental quality assessment. There are two types of weight determination methods: objective weight determination methods (OW) based on data mining and statistics [18], including entropy weight method (EWM) [19], random forest (RF) [20], and support vector machine (SVM) [21], and subjective weight determination methods (SW) based on prior knowledge or expert opinions, including analytic hierarchy process (AHP) [22], multi-criteria decision analysis (MCDM) [23], and fuzzy mathematics (FM) [24], among others. The subjective weight method is influenced by prior knowledge, leading to higher subjectivity in the final evaluation results, while using only PCA for each indicator's weight assignment is too objective and may cause biased results due to the large amount of information [25]. Game theory (GT) is a mathematical model of strategic interaction between rational and irrational agents, which can effectively use SW and OW weight information to obtain combined weights (CW) [26]. Therefore, this study uses game theory (GT) to calculate weights, with PCA used as the objective weight assignment method and AHP as the subjective weight assignment method, to obtain optimal weights by balancing subjective and objective weights.

The Lhasa–Nyingchi Motorway from Nyingchi to Gongbo'gyamda section was put into use in September 2015, meeting the requirements for ecological environment assessment of highways (generally 3 to 5 years after completion and acceptance). Therefore, this study takes the Lhasa–Nyingchi Motorway from Nyingchi to Gongbo'gyamda section as an example, using a modified remote sensing ecological index to evaluate the ecological environmental quality within a 5 km range on both sides of the highway from 2012 to 2020, studying its change trends and analyzing the reasons for the change in order to provide reference and basis for the ecological environment restoration work after highway construction in the Xizang region.

2. Study Area and Materials

2.1. Study Area

The Lhasa–Nyingchi Motorway, also known as the Lhoka–Lhasa Expressway, has a total length of 409.2 km and is the first phase project of the "12th Five-Year Plan" key project in the Xizang region of China. The Nyingchi to Gongbo'gyamda section of the highway is 114.3 km long and runs parallel to the Lhasa to Nyingchi section of National Highway 318. It is primarily designed as a dual-carriageway with four lanes, and construction began in late May 2013. The section was officially opened to traffic on 15 September 2015. This section is located in the southeastern part of the Xizang region, in the northwest of Nyingchi Prefecture, with coordinates ranging from 29°31′N to 29°54′N and 93°10′E to 94°27′E. It is situated in the central part of Bayi District and the southeastern part of Gongbo'gyamda (Figure 1). Vegetation is the main component of the ecosystem in this region. The Nyang River valley area is mostly wetlands and water bodies, while the land use types around the highways are mainly forests and shrubs. Influenced by the warm and humid air currents of the Indian Ocean, the vegetation in the study area belongs to the semi-humid temperate mountain vegetation and has a clear vertical distribution pattern. The land use of the residential area is mainly farmland, with no obvious impervious areas. The region has a large range of elevation, a complex and diverse climate, and significant local climate differences, making it prone to natural disasters such as hailstorms, drought, or floods [27–30]. The highways mainly traverse the middle and lower reaches of the Nyang River valley, which is characterized by a wide valley landform with flat terrain, facilitating the implementation of linear engineering. However, there are risks of collapse and debris flows on the mountainsides on both sides of the Nyang River valley [31].

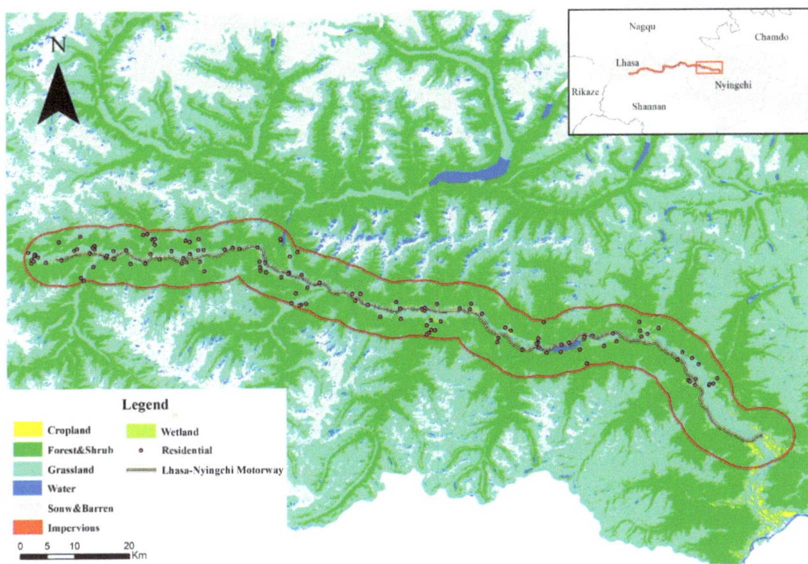

Figure 1. Location of Nyingchi to Gongbo'gyamda section of Lhasa–Nyingchi Motorway.

2.2. Data Source and Preprocessing

The remote sensing data used in this study mainly come from the MODIS (moderate-resolution imaging spectroradiometer) data provided by the Google Earth Engine (GEE) platform. MODIS is a remote sensing product carried by the Terra and Aqua satellites; it is used to observe global climate change and biological activities. The MODIS data products used in this study are processed and analyzed synthetic images, with imaging time ranging from July to September between 2012 and 2020. This period represents the peak growing season for vegetation in the Xizang region and helps distinguish vegetation areas from bare land and snow-covered areas. Five types of MODIS data products (Table 1) were used in this study. These data products have high quality and have undergone radiometric and atmospheric corrections. However, due to geometric distortions, geometric correction was performed using the reproject and resample functions in the GEE platform. Other data processing tasks, such as mosaicking, mask extraction, and uniform resolution, were also implemented by coding in the GEE platform. Finally, the indicator data for the study area could be calculated.

Table 1. Data details.

Index	Product	Spatial Resolution (m)	Time Resolution (d)	Time
NDVI/FVC	MOD13Q1	250	16	2012–2020
LAI	MCD15A3H	500	8	2012–2020
GPP	MOD17A2H	500	8	2012–2020
LST	MOD11A2	1000	8	2012–2020
Wet	MOD09A1	500	8	2012–2020

2.3. Method

In recent years, vegetation changes have dominated ecological research in the Xizang region of China. The original RSEI (remote sensing ecological index) [6] only used vegetation indices (VI) to represent vegetation. However, in this study, considering the extremely limited ecological carrying capacity and increased sensitivity to greenhouse effects in the Qinghai-Xizang Plateau, detecting short-term vegetation changes becomes particularly important [32,33]. Therefore, this study incorporates the leaf area index (LAI), which reflects vegetation growth quality, and the gross primary productivity (GPP) index, which reflects vegetation carbon sequestration capacity, to improve the vegetation index in the evaluation system. Soil wetness can be used as an important indicator for monitoring soil degradation. Additionally, Yijin Wu [34] found that the land surface temperature (LST) in the Xizang region is positively correlated with vegetation growth; specifically, this abnormal phenomenon is manifested as follows: the larger the aridity index, the better the ecological environment. Therefore, this study uses the LST and the wetness (Wet) to replace the original TVDI(Temperature Vegetation Dryness Index). Finally, the five indicators of greenness (FVC, LAI, and GPP), warmth (LST), and wetness (Wet) were used to construct an improved RSEI [7] (MRSEI).

2.3.1. Indicator Calculation

(1) Greenness

FVC (Fractional Vegetation Cover) is the percentage of the area covered by vegetation in relation to the total surface area. It is a simple measure of vegetation coverage and growth and a key factor in soil and water conservation and ecological assessment [35]. FVC can be used to estimate shrub biomass in high mountain or subalpine environments [36]. In this study, FVC is derived using the normalized difference vegetation index ($NDVI$, and pixel-based binary method [37]. The $NDVI$ is primarily derived from MODIS data products, as shown in the following formula:

$$NDVI = NDVI_{veg} \times FVC + NDVI_{soil} \times (1 - FVC) \tag{1}$$

$NDVI_{veg}$ and $NDVI_{soil}$ represent the $NDVI$ values in areas with complete vegetation coverage and areas with complete bare soil or no vegetation coverage, respectively. The calculation formulas are as follows:

$$NDVI_{soil} = \frac{FVC_{max} \times NDVI_{min} - FVC_{min} \times NDVI_{max}}{FVC_{max} - FVC_{min}} \tag{2}$$

$$NDVI_{veg} = \frac{(1 - FVC_{min}) \times NDVI_{max} - (1 - FVC_{max}) \times NDVI_{min}}{FVC_{max} - FVC_{min}} \tag{3}$$

Since the MODIS data used in this study were acquired from July to September, which is the period of the year with the best vegetation growth in the study area, we can approximate that FVC_{max} = 100% and FVC_{min} = 0%. Based on the pixel-based binary method, the vegetation coverage model is defined as follows:

$$FVC = \frac{NDVI - NDVI_{min}}{NDVI_{max} - NDVI_{min}} \tag{4}$$

In the equation, $NDVI_{max}$ represents the normalized difference vegetation index value of pixels with vegetation coverage, while $NDVI_{min}$ represents the normalized difference vegetation index value of pixels with complete bare soil. Since the study area is mostly inaccessible and obtaining extensive field measurements is challenging, this study uses the confidence interval values of the cumulative frequency of $NDVI$ at 5% and 95% as the $NDVI_{min}$ and $NDVI_{max}$ values.

The leaf area index (LAI) is a dimensionless parameter that measures the amount of foliage in the canopy and is used to reflect the vegetation growth quality in the study area [38]. In this study, the MODIS data product MOD15A3H is used to obtain LAI data for the study area. The calculation formula is as follows [39]:

$$LAI = 0.1 \times DN_A \tag{5}$$

DN_A represents the gray value of the leaf area index image except for bare land, glaciers, or water areas.

Gross primary productivity (GPP) refers to the total amount of organic carbon fixed by vegetation through photosynthesis in a unit of time. The accumulation of organic carbon by vegetation is a driving force for multiple ecosystems and a key component of land carbon balance, mainly reflecting the carbon source-sink status of vegetation in the study area [40]. In this study, the MOD17A2H product of GPP is extracted at the grid level and dimensionless processing is applied to obtain the distribution of GPP in the study area.

(2)   Heat

The thermal index utilizes land surface temperature (LST), which refers to the surface temperature of the Earth's land surface that has been corrected for emissivity. It is measured in degrees Celsius (°C) and is an important surface parameter that controls the energy balance between the atmosphere and the land surface. The process of extracting LST involves converting the digital number (DN) values from the MOD11A2 product data into Celsius to represent the distribution of land surface temperature in the study area. The calculation formula is as follows:

$$LST = 0.02 \times DN_S - 273.15 \tag{6}$$

(3)   Wet

This study utilizes a modified tasseled cap transformation (Kauth–Thomas, K–T) formula [41] to extract the wetness index. The K–T transformation is a method based on image physical characteristics, which uses a linear orthogonal transformation to project and transform multi-spectral remote sensing images into three-dimensional space. This method is used to reflect the degree of plant growth and withering, as well as changes in land information.

We use the MOD09A1 surface reflectance product, which is sensitive to land moisture information in certain bands, to calculate the humidity of the study area. The formula is as follows:

$$Wet = 0.1147\rho_1 + 0.2489\rho_2 + 0.2408\rho_3 + 0.3132\rho_4 - 0.3122\rho_5 - 0.6416\rho_6 - 0.5087\rho_7 \tag{7}$$

$\rho_1 \sim \rho_7$ The seven bands of reflectance in the MOD09A1 surface reflectance product are red, NIR1, Blue, Green, NIR2, SWIR1, and SWIR2.

2.3.2. Weight Determination

(1)   Principal component analysis (PCA)

Principal component analysis (PCA) is a commonly used statistical analysis method that can combine several correlated indicators into a mutually independent composite index through a linear transformation. The advantage of PCA is that it can reduce the dimensionality of the indicators while retaining the maximum principal components, and the determination of weights is objective. The steps for determining weights based on PCA are as follows: ① Standardize the original data. ② Calculate the correlation coefficient matrix of each indicator. ③ Calculate the eigenvalues, contribution rates, and cumulative contribution rates of each principal component. ④ Calculate the loading values of each principal component. ⑤ Obtain the scores of each principal component. ⑥ Calculate the weight of each indicator; the formula is as follows:

$$PC = \sum_{i=1}^{i} \frac{pc_i \times e_i}{\sqrt{k_i} \times E} \tag{8}$$

where PC represents the score coefficient of each indicator, pc_i represents the load of each principal component, e_i represents the variance contribution rate of each principal component, k_i represents the eigenvalues of each principal component, and E represents the cumulative contribution rate of the extracted principal components. Finally, the weights of each indicator can be obtained by normalizing the PC.

(2) Combination weight method based on game theory

Game theory (GT) can effectively utilize the combined information of subjective weights (SW) and objective weights (OW) and obtain the most balanced weights, also known as combined weights (CW). Supposing that the weight set generated by using a weighting method for m basic indicator vectors is ω_1, the weight vector obtained by n empowerment methods is then ω_i:

$$\omega_1 = [\omega_{11}, \omega_{12}, \omega_{13}, \ldots, \omega_{1m}] \tag{9}$$

$$\omega_i = [\omega_{n1}, \omega_{n2}, \omega_{n3}, \ldots, \omega_{nm}] = [\omega_1, \omega_2, \omega_3, \ldots, \omega_n], i = 1, 2, \ldots, n \tag{10}$$

The arbitrary linear combination W of the n vectors is as follows:

$$W = \sum_{i=1}^{n} \gamma_i \omega_i^T, \; i = 1, 2, \ldots, n \tag{11}$$

Solving the optimal combination coefficient γ_i to minimize the deviation between the combination weight and the weight involved in optimization can be achieved through the following function:

$$\min || \sum_{i=1}^{n} \gamma_i \omega_i^T - \omega_i^T ||, i = 1, 2, \ldots, n \tag{12}$$

Solving the first derivative of the above equation for optimization yields the following equivalent linear equations:

$$\sum_{i=n}^{n} \gamma_i \omega_i \omega_i^T = \begin{bmatrix} \gamma_1 \\ \gamma_2 \\ \cdot \\ \cdot \\ \cdot \\ \gamma_n \end{bmatrix} \times \begin{bmatrix} \omega_1 \times \omega_1^T & \omega_1 \times \omega_2^T & \cdots & \omega_1 \times \omega_n^T \\ \omega_2 \times \omega_1^T & \omega_2 \times \omega_2^T & \cdots & \omega_2 \times \omega_n^T \\ \cdot & \cdot & \cdot & \cdot \\ \cdot & \cdot & \cdot & \cdot \\ \cdot & \cdot & \cdot & \cdot \\ \omega_n \times \omega_1^T & \omega_n \times \omega_2^T & \cdots & \omega_n \times \omega_n^T \end{bmatrix} = \begin{bmatrix} \omega_1 \times \omega_n^T \\ \omega_2 \times \omega_2^T \\ \cdot \\ \cdot \\ \cdot \\ \omega_n \times \omega_n^T \end{bmatrix} \tag{13}$$

The process of calculating the combined weight $(\gamma_1, \gamma_2, \ldots, \gamma_n)$, and then standardizing it to obtain the optimal combination weight ω^*, can be expressed as follows:

$$\omega^* = \sum_{i=1}^{n} \gamma_i \times \omega_i^T, i = 1, 2, \ldots, n \tag{14}$$

Finally, we use the comprehensive index method to calculate the comprehensive value of the ecological environment, the *MRSEI*, which can be expressed by the following formula:

$$MRSEI = \sum_{i=1}^{P} \omega_i^* * x_i \tag{15}$$

where x_i represents the standardized value of the i-th indicator, ω_i^* represents the optimal combined weight of the i-th indicator, and P represents the number of evaluation indicators.

3. Results and Analysis

3.1. PCA and Combination Weights

This study utilized IBM SPSS Statistics 25.0 to perform a principal component analysis on the indicator data. Table 2 presents the results of the principal component analysis for the years 2012 and 2020. It can be observed that the cumulative contribution rates of the first three principal components are all greater than 85%, indicating that these three components capture the majority of the characteristics of the five indicators. Specifically, PC1 has high loading values for FVC, LAI, and GPP, PC2 has a high loading value for Wet, and PC3 has a high loading value for LST. Therefore, we can summarize the extracted principal components as follows: PC1 represents the vegetation factor, represented by FVC, LAI, and GPP; PC2 represents the humidity factor, represented by Wet; and PC3 represents the thermal factor, represented by LST.

Table 2. Results of PCA in 2012 and 2020.

Index	2012			2020		
	PC1	PC2	PC3	PC1	PC2	PC3
FVC	0.847	−0.201	−0.066	0.858	−0.241	0.056
LAI	0.909	−0.005	−0.002	0.894	−0.134	−0.128
GPP	0.86	−0.095	0.25	0.882	−0.103	−0.24
Wet	0.081	0.806	0.572	0.226	0.88	−0.404
LST	0.293	0.65	−0.693	0.496	0.439	0.745
Eigenvalues	2.375	1.123	0.874	2.61	1.053	0.794
Variance Contribution Rate (%)	47.5	22.454	17.479	52.201	21.059	15.884
Total Contribution Rate (%)	47.5	69.954	87.433	52.201	73.26	89.144

Based on the principal component analysis results, the weights for each indicator can be calculated for each year from 2012 to 2020. These weights are then combined with the weights obtained from the analytic hierarchy process (AHP) to calculate the composite weight coefficients. Taking 2012 and 2020 as examples, the final weights are shown in Table 3.

Table 3. Index weight in 2012 and 2020.

Index	2012			2020		
	OW	SW	CW	OW	SW	CW
FVC	0.1750	0.3178	0.2412	0.1884	0.3178	0.2575
LAI	0.2367	0.1761	0.2086	0.1890	0.1761	0.1821
GPP	0.2477	0.1761	0.2145	0.1752	0.1761	0.1757
Wet	0.2570	0.165	0.2143	0.1439	0.165	0.1552
LST	0.0836	0.165	0.1214	0.3036	0.165	0.2295

3.2. Analysis on the Overall Change Trend of Ecological Quality

As shown in Table 4, the MRSEI values for the Lhasa–Nyingchi Motorway segment from Nyingch to Gongbo'gyamda in the years 2012–2020 are as follows: 0.5885, 0.5951, 0.5296, 0.6202, 0.59, 0.5777, 0.5898, 0.5703, and 0.5987. Overall, compared to before the construction of the highway in 2012, the MRSEI value in 2020 has slightly increased, but there is significant fluctuation in the MRSEI values each year. The MRSEI values showed a noticeable decrease from 2013 to 2014, which is closely related to highway construction. The MRSEI values then increased significantly from 2014 to 2015, mainly due to the implementation of ecological restoration projects. From 2015 to 2017, there was a slow decline in the MRSEI values, but from 2017 to 2020, the MRSEI values showed a stair-step growth trend, indicating that the ecological restoration work was challenging but that the ecological environmental quality was improving overall.

Table 4. MRSEI from 2012 to 2020.

	2012	2013	2014	2015	2016	2017	2018	2019	2019
MRSEI	0.5885	0.5951	0.5296	0.6202	0.59	0.5777	0.5898	0.5703	0.5987

Figure 2 shows the spatiotemporal changes in the ecological environmental quality of the study area. The MRSEI values were divided into five categories: poor, fair, moderate, good, and excellent, with intervals of 0.2. It can be observed that in 2015, the ecological environmental quality was rated as good, while the rest of the years were rated as moderate. In terms of the spatial distribution of the MRSEI values in the nine periods, the ecological environmental quality gradually decreased from the eastern segment to the western segment. The areas with good and excellent ecological quality were mainly located in the southeastern part of the study area, where the humidity and temperature conditions were favorable and the vegetation coverage was relatively high. The areas with poor, fair, and moderate ecological quality were primarily located in the road area and the peripheral areas of the central and western segments. This is mainly due to the higher altitude and colder, drier climate in these areas, resulting in relatively lower vegetation coverage. The relatively poor ecological environmental quality in the western segment may also be related to the denser distribution of local residential areas.

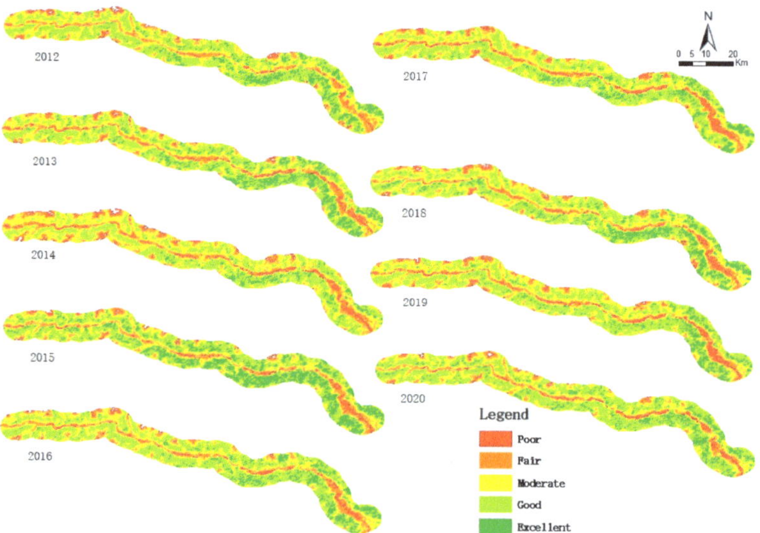

Figure 2. Spatial distribution of MRSEI values of the Nyingchi–Gongbo'gyamda section of the Lhasa–Nyingchi Motorway from 2012 to 2020.

The area and proportion of the ecological environmental quality in the study area for the nine periods were statistically analyzed (Table 5, Figure 3). It can be observed that the proportion of good ecological environmental quality was the largest, followed by the moderate category, from 2012 to 2020. The ecological environmental quality structure in 2012–2013 was similar, with the most significant variation occurring in 2014; the proportions of poor, fair, and moderate ecological quality were the highest in the nine-year period, and the MRSEI value was the lowest. Although the MRSEI value in 2015 was the highest in the nine years, there was no significant change in the area of good ecological quality compared to the previous year, and the area of moderate quality was the lowest in the nine years, decreasing by 9.72% compared to 2014. It is worth noting that in 2015, the area of excellent ecological quality reached 21.48%, which was nearly 10% higher than the average proportion of excellent quality in the other seven years, excluding 2014. This can be attributed to the construction period of the highway from 2013 to 2015, with full construction taking place in 2014, stripping the original vegetation and damaging the ecological environment along the route, resulting in a sudden decrease in the MRSEI value in 2014. However, in 2015, thanks to the smooth implementation of the highway ecological environment project, the MRSEI value rapidly increased. After the construction was completed, the construction facilities were removed, construction waste was cleared, depressions were leveled, and the original vegetation and turf were replanted, leading to a significant improvement in the overall ecological environmental quality in a short period of time. Clearly, the ecological environmental quality in the road area and the peripheral areas of the study area has been significantly improved. The proportions of excellent, good, and moderate levels showed a declining trend in 2016 and 2017, indicating a slight rebound in the ecological restoration effect, mainly due to the higher requirements for transplanting vegetation in the high-altitude ecological environment. The ecological environmental quality structure in 2018 was similar to that in 2017, with similar proportions of excellent and good levels, but there was a noticeable increase in the proportion of excellent quality areas in 2018. Looking at Figure 2, it can be further observed that these increased areas were mainly located in the southern part of the eastern segment of the study area. Compared to 2018, the MRSEI value decreased in 2019, and the area of excellent quality decreased by 6.12%. Although there was a slight increase in the proportion of good-quality areas, the combined proportion of excellent and good levels was lower than in 2018. In 2020, the overall ecological quality level was good, with a 1.01% decrease in the area of excellent quality compared to 2012, while the area of good quality increased by 5.46% and had the highest proportion in the nine years. The area of moderate quality decreased by 2.74%, and the combined area of poor and fair levels was the smallest in the nine years. In conclusion, the ecological environmental quality in the study area in 2020 did not show significant changes compared to before the construction of the highway.

Table 5. Statistics of MRSEI area proportions in different years.

	Excellent		Good		Moderate		Fair		Poor	
	S/km^2	Pct./%	S/km^2	Pct./%	S/km^2	Pct./%	S/km^2	Pct./%	S/km^2	Pct./%
2012	172.8125	10.51	728.3125	44.3	434.4375	26.43	237.75	14.46	70.6875	4.3
2013	232.0625	14.12	688.25	41.86	405.0625	24.64	235.75	14.34	82.875	5.04
2014	72.5	4.41	640.0625	38.93	511.3125	31.1	292.6875	17.8	127.4375	7.75
2015	353.1875	21.48	652.6875	39.7	351.4375	21.38	211.5	12.86	75.1875	4.57
2016	147.8125	8.99	765.6875	46.57	428.5	26.06	234.125	14.24	67.875	4.13
2017	146.0625	8.88	740.3125	45.03	423.5	25.76	230.5	14.02	103.625	6.3
2018	221.125	13.45	676.1875	41.13	424.0625	25.79	234.75	14.28	87.875	5.35
2019	120.5	7.33	742.875	45.19	439.125	26.71	237.75	14.46	103.75	6.31
2020	156.1875	9.5	818.0625	49.76	389.4375	23.69	204.5	12.44	75.8125	4.61

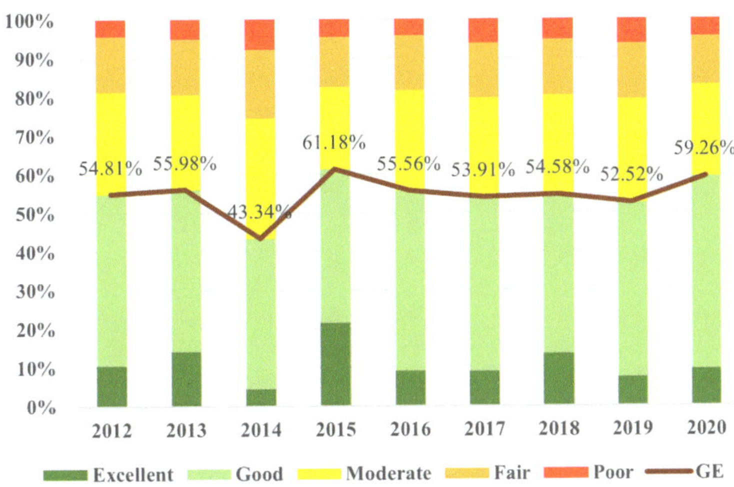

Figure 3. Changes in the proportion of MRSEI levels in the Nyingchi–Gongbo'gyamda section of the Lhasa–Nyingchi Motorway from 2012 to 2020.

Furthermore, the sums of the proportions of good and excellent quality (GE%) were separately calculated. The GE% for the years 2012–2020 are as follows: 54.81%, 55.98%, 43.34%, 61.18%, 55.56%, 53.91%, 54.58%, 52.52%, and 59.26%. Obviously, the GE% initially decreased due to road construction, then increased due to the implementation of ecological environment projects and ecological restoration, followed by a stepwise increase.

3.3. Analysis of the Spatial and Temporal Evolution of Ecological Quality

In order to further analyze the dynamic changes in the ecological environmental quality of the Nyingchi–Gongbo'gyamda section from 2012 to 2020, Table 6 presents the calculation of the difference in the modified remote sensing ecological index (MRSEI) for every two years. The difference values are categorized into five levels: significantly deteriorated (-2), deteriorated (-1), essentially unchanged (0), improved ($+1$), and significantly improved ($+2$). Figure 4 provides a visual representation of the proportions of each category and their changes over time.

Table 6. Monitoring of the level of MRSEI changes from 2012 to 2020.

Years	The Ratio and Area of Changes	Significantly Worse	Worse	Invariability	Improved	Significantly Improved
2012–2014	Change Area (km^2)	0.3125	545.375	990.0625	107.4375	0.8125
	Percentage (%)	0.02	33.17	60.22	6.54	0.05
2014–2016	Change Area (km^2)	0	86.0625	1028.125	529.1875	0.625
	Percentage (%)	0	5.23	62.54	32.19	0.04
2016–2018	Change Area (km^2)	1.125	199.25	1243.75	199.875	0
	Percentage (%)	0.07	12.12	75.65	12.16	0
2018–2020	Change Area (km^2)	0.4375	193.0625	1192.875	256.75	0.875
	Percentage (%)	0.03	11.74	72.56	15.62	0.05

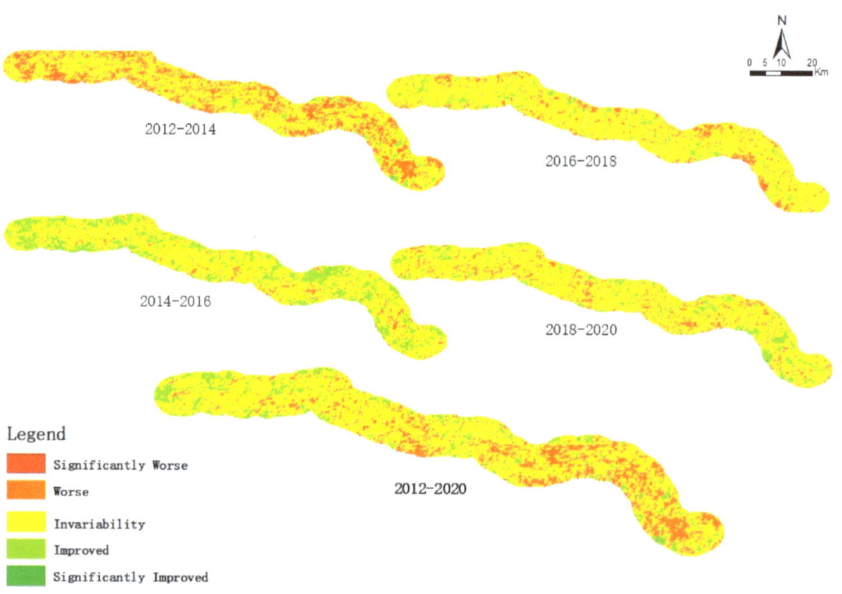

Figure 4. Dynamic changes of the MRSEI in the Nyingchi–Gongbo'gyamda section of the Lhasa–Nyingchi Motorway from 2012 to 2020.

Combining Table 7 and Figure 5, it can be observed that the ecological quality remained relatively stable across the four periods. The proportion of areas with deteriorated ecological quality was highest in the 2012–2014 period, accounting for 33.17% of the total area. The proportions of areas with unchanged and improved ecological quality were the lowest in the four periods. Clearly, the construction of the highway had a significant impact on the local ecological environmental quality. Figure 4 reflects that the regions with deteriorated ecological environments were mainly located along the road and at the edges of the study area. During the 2014–2016 period, only 5.23% of the area experienced deteriorated MRSEI, while the proportion of areas with improved ecological quality reached 32.19%. This improvement can be attributed to the restoration effects of ecological environment projects. Overall, the ecological quality of the study area showed a positive trend. From the MRSEI changes in the 2016–2018 period, it can be seen that the ecological environmental quality in the study area exhibited a rebound trend. Looking at Figure 2, it is evident that the areas with deteriorated ecological quality were mainly located along the road and at the edges of the study area, indicating the high ecological sensitivity and the difficulty of ecological restoration in the alpine region. The changes in the 2018–2020 period were similar to the previous period. Looking at the 2016–2018 period, it can be observed that the proportion of areas with unchanged ecological quality tended to stabilize. With the continued efforts in ecological governance, the transplanted vegetation can better adapt to the local climate and environment.

Table 7. Monitoring of MRSEI changes in the Nyingchi–Gongbo'gyamda section of the Lhasa–Nyingchi Motorway from 2012 to 2020.

	Change Level	Change Area (km²)	Percentage (%)	Total Change (km²)	Total Percentage (%)
Improved	Significantly Improved	1.9375	0.12	289	17.58
	Improved	287.0625	17.46		
Invariability	Invariability	1142.5	69.5	1142.5	69.5
Degenerate	Worse	212.375	12.92	212.5	12.93
	Significantly Worse	0.1215	0.01		

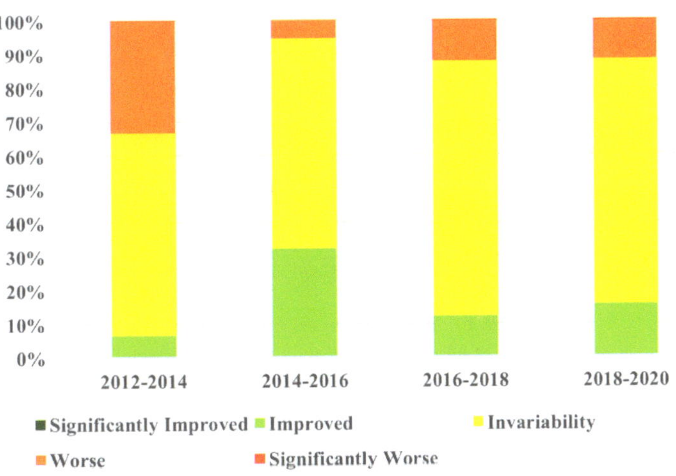

Figure 5. Changes in the level of the MRSEI from 2012 to 2020.

Table 7 presents the calculation of the MRSEI difference between the year before road construction in 2012 and the fifth year after road implementation in 2020. It is evident that the overall ecological quality in the study area remained relatively stable from 2012 to 2020, with a value of 69.5%. The majority of this ecological quality was concentrated in the western and central parts of the study area. Additionally, a significant proportion (17.58%) of the area experienced improvement in ecological environmental quality, mainly located in the central and western sections, as well as the periphery of the entire study area. Lastly, there were areas where ecological quality degraded, primarily in the eastern-central part of the study area. Table 8 analyzes the overall changes in ecological quality using a transition matrix. With the exception of the "excellent" and "poor" levels, the proportion of each ecological quality level that remained unchanged was the largest. Additionally, the proportion of transitions between two or more levels was relatively small, regardless of whether the changes were "improved" or "degenerate". For each level, except for the "excellent" level, the proportion of improvements was greater than that of degradations.

Table 8. Level transfer matrix of the MRSEI for the Nyingchi–Gongbo'gyamda section of the Lhasa-Nyingchi Motorway from 2012 to 2020 (%).

		2012				
	Level	Excellent	Good	Moderate	Fair	Poor
	Excellent	33.35	12.88	0.95	0.21	0.18
	Good	65.39	75.98	32.10	4.57	1.95
2020	Moderate	1.27	10.36	54.22	29.60	8.31
	Fair	0.00	0.71	11.57	49.21	45.36
	Poor	0.00	0.07	1.17	16.40	44.21
	Rate of change	9.62	−12.32	10.36	13.99	−7.25

From the perspective of MRSEI level transitions, the proportions and areas of land that did not undergo transitions from the "excellent" to "poor" ecological quality levels are as follows: 33.35%/57.625 km^2, 75.98%/553.375 km^2, 54.22%/235.5625 km^2, 49.21%/117 km^2 and 44.21%/31.25 km^2. Among them, the land with "good" and "fair" ecological quality levels had a larger area that did not undergo any transitions. Therefore, the land with unchanged ecological environmental quality from 2012 to 2020 was mainly dominated by "good" and "fair" levels.

In terms of improvements in ecological environmental quality levels, the proportions and areas of land that transitioned from "poor", "fair", "moderate", and "good" to the next higher level are as follows: 45.36%/32.0625 km^2, 29.61%/70.375 km^2, 32.1%/139.4375 km^2, and 12.88%/93.8125 km^2, respectively. It is evident that there was a significant increase in the areas for all categories. Therefore, improvements in ecological environmental quality were evident in all four levels: "poor", "fair", "moderate", and "good". In terms of degradation types, the proportions and areas of land that transitioned from "excellent", "good", "moderate", and "fair" to the next lower level are as follows: 65.39%/113 km^2, 10.36%/75.4375 km^2, 11.57%/50.25 km^2, and 16.4%/39 km^2, respectively. It can be seen that there was a higher proportion of degradation in the "excellent" and "good" categories. Therefore, the degradation of ecological quality mainly occurred in the transition from "excellent" to "good" and "good" to "fair". In summary, the ecological environmental quality in the study area was slightly better in 2020 than in 2012, with more obvious improvements in the central and western sections. The areas with significant ecological degradation were mainly located in the eastern part of the study area, where the "excellent" and "good" ecological quality levels were degraded to "good" and "moderate", respectively.

4. Conclusions

This study mainly utilized MODIS data from the GEE platform and combined it with the characteristics of the Xizang region. By introducing a combination weighting method based on GT, the RSEI was improved to evaluate the ecological environmental quality of the Lhasa–Nyingchi Motorway's Nyingchi to Gongbo'gyamda section. The main conclusions are as follows:

(1)　The overall ecological environmental quality of the Nyingchi to Gongbo'gyamda section has significant regional differences. The quality of the ecological environment decreases from east to west. In the central and eastern sections, the ecological environmental quality is better on the southern side compared to the northern side. The areas with better ecological quality are mainly located in the southeastern part, while areas with poor, fair, and moderate ecological quality are mainly located in the roadside areas and the peripheral regions of the central and western sections.

(2) During the process of highway construction and operation, the ecological environmental quality shows a trend of "initial decline, subsequent improvement, and then a stair-step increase". The construction of the highway is the primary driver of ecological degradation in the study area, significantly reducing its ecological environmental quality. However, short-term ecological restoration and compensation efforts have allowed for the recovery of the ecological environmental quality, followed by a stair-step growth. This is mainly due to the strong ecological sensitivity of the Xizang region, as well as its unique geographical and climatic characteristics such as high altitude and large diurnal temperature differences, which increase the difficulty of ecological restoration.

(3) The ecological environmental quality in the study area from 2012 to 2020 mainly remained unchanged; with the area of improved ecological quality being greater than the degraded area, the ecological restoration project of the Lhasa–Nyingchi Motorway has shown significant effectiveness. The unchanged rate of ecological environmental quality is 69.5%, mainly occurring in good and moderate levels. In the area where the quality of the ecological environment has changed, there were significant improvements in all four levels of ecological quality: poor, fair, moderate, and good, mainly occurring in the middle and western sections. The types of ecological quality degradation mainly transformed from excellent to good and good to moderate, with the transformation being more pronounced in the central and eastern parts of the study area.

Overall, the assessment results validate the importance of ecological restoration projects. However, it is worth noting that due to the high cloud cover in the Nyingchi region from July to September, both Landsat and Sentinel data were not available for use in this study. As a result, the assessment primarily relied on MODIS imagery data and lacked consideration of factors such as meteorology and economic development. In future research, the focus will be on integrating high-temporal-resolution land satellite data, meteorological data, and other relevant factors for a more in-depth analysis. This aims to further enhance the effectiveness of ecological restoration in the Xizang region.

Author Contributions: Methodology, P.J.; Data curation, J.Y. and S.S.; Writing—original draft, X.W.; Project administration, X.H.; Funding acquisition, Q.L. and Z.M. All authors have read and agreed to the published version of the manuscript.

Funding: This study's research was funded by the National Natural Science Foundation of China (grant number 42101086, 42371086), a study on flash flood risk assessment method based on ensemble learning (grant number IWHR-SKL-KF202310), a flash flood warning method coupled with disaster-causing mechanism (grant number CKWV2021885/KY), and research on key technologies for flood disaster defense based on machine learning multi-models (grant number 2019KJ086).

Data Availability Statement: The data presented in this study are available on request from the corresponding author.

Conflicts of Interest: Author Pengfei Jia was employed by the company Citic Construction Co., Ltd. The remaining authors declare that the research was conducted in the absence of any commercial or financial relationships that could be construed as a potential conflict of interest.

References

1. Sheng, Y.; Cao, Y.B.; Li, J.; Wu, J.C.; Chen, J.; Feng, Z.L. Characteristics of permafrost along Highway G214 in the Eastern Qinghai-Tibet Plateau. *J. Mt. Sci.* **2015**, *12*, 1135–1144. [CrossRef]
2. Wu, Q.; Li, X.; Li, W. The Prediction of Permafrost Change along Qinghai–Tibet Highway, China. *Permafr. Periglac. Process.* **2015**, *11*, 371–376.

3.   Jin, H.; Zhao, L.; Wang, S.; Jin, R. Thermal regimes and degradation modes of permafrost along the Qinghai-Tibet Highway. *Sci China* **2006**, *49*, 1170–1183. [CrossRef]

4.   Lin, Z.J.; Niu, F.J.; Luo, J.; Lu, J.H.; Liu, H. Changes in permafrost environments caused by construction and maintenance of Qinghai-Tibet Highway. *J. Cent. South Univ. Technol.* **2011**, *18*, 1454–1464. [CrossRef]

5.   Yang, D.; Qiu, H.; Hu, S.; Pei, Y.; Wang, X.; Du, C.; Long, Y.; Cao, M. Influence of successive landslides on topographic changes revealed by multitemporal high-resolution UAS-based DEM. *Catena* **2021**, *202*, 105229. [CrossRef]

6.   Nip, M.J.; Haes, H. Ecosystem approaches to environmental quality assessment. *Environ. Manag.* **1995**, *19*, 135–145. [CrossRef]

7.   Xu, H.Q. A remote sensing index for assessment of regional ecological changes. *China Environ. Sci.* **2013**, *33*, 889–897.

8.   Wu, J.; Wang, X.; Zhong, B.; Yang, A.; Jue, K.; Wu, J.; Zhang, L.; Xu, W.; Wu, S.; Zhang, N.; et al. Ecological environment assessment for Greater Mekong Subregion based on Pressure-State-Response framework by remote sensing. *Ecol. Indic.* **2020**, *117*, 106521. [CrossRef]

9.   Shan, W.; Jin, X.; Ren, J.; Wang, Y.; Xu, Z.; Fan, Y.; Gu, Z.; Hong, C.; Lin, J.; Zhou, Y. Ecological environment quality assessment based on remote sensing data for land consolidation. *J. Clean. Prod.* **2019**, *239*, 118126. [CrossRef]

10. Xiong, Y.; Xu, W.; Lu, N.; Huang, S.; Wu, C.; Wang, L.; Dai, F.; Kou, W. Assessment of spatial–temporal changes of ecological environment quality based on RSEI and GEE: A case study in Erhai Lake Basin, Yunnan province, China—ScienceDirect. *Ecol Indic.* **2021**, *125*, 107518. [CrossRef]

11. Kamara, D.M.; Yang, Z.; Alhaji, L.Y. Ecological Geospatial Monitoring and Assessment of Surface Water Environment Using Remote Sensing Ecological Index Model (RSEI) in Freetown, Sierra Leone, from 2010 to 2018. *Glob. Sci. J.* **2020**, *8*, 45978.

12. Boori, M.S.; Choudhary, K.; Paringer, R.; Kupriyanov, A. Eco-environmental quality assessment based on pressure-state-response framework by remote sensing and GIS. *Remote Sens. Appl. Soc. Environ.* **2021**, *23*, 100530. [CrossRef]

13. Indrawati, L.; Murti, B.; Rachmawati, R. Integrated ecological index (IEI) for urban ecological status based on remote sensing data: A study at Semarang—Indonesia. *IOP Conf. Ser. Earth Environ. Sci.* **2020**, *500*, 012074. [CrossRef]

14. Song, M.J.; Luo, Y.Y.; Duan, L.M. Evaluation of Ecological Environment in the Xilin Gol Steppe Based on Modified Remote Sensing Ecological Index Model. *Arid Zone Res.* **2019**, *36*, 1521–1527.

15. Gou, R.; Zhao, J. Eco-Environmental Quality Monitoring in Beijing, China, Using an RSEI-Based Approach Combined with Random Forest Algorithms. *IEEE Access* **2020**, *8*, 196657–196666. [CrossRef]

16. Firozjaei, M.K.; Fathololoumi, S.; Kiavarz, M.; Biswas, A.; Homaee, M.; Alavipanah, S.K. Land Surface Ecological Status Composition Index (LSESCI): A novel remote sensing-based technique for modeling land surface ecological status. *Ecol. Indic.* **2021**, *123*, 107375. [CrossRef]

17. Wang, X.; Liu, C.; Fu, Q.; Yin, B. Construction and Application of Enhanced Remote Sensing Ecological Index. *Int. Arch. Photogramm. Remote Sens. Spat. Inf. Sci.* **2018**, *42*, 1809–1813. [CrossRef]

18. Lai, C.; Chen, X.; Wang, Z.; Yu, H.; Bai, X. Flood Risk Assessment and Regionalization from Past and Future Perspectives at Basin Scale. *Risk Anal.* **2020**, *40*, 1399–1417. [CrossRef]

19. Huang, S.; Ming, B.; Huang, Q.; Leng, G.; Hou, B. A Case Study on a Combination NDVI Forecasting Model Based on the Entropy Weight Method. *Water Resour. Manag. Int. J. Publ. Eur. Water Resour. Assoc.* **2017**, *31*, 3667–3681. [CrossRef]

20. Liaw, A.; Wiener, M. Classification and Regression by random Forest. *R News* **2002**, *23*, 18–22.

21. Foody, G.M.; Mathur, A. Toward intelligent training of supervised image classifications: Directing training data acquisition for SVM classification. *Remote Sens. Environ.* **2004**, *93*, 107–117. [CrossRef]

22. Orencio, P.M.; Fujii, M. A localized disaster-resilience index to assess coastal communities based on an analytic hierarchy process (AHP). *Int. J. Disaster Risk Reduct.* **2013**, *3*, 62–75. [CrossRef]

23. Stewart, T.J. Multicriteria Decision Analysis. In *International Encyclopedia of Statistical Science*; Lovric, M., Lovric, M., Eds.; Springer: Berlin/Heidelberg, Germany, 2011; pp. 872–875.

24. Bahar Gogani, M.; Douzbakhshan, M.; Shayesteh, K.; Ildoromi, A.R. New Formulation of Fuzzy Comprehensive Evaluation Model in Groundwater Resources Carrying Capacity Analysis. *Ecopersia* **2018**, *6*, 79–89.

25. Zou, Q.; Liao, L.; Qin, H. Fast Comprehensive Flood Risk Assessment Based on Game Theory and Cloud Model Under Parallel Computation (P-GT-CM). *Water Resour. Manag.* **2020**, *34*, 1625–1648. [CrossRef]

26. Wu, T.Y.; Lee, W.T.; Guizani, N.; Wang, T.M. Incentive mechanism for P2P file sharing based on social network and game theory. *J. Netw. Comput. Appl.* **2014**, *41*, 47–55. [CrossRef]

27. Zhou, W.; Qiu, H.; Wang, L.; Pei, Y.; Tang, B.; Ma, S.; Yang, D.; Cao, M. Combining rainfall-induced shallow landslides and subsequent debris flows for hazard chain prediction. *Catena* **2022**, *213*, 106199. [CrossRef]

28. Ma, S.; Qiu, H.; Yang, D.; Wang, J.; Zhu, Y.; Tang, B.; Sun, K.; Cao, M. Surface multi-hazard effect of underground coal mining. *Landslides* **2022**, *20*, 39–52. [CrossRef]

29. Liu, Z.; Qiu, H.; Zhu, Y.; Liu, Y.; Yang, D.; Ma, S.; Zhang, J.; Wang, Y.; Wang, L.; Tang, B. Efficient Identification and Monitoring of Landslides by Time-Series InSAR Combining Single- and Multi-Look Phases. *Remote Sens.* **2022**, *14*, 1026. [CrossRef]

30. Wang, L.; Qiu, H.; Zhou, W.; Zhu, Y.; Liu, Z.; Ma, S.; Yang, D.; Tang, B. The post-failure spatiotemporal deformation of certain translational landslides may follow the pre-failure pattern. *Remote Sens.* **2022**, *14*, 2333. [CrossRef]

31. Li, T.; Sun, G.; Huang, J. Analysis on Typical Route Selection along River of High-grade Lhasa-Nyingchi Highway. *Urban Roads Bridges Flood Control* **2018**, *8*, 27–30+8.

32. Piao, S.L.; Fang, J.Y. Seasonal changes in vegetation activity in response to climate changes in China between 1982 and 1999. *Acta Geogr. Sin.* **2003**, *58*, 119–125.
33. Zhong, L.; Ma, Y.M.; Salama, M.S.; Su, B. Assessment of vegetation dynamics and their response to variations in precipitation and temperature in the Tibetan Plateau. *Clim. Chang.* **2010**, *103*, 519–535. [CrossRef]
34. Wu, Y.J.; Zhao, X.; Xi, Y.; Liu, H.; Li, C. Comprehensive evaluation and spatial-temporal changes of eco-environmental quality based on MODIS in Tibet during 2006–2016. *Acta Geogr. Sin.* **2019**, *74*, 1438–1449.
35. Bing, G.; Fang, H.; Lin, J. An Improved Dimidiated Pixel Model for Vegetation Fraction in the Yarlung Zangbo River Basin of Qinghai-Tibet Plateau. *J. Indian Soc. Remote Sens.* **2017**, *46*, 219–231.
36. Zhang, L.; Cui, G.; Shen, W.; Liu, X. Cover as a simple predictor of biomass for two shrubs in Tibet. *Ecol. Indic.* **2016**, *64*, 266–271. [CrossRef]
37. Leprieur, C.; Verstraete, M.; Pinty, B. Evaluation of the performance of various vegetation indices to retrieve vegetation cover from AVHRR data. *Remote Sens. Rev.* **1994**, *10*, 265–284. [CrossRef]
38. Asner, G.P.; Scurlock, J.; Hicke, J.A. Global synthesis of leaf area index observations: Implications for ecological and remote sensing studies. *Glob. Ecol. Biogeogr.* **2003**, *12*, 191–205. [CrossRef]
39. Jensen, J.L.; Humes, K.S.; Hudak, A.T.; Vierling, L.A.; Delmelle, E. Evaluation of the MODIS LAI product using independent lidar-derived LAI: A case study in mixed conifer forest. *Remote Sens. Environ.* **2011**, *115*, 3625–3639. [CrossRef]
40. Schaefer, K.; Schwalm, C.R.; Williams, C.; Arain, M.A.; Barr, A.; Chen, J.M.; Davis, K.J.; Dimitrov, D.; Hilton, T.W.; Hollinger, D.Y.; et al. A model-data comparison of gross primary productivity: Results from the North American Carbon Program Site Synthesis. *J. Geophys. Res. Biogeosci.* **2012**, *117*, G03010. [CrossRef]
41. Lobser, S.E.; Cohen, W.B. MODIS tasselled cap: Land cover characteristics expressed through transformed MODIS data. *Int. J Remote Sens.* **2007**, *28*, 5079–5101. [CrossRef]

Article

Landslide Hazard Assessment in Highway Areas of Guangxi Using Remote Sensing Data and a Pre-Trained XGBoost Model

Yuze Zhang [1], Lei Deng [2,*], Ying Han [3], Yunhua Sun [1], Yu Zang [1] and Minlu Zhou [4]

1   National Engineering Research Center for Transportation Safety and Emergency Informatics, China Transport Telecommunications & Information Center, Beijing 100028, China; zhangyz@igsnrr.ac.cn (Y.Z.); sunyunhua@cttic.cn (Y.S.); zangyu@cttic.cn (Y.Z.)
2   School of Traffic and Transportation, Beijing Jiaotong University, Beijing 100044, China
3   State Grid Siji Location Service Co., Ltd., Beijing 102200, China; hanying12@mails.ucas.ac.cn
4   GuangXi Communications Design Group Co., Ltd., Nanning 530012, China
*   Correspondence: denglei@cttic.cn

Abstract: This study presents a novel method for assessing landslide hazards along highways using remote sensing and machine learning. We extract geospatial features such as slope, aspect, and rainfall over Guangxi, China, and apply an extreme gradient boosting model pre-trained on contiguous United States datasets. The model produces susceptibility maps that indicate landslide probability at different scales. However, the lack of accurate data on historical landslides in Guangxi challenges the model evaluation and comparison between regions. To overcome this, we calibrate the model to fit the local conditions in Guangxi. The calibrated model agrees with the observed landslide locations, implying its capability to capture regional variations in landslide mechanisms. We apply the model at a 30 m resolution along the Heba Expressway and validate it against reports from July 2021 to March 2022. The model correctly predicts five of seven landslide events in this period with a reasonable alarm rate. This framework has the potential for large-scale landslide risk management by informing transportation planning and infrastructure maintenance decisions. More data on landslide timing and human disturbance events may improve the model's accuracy across diverse geographical areas and terrains.

Keywords: landslide hazard assessment; remote sensing data; XGBoost algorithm; highway; pre-trained model

Citation: Zhang, Y.; Deng, L.; Han, Y.; Sun, Y.; Zang, Y.; Zhou, M. Landslide Hazard Assessment in Highway Areas of Guangxi Using Remote Sensing Data and a Pre-Trained XGBoost Model. *Remote Sens.* **2023**, *15*, 3350. https://doi.org/10.3390/rs15133350

Academic Editor: Magaly Koch

Received: 9 May 2023
Revised: 16 June 2023
Accepted: 28 June 2023
Published: 30 June 2023

1. Introduction

Landslides are a common natural disaster that pose a significant threat to human society and the environment. They can be triggered by various factors, such as intense rainfall, earthquake shaking, water level change, or human activities [1]. The occurrence of landslides is often accompanied by secondary disasters such as mudflows, which can endanger lives and property [2]. In recent years, the frequency and scale of landslides have increased due to factors like climate change and human activities [3,4].

Guangxi Zhuang Autonomous Region in China is a region prone to landslides due to its complex terrain and landform, which are affected by natural hazards such as typhoons and heavy rains. Landslides in Guangxi often cause damage to infrastructure and agricultural land, and even lead to casualties. For instance, the Jingxi landslide in June 2010 resulted in 26 deaths and 4 missing. Therefore, it is important to conduct research on landslide hazard assessment techniques to better forecast and prevent landslides in Guangxi, which can enhance the protection of the local people and the sustainable development of the region. However, traditional field investigation approaches are often time-consuming, expensive, and difficult to implement for regional-scale analysis.

Landslide hazard assessment is a complex task that involves several aspects. It requires the identification of landslide-prone areas, the evaluation of landslide-triggering factors, the estimation of landslide frequency and magnitude, and the assessment of potential consequences and risks. Different methods have been developed and applied for landslide hazard assessment, depending on the scale, purpose, data availability, and complexity of the problem [5,6]. The main challenges of landslide hazard assessment are to select appropriate methods, integrate multiple factors and sources of uncertainty, and validate and communicate the results effectively. Some of the commonly used methods for landslide hazard assessment are heuristic, statistical, deterministic, probabilistic, and multi-criteria decision-making methods [7]. Heuristic methods are based on expert knowledge and experience, and they assign qualitative ratings or weights to different factors influencing landslides [8]. Statistical methods are based on empirical relationships between landslide occurrence and explanatory variables, and they use mathematical models to calculate the probability of or susceptibility to landslides [2]. Deterministic methods are based on physical laws and principles, and they evaluate the stability or instability of slopes under given conditions [9]. Probabilistic methods are based on stochastic models and Bayesian inference, and they estimate the likelihood or frequency of landslides considering uncertainties in input data and parameters [10]. Multi-criteria decision-making methods are based on analytical techniques and fuzzy logic, and they combine multiple criteria and preferences to rank or classify landslide hazards [11]. Each method has its advantages and limitations, and no single method can be universally applicable or optimal for all situations. Therefore, it is important to choose a suitable method according to the objectives, data availability, scale, and complexity of the study area. Moreover, it is also useful to compare different methods or combine multiple methods to improve the reliability and accuracy of landslide hazard assessment.

Recent advances in remote sensing and machine learning provide powerful tools for landslide hazard assessment over large areas [12–18]. Remote sensing techniques enable the acquisition of key biophysical indicators related to slope instability over vast areas. Machine learning algorithms can detect complex patterns from big data and generate predictive models. Studies have applied remote sensing and machine learning for landslide hazard assessment, but there are still challenges and opportunities for further development [19–21].

This study introduces a novel methodology for landslide hazard assessment along highways that integrates an extreme gradient boosting (XGBoost) model and geospatial analysis. The framework uses remote sensing imagery and geospatial data to capture the variations in topography, hydrology, soil properties, and vegetation conditions over Guangxi, China. XGBoost is an efficient and scalable end-to-end tree-boosting system that has achieved state-of-the-art results on many machine learning challenges [22–25]. It can capture complex and nonlinear relationships between landslide factors and susceptibility, and it can also handle imbalanced data and missing values effectively [26]. In our proposed framework, we adopt the XGBoost algorithm to first pre-train a base model on United States (U.S.) datasets and then calibrate it for local conditions in Guangxi using statistical approaches. The calibrated model is then applied to a 30 m resolution site along the landslide-prone Heba Expressway for validation with documented landslide events from 2021 to 2022 [27–31]. According to the results of this work, the framework has the potential to provide quantitative risk assessment over large areas to aid policymaking, reduce loss from disasters, and promote environmentally sustainable infrastructure in mountainous regions. Continued progress in remote sensing, machine learning, and computing power is poised to enable the wider application of this approach worldwide.

The rest of this paper is organized as follows: Section 2 introduces the study area and the data sources. Section 3 describes the methodology, including the data processing, the machine learning models, and the evaluation metrics. Section 4 presents the results and discussion. Section 5 concludes the paper and suggests some future work.

2. Study Area and Data Sources

2.1. Study Area

The study area of this paper is Guangxi, a coastal autonomous region in southern China that covers an area of about 237,600 square kilometers. The region has complex terrain and frequent landslide disasters, which not only pose a great threat to road traffic safety but also have a serious impact on surrounding residents and enterprises. Guangxi was chosen as the study area because it is one of the major areas prone to landslides in China. According to historical data from the catalog, over 15,000 geological hazards occurred in Guangxi in the past few decades (1950–2020), among which over 4200 landslides were predominantly located within a 1 km range along roads. Figure 1 shows the spatial distribution of all recorded landslides along roads in Guangxi from 1950 to 2020.

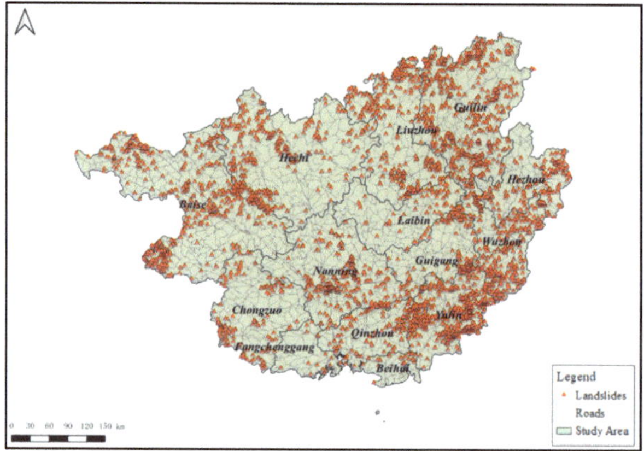

Figure 1. Spatial distribution of landslides along roads in Guangxi from 1950 to 2020.

2.2. Remote Sensing and Terrain Data

In general, landslide risk assessment requires the identification and analysis of various factors that influence the occurrence of landslides, such as geology, topography, soil type, vegetation cover, and rainfall patterns. Previous studies have shown that these factors vary in different regions and need to be selected based on local conditions. In this paper, we focus on the Guangxi area in China, where landslides are frequent and pose a serious threat to the local population and infrastructure. Based on the geological terrain, historical landslide patterns, and existing research in this area, we selected 12 landslide conditioning factors, namely, slope angle, slope aspect, three types of precipitation, soil moisture, lithology, vegetation index, two types of land cover, distance to faults, and distance to rivers. In addition, besides conditioning factors, the landslide catalog is also an important source of data for landslide susceptibility mapping. It can be used to identify areas prone to landslides and to develop effective landslide risk management strategies.

1. Slope (1) and aspect (2) are important terrain factors that affect landslide occurrence. Steep terrain and slopes with different orientations can influence soil stability and the risk of soil erosion.
2. Daily (3) and antecedent rainfall (4) (7 days in this paper) are important climate factors that affect soil saturation and pore water pressure. The amount and distribution of rainfall can both affect soil stability. In addition, to avoid overfitting the pre-trained models on the training set, this study further incorporated daily accumulated precipitation information for the past 10 years (5) in the region, which could help reduce biases caused by differences in regional precipitation distribution.
3. Soil moisture (6) is another important soil factor, and excessive moisture can make soil unstable and increase the risk of landslides.

4. The Normalized Difference Vegetation Index (NDVI) (7), as a vegetation index, can reflect the density of vegetation. Dense vegetation can reduce soil erosion and improve soil stability.

5. Land cover type is a key factor that influences soil stability and affects its physical and chemical properties. We used two kinds of classification from the Moderate Resolution Imaging Spectroradiometer (MODIS) in this paper. The first one is the International Geosphere-Biosphere Programme (IGBP) (8), which consists of 17 classes defined by the IGBP. These data indicate the dominant land cover type for each pixel every year. The second one is the annual Leaf Area Index (LAI) (9) based on land cover type, which is an important variable for estimating photosynthesis, evapotranspiration, and crop growth. Moreover, by combining annual LAI with the latest NDVI, we can capture the heterogeneity and dynamics of land surface properties in our model.

6. Lithology (10) is one of the important factors affecting soil stability, and different types of rocks can affect the properties and stability of the soil.

Details about all variables utilized in this work can be found in Tables 1 and 2.

Table 1. Selected landslide catalogs.

Landslide Catalog	Time Range	Landslide Events	Reference
United States	2009–2020	1963	https://www.gislounge.com/using-nasas-global-landslide-catalog-for-landslide-risk-management/ (accessed on 13 May 2023)
Guangxi	1950–2020	4229	/
Heba Expressway	2021.01–2022.03	7	/

Table 2. Selected indicators and data sources.

Indicator	Source	Dataset	Time Resolution	Spatial Resolution	Reference
Slope and Aspect	ASTGTM3 DEM	DEM	/	30 m	https://lpdaac.usgs.gov/products/astgtmv003/ (accessed on 13 May 2023)
Rainfall	GPM Level 3 IMERG Late Daily	Precipitation	1 day	10 km	https://gpm.nasa.gov/data/directory (accessed on 13 May 2023)
Soil Moisture	SMAP L4	Root zone soil moisture percentile	1 day	9000 m	https://grace.jpl.nasa.gov/data/get-data/ (accessed on 13 May 2023)
NDVI	MODIS Surface Reflectance Daily L2G	Surface Reflectance	1 day	500 m	https://lpdaac.usgs.gov/products/mod09gav061/ (accessed on 13 May 2023)
	Harmonized Landsat Sentinel-2 (HLS)		2–3 days	30 m	https://lpdaac.usgs.gov/products/hlsl30v002/ (accessed on 13 May 2023)
Land Cover Types	MODIS Land Cover Type	LC_Type1(IGBP) and 3(LAI)	1 year	250 m	https://lpdaac.usgs.gov/products/modis_products_table/mcd12q1 (accessed on 13 May 2023)
	China Land Cover Dataset (CLCD)	/		30 m	https://essd.copernicus.org/articles/13/3907/2021/ (accessed on 13 May 2023)
Lithology	Global Lithological Map (GLiM)	/	/	/	https://www.earthdoc.org/content/papers/10.3997/2214-4609.201802924 (accessed on 13 May 2023)

7. Distance to faults (11) and rivers (12) are also important factors influencing landslide occurrence. Faults can weaken the rock mass and make it vulnerable to sliding. Rivers can erode the toe of slopes and cause bank collapse. In addition, river erosion and undercutting on the outside bends of meandering rivers can also trigger landslides.

3. Methodology

3.1. Landslide Hazard Assessment (LHA) Framework

The steps of landslide hazard assessment may vary depending on the scale, purpose, and data availability of the study. However, a general framework can be summarized as follows:

1. Define the study area and objectives of the assessment.
2. Collect and review existing data on landslide inventory, causative factors, and consequences.
3. Select and apply appropriate methods and techniques for landslide susceptibility, hazard, and risk analysis.
4. Validate and evaluate the results of the analysis using historical records, field observations, or expert opinions.
5. Prepare and present landslide hazard maps and reports with recommendations for mitigation measures and further studies.

In this work, we chose the XGBoost algorithm together with 12 landslide conditioning factors for implementing the Landslide Hazard Assessment (LHA) model. One of the challenges we faced while collecting information on landslides in Guangxi was the lack of detailed geological disaster data. Specifically, the data often did not include important time information, which is essential for understanding the temporal patterns and trends of landslides and developing a suitable landslide risk assessment model. To overcome this challenge, we used a pre-trained model that was based on data from the U.S. with complete information. Then, we calibrated and transferred the pre-trained model to assess landslide hazards in the Guangxi area. Additionally, by considering the transferability and generality of the pre-trained model as well as the larger area and coarse resolution in training data, the grid cells were chosen as the mapping unit instead of the slope units [32–36].

3.2. XGBoost Algorithm

XGBoost is a powerful machine learning algorithm that has shown a lot of promise for landslide hazard mapping and assessment. It is an ensemble method that combines many decision trees to produce robust predictions, even with noisy and complex data. XGBoost also offers advantages like built-in cross-validation, handling of missing values, and computational scalability. However, XGBoost also has some significant limitations in this domain. It may face challenges in extrapolating to new regions. XGBoost is also prone to overfitting with small datasets and is not inherently interpretable, limiting transparency for end users. While XGBoost should definitely be considered as a tool for landslide mapping, it needs to be implemented carefully with these limitations in mind. With issues of scalability and interpretability adequately addressed, XGBoost can be a highly effective method for modeling landslide hazards over large areas [22,37–39].

3.2.1. Sample Preparation

Landslides are complex natural phenomenon that involve various factors and have high uncertainty and randomness. To predict the likelihood of landslides effectively, we needed to construct a reliable sample set that represented the conditions and characteristics of landslide occurrence. However, landslides are rare events both globally and regionally, so using all data in study areas as samples will create a severe imbalance between positive (landslide area) and negative samples (non-landslide area), which will impair the predictive performance of the model. Therefore, we applied a date-based sampling method; that is, on the day of a landslide event, we randomly selected 100 areas that did not experience landslides as negative samples, which, along with the landslide events as positive samples, formed a sample data set. This way, we could control the overall ratio of positive to negative samples in the data while also considering the temporal factors that affect landslides.

Finally, after removing invalid and missing data samples, we obtained a total of 197,431 model samples, including 1963 positive samples and 195,468 negative samples.

3.2.2. Model Training, Evaluation, and Validation

When training XGBoost, we usually need to divide the dataset into three parts: training set, test set, and validation set. The training set is used to train the model parameters, and the test and validation sets are used to evaluate the model's generalization ability, which is unknown during the training process to avoid overfitting or underfitting. There are many ways to create training sets, test sets, and validation sets. A common method is random splitting. That is, the dataset is shuffled randomly and then allocated to the training set, test set, and validation set according to a certain proportion.

In this paper, we used the data from 2009 to 2017 from the United States Geological Survey (USGS) NASA (National Aeronautics and Space Administration) Global Landslide Catalog and randomly split it into 80% training set and 20% test set. This can ensure that the distribution of the dataset is uniform and there will be no situation where there are too many or too few of a certain category or feature. In addition, we also used all the data after 2018 as a validation set to evaluate the model's generalization ability. Since the validation set comes from different time and geographic distributions, it can better test the model's adaptability to unknown data. Table 3 shows the three subsets used in this article for training the model.

Table 3. Summary of datasets for model training.

Dataset	Total	Time Range	Positive Samples
Training	145,738		1478
Test	36,435	2009–2017	335
Validation	15,258	2018–2020	150
Total	197,431	-	1963

Besides a high-quality dataset, the choice of model parameters is also very important for improving the performance and generalization ability of the model. In this article, we used the grid search method to determine the model parameters such as the learning rate, tree depth, and subsampling rate. Grid search is a technique that evaluates the performance of different combinations of parameters on a given dataset using cross-validation. The parameter value combined with the highest evaluation metric is finally selected as the optimal parameter combination. Parameter ranges selected for grid search are shown in Table 4.

Table 4. Parameters for grid search.

Parameter Name	Parameter Content	Value Range
Learning Rate	The rate at which the model learns	0.001, 0.05, 0.1
Max Depths	The maximum depth of the tree	3, 4, 5,
Subsamples	The subsample ratio of the training instances	0.5, 0.8, 1.0
Gamma	The minimum loss reduction. The larger it is, the more conservative the algorithm will be.	0, 0.01, 0.1
Minimum Child Weight	The minimum sum of instance weight (hessian) needed in a child	1, 2, 3
Estimators	Number of gradient-boosted trees	100, 200, 300

In terms of the evaluation of model results, we calculated the metrics on the test set and validation set with the optimal parameters of XGBoost, including overall accuracy (OA), kappa coefficients, F1-score, true-positive rate (TPR), false-positive rate (FPR), recall, and Area Under the Curve (AUC). Recall is the TPR, which is the ratio of true positives to the total number of positives. It assesses how well the model can identify the target conditions (in this case, landslides). Overall accuracy measures the total proportion of correct predictions, regardless of class. However, OA can be misleading when there is a class imbalance, so additional metrics are needed. The kappa coefficient measures the proportion of agreement beyond what would be expected by chance alone. It is considered a robust measure when there are multiple categories and high accuracy that could be due to class imbalance. The F1-score is the harmonic mean of recall and precision, providing a balanced measure of the model's accuracy on the positive class. Finally, the AUC measures the area under the Receiver Operating Characteristic (ROC) curve, which plots the true-positive rate against the false-positive rate at various threshold settings. The AUC ranges from 0 to 1, with higher values indicating better distinguishing ability between the target and non-target classes [40–42].

Equations of evaluation metrics are as follows:

$$OA = \frac{TP + TN}{TP + FP + FN + TN} \tag{1}$$

$$Recall = \frac{TP}{TP + FN} \tag{2}$$

$$Precision = \frac{TP}{TP + FP} \tag{3}$$

$$F1 - score = \frac{2 * Precision * Recall}{Precision + Recall} \tag{4}$$

$$Kappa = \frac{(OA - CA)}{(1 - CA)} \tag{5}$$

$$CA = \frac{(TP + FN)(TP + FP) + (FP + TN)(FN + TN)}{(TP + FPFN + TN)^2} \tag{6}$$

where *TP* is the number of true-positive predictions, which means the number of positive samples that are correctly predicted as positive. *FP* is the number of false-positive predictions, which means the number of negative samples that are incorrectly predicted as positive. *TN* is the number of true-negative predictions, which means the number of negative samples that are correctly predicted as negative. *FN* is the number of false-negative predictions, which means the number of positive samples that are incorrectly predicted as negative.

4. Results and Discussion

4.1. Evaluation of Pre-Trained Model in the Contiguous United States

In this section, we compare the performance of seven machine learning algorithms on three datasets to evaluate their applicability in a landslide hazards assessment task. The algorithms included XGboost-1, XGBoost-2 (less complex), MLP-1 (MultiLayer Perceptron with 8 × 8 layers), MLP-2 (16 × 16 layers), MLP-3 (32 × 32 layers), Support Vector Machine (SVM), and logistic regression (LR).

As mentioned in Section 3.2.2, the metrics used to evaluate the models were AUC, kappa, F1-score, recall, overall accuracy, TPR, and FPR, which could provide insights into how well the model is performing and help in comparing different models or algorithms. However, due to the serious imbalance problem in the landslide and non-landslide samples, we utilized the weighted recall and F1-score instead of the recall and F1-score for the landslide class. These weighted metrics were calculated with the classification report function provided by Scikit-learn in Python.

Finally, we summarize the metrics of seven models on three datasets in Table 5.

Table 5. Evaluation metrics of 7 pre-trained models.

Metrics	Datasets	XGBoost-1	XGBoost-2	MLP-1 (8 × 8)	MLP-2 (16 × 16)	MLP-3 (32 × 32)	SVM	LR
AUC (%)	Training	97	97	95	96	98	88	92
	Test	95	96	94	95	93	87	92
	Validation	93	93	94	91	90	86	91
Kappa (%)	Training	81	80	76	79	85	61	68
	Test	78	78	74	77	70	57	70
	Validation	68	63	72	68	64	55	67
Weighted F1-Score (%)	Training	94	95	93	93	94	91	91
	Test	94	95	93	94	94	91	91
	Validation	93	94	92	92	93	88	89
Weighted Recall (%)	Training	89	91	88	89	91	85	84
	Test	89	91	88	89	90	85	85
	Validation	88	90	86	88	89	79	82
Overall Accuracy (%)	Training	89	91	88	89	91	85	84
	Test	89	91	88	89	90	85	85
	Validation	88	90	86	88	89	79	82
FPR (%)	Training	11	9	12	11	10	15	16
	Test	11	9	12	11	10	15	15
	Validation	12	10	14	12	11	21	18
TPR (%)	Training	92	89	88	90	95	76	84
	Test	88	87	86	88	79	72	86
	Validation	80	73	86	80	75	75	85

The results above show that the performance of the seven models varies depending on the dataset and the metric used. To compare them more clearly, we highlight some key points as follows:

1. The XGBoost-2 model achieved the best performance in most of the metrics, especially in F1-score, recall, FPR, and overall accuracy, which represent the overall classification ability and precision. However, it also had a significant drop in kappa and TPR on the validation set, which means it was less consistent and sensitive in LHA.

2. MLP-3 showed similar performance to XGBoost-2, ranking second in most of the metrics. Compared to XGBoost-2, MLP-3 was slightly inferior, especially on the test and validation sets. There was a steep decline in metrics from training sets to test and validation sets, indicating a more serious overfitting problem.

3. XGBoost-1 and MLP-2 also share similar performances in most metrics. In general, XGBoost-1 was slightly better than MLP-2 as all metrics ranked 1–3 among seven models. As for MLP-2, most values ranked third but with small gaps (1–2%) with XGBoost-1. It should be noted that both XGBoost-1 and MLP-2 had an obvious improvement in TPR but a slight loss in FPR on validation sets compared to XGBoost-2 and MLP-3. A higher TPR with a relatively low FPR is more valuable for balancing the accuracy and management cost in engineering applications for LHA.

4. Finally, the SVM and LR models did not outperform XGBoost and MLP models. The classification performance and generalization ability were relatively poor. However, the LR model was the most robust one when tested on three datasets. The difference in metrics only ranged from 2 to 3%.

In summary, based on the overall ranks in metrics, we can infer that the XGBoost-1 model outperformed the other models in this evaluation scenario. The MLP-2 model ranked second and showed a stable trade-off between TPR and FPR. The XGBoost-2 and MLP-3 models had lower ranks but achieved higher scores in LHA, which could be preferable for applications that prioritize accuracy over sensitivity and specificity. Therefore, we finally chose the XGBoost-1 model as the best choice for this problem domain. We also report the optimal parameters of the XGBoost-1 model in Table 6 and visualize the feature importance and the ROC curves for this model in Figures 2 and 3. Feature importance is a technique that assigns a score to each input feature based on how useful it is for predicting the target variable. Feature importance scores can help us understand which features are most relevant for our prediction task, and which ones can be ignored or removed.

Table 6. Values of hyperparameters for pre-trained model.

Parameter Name	Values
Learning Rate	0.1
Max Depths	4
Subsamples	0.5
Gamma	0.1
Minimum Child Weight	2
Estimators	200
Tree Method	exact

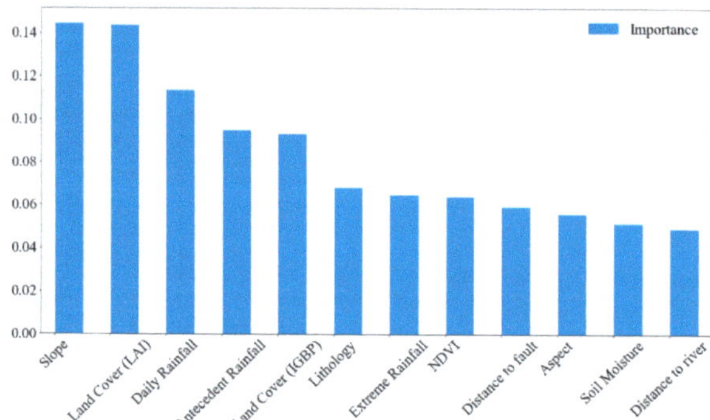

Figure 2. Feature importance of pre-trained model.

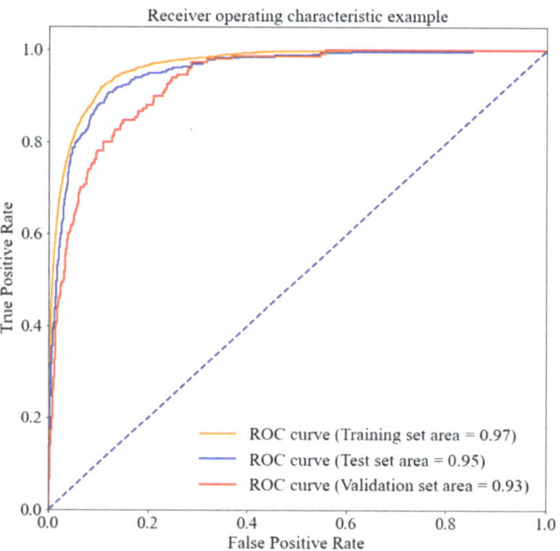

Figure 3. Plots of ROC curves with different sets.

The results in Figure 2 show that our model identified slope as the most important factor influencing landslide risk. However, the relationship between slope and landslide susceptibility is not linear, but rather complex and influenced by multiple factors. The slope is an important first-order control, but not the only one.

1. The slope is the most important factor affecting landslide occurrence in the region, according to the ranking analysis. Aspect also influences landslides but to a lesser degree. Generally, slope stability decreases when slope and aspect are unfavorable, as they increase the driving forces. However, when the slope exceeds a certain threshold, collapse becomes more likely than a landslide. Therefore, setting an optimal maximum slope value to distinguish between landslide and collapse could further improve the performance of the LHA model.

2. Climatic factors (daily rainfall, antecedent rainfall, and extreme rainfall) are also important features of the model. The importance of daily rainfall and antecedent rainfall (7 days) ranks third and fourth, respectively, indicating that the amount and distribution of rainfall are important reasons for soil instability in the region. For example, as rainwater infiltrates into the soil, it fills up the pore spaces and exerts pressure on soil particles. When the pore pressure builds up rapidly due to intense daily rainfall, it reduces the effective stress between soil particles that provide shear strength. This can quickly destabilize the slope and trigger landslides. In contrast, antecedent rainfall has already had time to drain from the soil, so it contributes less to elevated pore pressures on the specific day of landslide occurrence. In addition, sudden heavy rains will also increase the risk of landslides as excessive rainfall can increase soil saturation and pore water pressure, thus reducing soil shear strength and stability.

3. Land cover factors (LAI land cover and IGBP land cover) also account for a relatively large weight in the important features of the model. The importance of LAI and IGBP classification ranks second and fifth, respectively, reflecting that land use type has a great influence on slope stability. This is because different land covers have different impacts on vegetation coverage, soil structure, and hydrological processes, thus affecting slope stability.

4. Vegetation factors (NDVI) are also one of the features considered in the model. Their importance ranks ninth, indicating that vegetation cover is also a factor affecting landslide occurrence, but relatively less than other factors such as terrain, climate, and land use type. In general, dense vegetation can improve slope stability by reinforcing soil shear strength through its root system.

5. Geological factors (lithology) and distance to faults and rivers also account for a certain weight in the model features. The importance of rock type, fault distance, and river distance ranks 7th, 10th, and 12th, respectively. These factors can weaken the rock mass and trigger slope collapse.

6. Soil properties (soil moisture) matter because weaker, more porous soils with low shear strength saturate more easily and provide less resistance to sliding. Pore pressures rise quickly in these soils during intense rain, reducing stability.

In summary, the proposed LHA model comprehensively considers factors such as terrain, climate, land use, vegetation, geology, and so on. Among them, terrain, climate, and land use type are important features of the model, which contribute greatly to the assessment of landslide risk. The analysis results of the model are reasonable to some extent and also provide a reference for us to further study the mechanism of landslides and improve the assessment model.

4.2. Evaluation of Pre-Trained Model Performance in Guangxi

To assess the pre-trained XGBoost model's generalizability across spatial scales and regions, two evaluation schemes were deployed for LHA in Guangxi. The first scheme calibrated the pre-trained model using the local landslide inventory to evaluate landslide hazards along highways across Guangxi. By calibrating to regional conditions, the pre-trained model accounted for broad-scale differences influencing slope stability across Guangxi and generated an overview of landslide susceptibility over a large area. This examined the model's ability to adapt to variable controls for land sliding at a regional scale. The second scheme selected a subsection from 141 to 143 km of the Heba Expressway as the validation site to evaluate landslide hazard assessment at 30 m resolution using the calibrated pre-trained model. This evaluated the model's capability to detect triggers of landslides at the hillslope scale, which commonly operates at a higher resolution. Together the schemes assessed the pre-trained model's potential for generalization across spatial scales as well as regional adaptation.

4.2.1. Calibrating the Pre-Trained Model for LHA across Guangxi

Conventional calibration approaches for model transfer rely on accurate data on the precise timing of landslide events, which were lacking in this study. Thus, to evaluate the pre-trained model in Guangxi, an alternative method was developed based on the spatial consistency between the predicted landslide hazard map and the distribution of known landslides. Specially, we first used the pre-trained model to calculate the daily landslide hazard probability along Guangxi roads within the time range of the landslide catalog. After that, we aggregated these probabilities and extracted the 98th percentile value to represent the overall probability of a hazardous landslide occurring. Finally, to verify the performance of our probabilistic model, we overlaid the final hazard map with a catalog of landslide occurrence locations from the same period and obtained Figure 4.

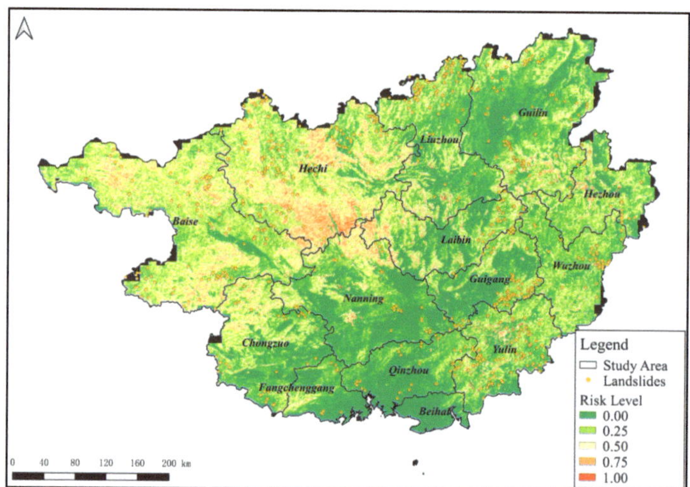

Figure 4. Overlay of landslide events and 98th probability map of predicted landslide hazard in Guangxi.

As shown in the probability map above, the pre-trained model developed in this paper generally produced relatively low-risk probability (less than 0.4 in most landslide points) when applied in the Guangxi region, making it difficult to determine whether an area is at risk. This phenomenon may be attributed to the fact that the pre-trained model is based on the regional characteristics of the United States, and applying it to a different region without calibration will lead to performance degradation due to feature distribution differences. To address this issue, this study adopted a statistical-based threshold-setting method and reclassified the regional risk probability. The threshold-setting method was as follows: (1) the probability values of all landslide points in the landslide catalog were extracted from the probability map; (2) percentiles were set according to the landslide areas of historical events, such as the {2, 39, 89, 99} calculated by counting landslides areas in the range of 100, 1000, 10,000, and 100,000 m^2; (3) the percentiles were used to reclassify the predicted probability into five categories of landslide hazard levels (from very low to very high as illustrated in Table 7); (4) urban and flat areas (<5 degree according to the minimum of slope that reported landslides) that rarely suffer landslides, and extreme steep areas (>58 degree, 98th percentile of slope that reported landslides) that are more prone to develop collapses instead of landslides, were mapped out.

Table 7. Thresholds for classifying the predicted probabilities.

Classified Number	Risk Level	Percentile Thresholds
0	Very Low	<2nd%
1	Low	[2nd%, 39th%)
2	Medium	[39th%, 89th%)
3	High	[89th%, 99th%)
4	Very High	≥99th%

After classification, we obtained the corresponding landslide risk as shown in Figure 5. Here, we only focused on highway roadsides, so only areas that were covered with highways are given in the figure.

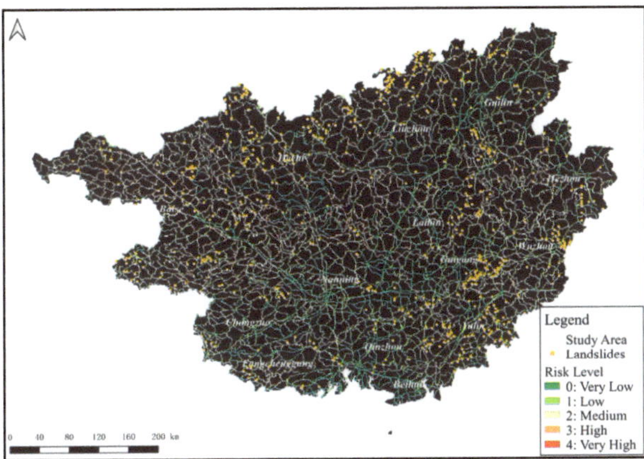

Figure 5. Landslide susceptibility map of highways in Guangxi Province.

The classified landslide hazard map that we produced is now clearer and consistent with the landslide catalog of Guangxi. The map shows that the regions with high landslide risk are mainly located in the mountainous areas of Guangxi, where the terrain is steep and the geology is complex. These areas are also prone to stone desertification due to the unique karst landscape. In contrast, the central and southern parts of Guangxi have a lower landslide risk because of their flat terrain and simple geology. Overall, the calibrated pre-trained model showed a high level of consistency in estimating the relative landslide hazard along the transportation network. This indicates that our pre-trained model can be transferred across different regions and still perform well.

4.2.2. Validating Localized Landslide Hazard Assessment at High Resolution

This section presents a daily evaluation of our predictions based on the date information of landslide events on the Heba Expressway, which was not available in the previous section. The Heba Expressway is a highway project in China that connects Hezhou and Bama. We collected the reports of seven landslides that occurred near sections 141–143 from July 2021 to March 2022 during the construction. We also improved the image resolution from 1 km to 30 m to further validate the performance of the pre-trained model. Specifically, we replaced the land cover data from the MODIS product (250 m) to the China Land Cover Dataset (CLCD) (30 m) with the mapping given in Table 8 and used the HLS images (30 m) for calculating the NDVI instead of the MODIS product (500 m). Moreover, due to the improvement in resolution, the set of percentiles for the probability convention was also updated according to the area of pixels. They are now 1, 2, 12, and 66 based on the percentage of landslide areas in the segments of $[5 \times 5, 10 \times 10, 20 \times 20, 50 \times 50]$ m^2.

Table 8. Mapping of MODIS land cover and ICLD surface types.

ICLD Values	IGBP Values	LAI Values
1	12	3
2	5	5
3	6	2
4	10	1
5	17	0
6	15	9
7	16	9
8	13	10
9	11	9

On the other hand, as the Harmonized Landsat Sentinel-2 (HLS) project is not a daily product, we utilized a new method for LHA in this section as follows:

1. First, we set an 8-day data collection cycle and fused the collected data of the same type to achieve a gap-filled and high-quality image for driving the model. According to the data illustration in Table 2, using 8 days as a cycle can ensure at least two images of the same type are obtained within the cycle for data fusion. The cycle length can be adjusted according to the needs, as long as the main data features of the model are not empty.

2. Next, we used the cycle-synthesized data to predict the landslide risk for each day within the cycle and took the maximum value as the output of the cycle's risk probability. The reason for this process is that we assume that the landslide risk is mainly influenced by the cumulative effects of rainfall and soil moisture. Therefore, taking the maximum value can capture the highest risk level within a cycle.

Table 9 shows the final predicted results for each prediction cycle covering the landslide and Figure 6 gives the distribution of predicted risk pixels. For comparison, we also provide the results of the MLP-2 and LR models.

Table 9. Landslide prediction results for each cycle.

Prediction Cycle	Start Date	End Date	Landslide Occurrence	Predicted Risk Pixels ($\geq$Medium)		
				XGBoost-1	MLP-2	LR
1	4 July 2021	11 July 2021	11 July 2021	3	1	0
	4 July 2021	11 July 2021	12 July 2021			
2	24 October 2021	31 October 2021	25 October 2021	0	0	0
3	17 November 2021	24 November 2021	24 November 2021	0	0	0
4	17 January 2022	24 January 2022	18 January 2022	1	2	1
	17 January 2022	24 January 2022	23 January 2022			
5	30 March 2022	6 April 2022	31 March 2022	14	30	3

The table and figures show that the XGBoost-1 model successfully predicted and warned of five landslide events in the first, fourth, and fifth prediction periods. In these periods, the model identified 3, 1, and 14 pixels with medium or higher risk, respectively. However, the model failed to detect any risk pixels in the second and third prediction periods, and only one pixel in the fourth period. This suggests that the model's performance is inconsistent and depends on various factors. Some possible reasons are the following:

(1) The input variables did not include human interference factors. The study area had some construction activities that may alter the soil's physical and hydrological properties and affect the occurrence of landslides. These factors were not considered in the model training, resulting in the model's inability to adapt to this change. (2) The model parameters and weights were not finely adjusted. Due to the limited number of samples in the study area, this paper used a statistical threshold-setting method to determine the predicted risk levels. This method is simple and easy to implement, but it cannot fully utilize the data, nor can it effectively handle nonlinearity and complexity problems. (3) The data resolution was mismatched. The model was trained based on 1 km resolution data, but 30 m resolution data were used for testing in this section. This may have led to a decline in data accuracy and quality and also affected the model's generalization ability.

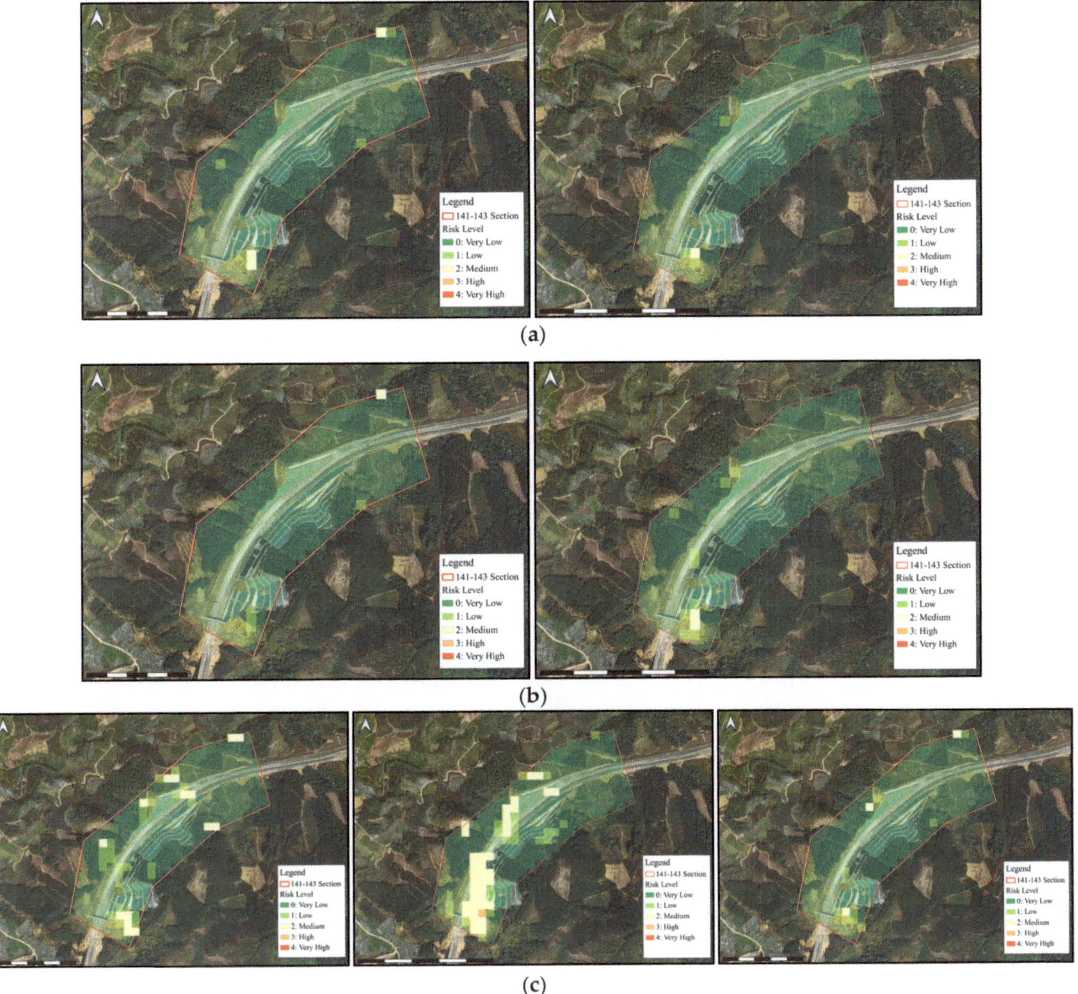

Figure 6. Predicted hazard pixels in study areas: (**a**) 2021-07-04 (XGBoost-1/MLP-2); (**b**) 2022-01-17 (XGboost-1/MLP-2); (**c**) 2022-03-30 (XGBoost-1/MLP-2/LR).

The MLP-2 model showed a similar overall performance to the XGBoost-1 model, even with a higher resolution, while the LR model performed poorly. The LR model labeled only 1 and 3 pixels as risky in periods 4 and 5, respectively, while the MLP-2 model labeled

1, 2, and 30 pixels as risky in periods 1, 4, and 5, respectively. Based on these results, we can infer that the MLP-2 model was slightly better than the XGBoost-1 model for the Heba Expressway. However, as we discussed earlier, the ability of a model to effectively control the false-alarm rate is also critical for real applications, which can avoid unnecessary alerts and interventions that save costs in landslide prevention management. To illustrate this, we calculated the 98th percentile of daily probabilities to count the risk frequency.

Figure 7 shows the distribution of the final predicted risk levels along the Heba Expressway with these three models.

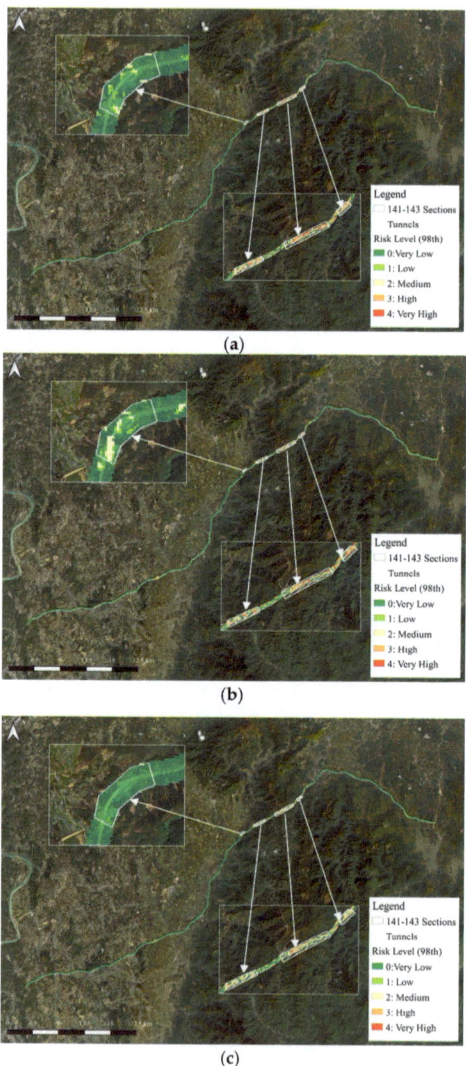

Figure 7. Landslide hazard levels along Heba Expressway: (**a**) XGBoost-1; (**b**) MLP-2; (**c**) LR.

As seen from the figure, the LR model was the most conservative, classifying fewer pixels as risky than the other two models. This contradicts the conclusions in Section 4.1 and suggests that the LR model might have poor generative ability when applied to a different region or resolution. Only 8.8% of the pixels were classified as a higher than medium risk level by the LR model, compared to about 20% by the pre-trained model. The MLP-2 model was the least conservative, classifying 14.7% of the pixels as risky, which was similar to the pre-trained model. The XGBoost-1 model classified 12.5% of the pixels as risky, which was also close to the pre-trained model. However, many of the risk pixels identified by these three models were located on the segments of tunnels (see zoomed inset), which were not suitable for our method since we focused on the roadside slopes that could be observed by satellites. After omitting these segments, the percentages of risk pixels were 4.2% for LR, 6% for XGBoost-1, and 9.5% for MLP-2. Based on these results, we conclude that all three models could work along the Heba Expressway under a 30 m resolution, but the LR model is not recommended for real applications unless further validation is completed. Both the XGBoost-1 and MLP-2 models showed good agreement with the results in Section 4.1, but the XGBoost-1 model could balance the effectiveness of landslide identification and the ratio of high-risk pixels better than the MLP-2 model. Therefore, we recommend the XGBoost-1 model for LHA applications.

In conclusion, this paper demonstrates the feasibility and effectiveness of using a pre-trained model to evaluate landslide hazards along a highway. The model can identify high-risk areas and provide useful information for mitigation and prevention measures. Despite some limitations, we can further refine and improve the model to enhance its accuracy and efficiency in practice, especially with more landslide samples. In addition, the results in Figure 7 indicate that risk levels vary with different models, suggesting that a further study on the integration of models would also improve the performance of pre-trained models in LHA.

5. Conclusions

Landslide disasters pose a serious threat to the safety of people and property along highways, especially in mountainous areas like Guangxi, China. To effectively prevent and mitigate landslide hazards, it is crucial to assess landslide risk and identify vulnerable locations. Remote sensing techniques have been widely used for landslide detection and monitoring because they provide large-scale, high-resolution, and multi-temporal data on various factors influencing landslide occurrence.

In this study, we propose a novel method that extracts these features from remote sensing data and feeds them into a pre-trained XGBoost model. The model can produce landslide hazard maps at different scales (1 km and 30 m in this work) and indicate the risk level (0–4) of landslide occurrence for each pixel. We compare the performance of the XGBoost model with other models such as MLP and LR on the test and validation datasets. The results show that the XGBoost model outperformed other models on most metrics, such as the AUC, F1-score, and kappa, indicating that it can avoid overfitting and achieve better trade-offs between TPR and FPR. When validated on the Heba Expressway, our method successfully recognized five out of seven events, with an alert rate of 6%. Compared to the MLP and LR models, the XGBoost model could provide a good balance between alert rate and landslide identification. Our method shows promise as a useful landslide hazard assessment tool that can provide valuable information for the preliminary risk evaluation of transportation networks and help prioritize areas needing further investigation or intervention.

Nevertheless, our method also faces some limitations. One major challenge is the scarcity of data on landslide occurrence in Guangxi. This hinders our ability to evaluate and compare model performance across different locations and to update the model with new data. Future research should gather more data on the temporal and spatial patterns of landslides in Guangxi and similar regions to better calibrate and validate the pre-trained model. Moreover, exploring the combinations of different types of models could also be valuable to enhance the performance of pre-trained models in LHA. Despite these limitations, the results on the Heba Expressway illustrate the potential of this method as a decision support tool for broad-scale risk assessment and management in the absence of timing data.

To conclude, this study proposes an innovative framework for landslide hazard assessment along highways based on remote sensing data and a pre-trained XGBoost model. This framework can help pinpoint areas susceptible to landslides at various scales and provide useful information for the preliminary risk assessment of transportation networks. This can potentially benefit transportation planning and emergency response by improving the resilience of critical infrastructure against landslide hazards.

Author Contributions: Conceptualization, Y.Z. (Yuze Zhang) and L.D.; methodology, Y.Z. (Yuze Zhang), L.D., and Y.H.; validation, Y.Z. (Yuze Zhang), Y.S., Y.Z. (Yu Zang) and M.Z.; formal analysis, Y.Z. (Yuze Zhang), L.D., and Y.H.; investigation, Y.S. and M.Z.; resources and data curation, Y.Z. (Yuze Zhang), Y.S., Y.H. and M.Z.; writing—original draft preparation, Y.Z. (Yuze Zhang) and Y.Z. (Yu Zang); writing—review and editing, Y.Z. (Yuze Zhang) and L.D.; visualization, Y.Z. (Yuze Zhang) and Y.H. All authors have read and agreed to the published version of the manuscript.

Funding: This work is supported by the National Natural Science Foundation of China (No. 42001301) and the Guangxi Science and Technology Project (AB20159034).

Conflicts of Interest: The authors declare no conflict of interest.

References

1. Schuster, R.L. Socioeconomic Significance of Landslides. In *Landslides: Investigation and Mitigation*; Turner, A.K., Schuster, R.L. Eds.; Transportation Research Board Special Report 247; National Academy Press: Washington, DC, USA, 1996.
2. Dai, F.C.; Lee, C.F.; Ngai, Y.Y. Landslide Risk Assessment and Management: An Overview. *Eng. Geol.* **2002**, *64*, 65–87. [CrossRef]
3. Crozier, M.J. Deciphering the Effect of Climate Change on Landslide Activity: A Review. *Geomorphology* **2010**, *124*, 260–267. [CrossRef]
4. Petley, D.N. Global Patterns of Loss of Life from Landslides. *Geology* **2012**, *40*, 927–930. [CrossRef]
5. Varnes, D.J. *Landslide Hazard Zonation: A Review of Principles and Practice*; UNESCO: Paris, France, 1984.
6. Guzzetti, F. Landslide Hazard and Risk Assessment. PhD. Thesis, University of Bonn, Bonn, Germany, 2006.
7. Pardeshi, S.D.; Autade, S.E.; Pardeshi, S.S. *Landslide Hazard Assessment: Recent Trends and Techniques*; Springer Plus: Berlin/Heidelberg, Germany, 2013; Volume 2, p. 523.
8. Lee, S.; Pradhan, B. Landslide hazard mapping at Selangor, Malaysia using frequency ratio and logistic regression models. *Landslides* **2007**, *4*, 33–41. [CrossRef]
9. Hungr, O. A model for the runout analysis of rapid flow slides, debris flows, and avalanches. *Can. Geotech. J.* **1995**, *32*, 610–623. [CrossRef]
10. Baecher, G.B.; Christian, J.T. *Reliability and Statistics in Geotechnical Engineering*; Wiley: Chichester, UK, 2003.
11. Malczewski, J. GIS-based multicriteria decision analysis: A survey of the literature. *Int. J. Geogr. Inf. Sci.* **2006**, *20*, 703–726. [CrossRef]
12. Ado, M.; Wang, R.-Y.; Lv, G.-A.; Jiao, L. Landslide Susceptibility Mapping Using Machine Learning: A Literature Survey. *Remote Sens.* **2022**, *14*, 3029. [CrossRef]
13. Yu, H.; Li, S.; Ruan, W.; Yao, J.; Liu, Y.; Zhang, L. Landslide Susceptibility Mapping and Driving Mechanisms in a Vulnerable Region Based on Multiple Machine Learning Models. *Remote Sens.* **2023**, *15*, 1886. [CrossRef]
14. Chen, S.; Abd Razak, K.I.; Shi, X.; Huang, R. Landslide susceptibility mapping using machine learning algorithms and multi-source remote sensing data. *J. Mt. Sci.* **2020**, *17*, 1897–1914.
15. Chen, Z.; Shahabi, H.; Shirzadi, A.; Chien, S.-F.; Koufos, G.D.; Yu, M.; Alipour, S.; Zhang, Y.; Yang, T.; Xu, C.; et al. Landslide susceptibility mapping using multi-source remote sensing data and an ensemble machine learning algorithm. *Remote Sens. Environ.* **2020**, *246*, 111853.
16. Alvioli, M.; Melillo, M.; Baum, R.L. A comparison of machine learning algorithms for regional landslide susceptibility mapping. *Landslides* **2020**, *17*, 1059–1078.
17. Hong, S.-H.; Park, J.; Lee, J.-S.; Kim, K.-S.; Yi, M.-J. Landslide susceptibility mapping using deep learning-based convolutional neural networks with high-resolution satellite imagery. *Remote Sens. Lett.* **2020**, *11*, 725–734.
18. Maji, A.K.; Martha, T.R.; Kerle, N.; van Westen, C.J. Landslide susceptibility mapping using deep convolutional neural network with multi-source remote sensing data. *Geomat. Nat. Hazards Risk* **2020**, *11*, 2336–2355.
19. Lissak, C.; Bartsch, A.; Michele, M.D.; Gomez, C.; Maquaire, O.; Raucoules, D.; Roulland, T. Remote Sensing for Assessing Landslides and Associated Hazards. *Surv. Geophys.* **2020**, *41*, 1391–1435. [CrossRef]
20. Casagli, N.; Intrieri, E.; Tofani, V.; Gigli, G.; Raspini, F. Landslide detection, monitoring and prediction with remote-sensing techniques. *Nat. Rev. Earth Environ.* **2023**, *4*, 51–64. [CrossRef]
21. Li, Z.; Chen, W.; Wang, J. Landslide identification using machine learning. *J. Mt. Sci.* **2020**, *17*, 1379–1392.
22. Chen, T.; Guestrin, C. XGBoost: A Scalable Tree Boosting System. In Proceedings of the 22nd ACM SIGKDD International Conference on Knowledge Discovery and Data Mining, San Francisco, CA, USA, 13–17 August 2016.

23. Althuwaynee, M.A.; Pradhan, B.; Lee, S.; Buchroithner, M.F. Landslide susceptibility mapping using XGBoost machine learning algorithm. *Geomat. Nat. Hazards Risk* **2020**, *11*, 2829–2854.

24. Pradhan, S.; Devkar, G.; Singh, U.K.; Kumari, P.; Singh, R. Landslide susceptibility assessment using XGBoost machine learning model: A case study of Uttarakhand state in India. *Geocarto Int.* **2020**, *35*, 1788–1813.

25. Rahmati, S.; Moumenifar, U.; Monavari, S.M.; Jolaei, S.A.; Shahabi, H. A novel hybrid machine learning model based on XGBoost and MARS for landslide susceptibility assessment. *Catena* **2020**, *187*, 104352.

26. Zhang, Y.; Takara, K.; Tachikawa, T. Landslide susceptibility mapping using an improved XGboost algorithm: A case study in the Kii Peninsula, Japan. *Remote Sens.* **2020**, *12*, 3413.

27. Breiman, L. Bagging predictors. *Machine Learn.* **1996**, *24*, 123–140. [CrossRef]

28. Paudel, U.; Oguchi, T.; Hayakawa, Y. Multi-resolution landslide susceptibility analysis using a DEM and random forest. *Int. J. Geosci.* **2016**, *7*, 726–743. [CrossRef]

29. Xu, Q.; Huang, R.Q.; Xiang, X.Q. Time and Spacial Predicting of Geological Hazards Occurrence. *J. Mt. Sci.* **2000**, *S1*, 112–117.

30. Youssef, A.M.; Pourghasemi, H.R.; Pourtaghi, Z.S.; Al-Katheeri, M.M. Landslide susceptibility mapping using random forest, boosted regression tree, classification and regression tree, and general linear models and comparison of their performance at Wadi Tayyah Basin, Asir Region, Saudi Arabia. *Landslides* **2016**, *13*, 839–856. [CrossRef]

31. Chen, T.; Zhong, Z.Y.; Niu, R.Q.; Liu, T.; Chen, S.Y. Mapping landslide susceptibility based on deep belief network. *Geomatics Inf. Sci. Wuhan Univ.* **2020**, *45*, 1809–1817.

32. Guzzetti, F.; Carrara, A.; Cardinali, M.; Reichenbach, P. Landslide hazard evaluation: A review of current techniques and their application in a multi-scale study, Central Italy. *Geomorphology* **1999**, *31*, 181–216. [CrossRef]

33. Feizizadeh, B.; Blaschke, T. An uncertainty and sensitivity analysis approach for GIS-based multicriteria landslide susceptibility mapping. *Int. J. Geogr. Inf. Sci.* **2014**, *28*, 610–638. [CrossRef]

34. Van Den Eeckhaut, M.; Vanwalleghem, T.; Poesen, J.; Govers, G.; Verstraeten, G.; Vandekerckhove, L. Prediction of landslide susceptibility using rare events logistic regression: A case-study in the Flemish Ardennes (Belgium). *Geomorphology* **2006**, *76*, 392–410. [CrossRef]

35. Blahut, J.; Van Westen, C.J.; Sterlacchini, S. Analysis of landslide inventories for accurate prediction of debris-flow source areas. *Geomorphology* **2010**, *119*, 36–51. [CrossRef]

36. Felicísimo, Á.M.; Cuartero, A.; Remondo, J.; Quirós, E. Mapping landslide susceptibility with logistic regression, multiple adaptive regression splines, classification and regression trees, and maximum entropy methods: A comparative study. *Landslides* **2012**, *10*, 175–189. [CrossRef]

37. Chen, T.; He, T. Higgs boson discovery with boosted trees. In Proceedings of the NIPS 2014 Workshop on High-energy Physics and Machine Learning 2015, Montreal, QC, Canada, 8–13 December 2014; PMLR: New York, NY, USA, 2015; Volume 42.

38. James, G.M.; Witten, D.; Hastie, T.; Tibshirani, R. *An Introduction to Statistical Learning*; Springer: New York, NY, USA, 2018; Volume 103, p. 18.

39. Molnar, C. *Interpretable Machine Learning*; Lulu Press: Morrisville, NC, USA, 2019.

40. Lv, L.; Chen, T.; Dou, J.; Plaza, A. A hybrid ensemble-based deep-learning framework for landslide susceptibility mapping. *Int. J. Appl. Earth Obs. Geoinf.* **2022**, *108*, 102713. [CrossRef]

41. He, Y.; Zhao, Z.; Yang, W.; Yan, H.; Liu, T. A unified network of information considering superimposed landslide factors sequence and pixel spatial neighbourhood for landslide susceptibility mapping. *Int. J. Appl. Earth Obs. Geoinf.* **2021**, *104*, 102508. [CrossRef]

42. Gao, X.; Chen, T.; Niu, R.; Plaza, A. Recognition and mapping of landslide using a fully convolutional densenet and influencing factors. *IEEE J. Sel. Top. Appl. Earth Obs. Remote Sens.* **2021**, *14*, 7881–7894. [CrossRef]

remote sensing

MDPI

Article

Air Pollution and Human Health: Investigating the Moderating Effect of the Built Environment

Chenglong Wang [1], Yunliang Sheng [1], Jiaming Wang [2], Yiyi Wang [2], Peng Wang [1,2,*] and Lei Huang [2]

[1] Faculty of Civil Engineering and Mechanics, Jiangsu University, Zhenjiang 212013, China; upesawcl@stmail.ujs.edu.cn (C.W.); upesasyl@stmail.ujs.edu.cn (Y.S.)
[2] State Key Laboratory of Pollution Control and Resource Reuse, School of the Environment, Nanjing University, Nanjing 210023, China; jiamingwang@smail.nju.edu.cn (J.W.); wangyiyi@nju.edu.cn (Y.W.); huanglei@nju.edu.cn (L.H.)
* Correspondence: upeswp@ujs.edu.cn

Abstract: Air pollution seriously threatens human health and even causes mortality. It is necessary to explore effective prevention methods to mitigate the adverse effect of air pollution. Shaping a reasonable built environment has the potential to benefit human health. In this context, this study quantified the built environment, air pollution, and mortality at 1 km $\times$ 1 km grid cells. The moderating effect model was used to explore how built environment factors affect the impact of air pollution on cause-specific mortality and the heterogeneity in different areas classified by building density and height. Consequently, we found that greenness played an important role in mitigating the effect of ozone (O_3) and nitrogen dioxide (NO_2) on mortality. Water area and diversity of land cover can reduce the effect of fine particulate matter ($PM_{2.5}$) and NO_2 on mortality. Additionally, gas stations, edge density (ED), perimeter-area fractal dimension (PAFRAC), and patch density (PD) can reduce the effect of NO_2 on mortality. There is heterogeneity in the moderating effect of the built environment for different cause-specific mortality and areas classified by building density and height. This study can provide support for urban planners to mitigate the adverse effect of air pollution from the perspective of the built environment.

Keywords: built environment; human health; air pollution; GIS; moderating effect

Citation: Wang, C.; Sheng, Y.; Wang, J.; Wang, Y.; Wang, P.; Huang, L. Air Pollution and Human Health: Investigating the Moderating Effect of the Built Environment. *Remote Sens.* **2022**, *14*, 3703. https://doi.org/10.3390/rs14153703

Academic Editors: Peng Jia, Shihong Du, Franz W. Gatzweiler, Xinhu Li and Jinchao Song

Received: 11 June 2022
Accepted: 29 July 2022
Published: 2 August 2022

Publisher's Note: MDPI stays neutral with regard to jurisdictional claims in published maps and institutional affiliations.

1. Introduction

Air pollution has become increasingly severe in recent years, which profoundly affects public health and is of concern to health organizations and governments [1,2]. About 90% of the population of the world lives in areas where air quality levels exceed the standard limited by the World Health Organization (WHO) [3]. Mitigating the effect of air pollution on human health has become a key issue to public health. During the last three decades, the effects of air pollution on human health have been documented, especially from particulate matter (PM), O_3, and NO_2 [4]. The global burden of disease caused by air pollution has barely declined since the 1990s [5]. In particular, an association with cardiovascular and respiratory causes of death has been reported [6,7].

The built environment is defined as human-made physical environment surroundings and conditions [8,9]. Urban spatial conditions can change the wind direction and velocity, which affect the air pollution dispersion [10]. Built environment characteristics can shape our activity patterns, which play a great role in health [11]. However, the health outcomes from physical activity could be offset by adverse exposure to air pollution [12]. Therefore, it is important to understand which built environment factors can alleviate the adverse effects of air pollution on human health. In addition, the definition of some built environment indicators does not adequately reflect local information [13]. Existing research on the built environment tends to select greenness, building density, and road density as major objects [14,15]. The use of multi-source data such as points of interest (POIs) can build a more comprehensive built environment framework [13].

Human health is an important landscape benefit and concern of many landscape users [16]. Although some studies have assessed the effect of the built environment on health, the results are inconsistent. As far as greenness is concerned, the reduction in mortality is associated with the increase in greenness [17]. Greenness was confirmed to link with the mortality of non-accident, cardiovascular, and respiratory diseases [3]. Other studies have pointed out a non-linear correlation between residential greenness and health [18,19]. Nevertheless, some studies point out that greenness is not associated with mortality [20,21]. The effect of the built environment on health is unclear and heterogeneity may exist across different areas. By incorporating building density and height into the framework of the built environment, the moderating mechanism of the built environment can be fully explored from a three-dimensional perspective.

Over the years, the study of the built environment, air pollution, and human health has shown a pairwise relationship, especially the effect of air pollution on health [2,11,22]. A spectrum of the key mechanisms link the built environment to human health, including air pollution and health behaviors [23,24]. The relationship between the built environment and health has been discussed from the perspectives of sleep, resilience, physicochemical environmental factors, movement, relationships, spirituality, and nutrition [25]. However, there is a lack of research linking these three fields.

Therefore, taking Nanjing city of Jiangsu Province in China as the study area, the main objectives of this study were (i) to determine which built environment can mitigate the adverse health effect of air pollution, (ii) to compare the moderating effect of the built environment on the relationship between air pollution (O_3, $PM_{2.5}$, and NO_2) and cause-specific mortality, and (iii) to analyze the heterogeneity of the built environment moderating effect in different areas classified by building density and height. This study quantified annual O_3, $PM_{2.5}$, NO_2, non-accident deaths, cardiovascular disease deaths, and respiratory disease deaths at a 1 km × 1 km grid scale from 2007 to 2015. Built environment factors were assumed to remain unchanged in this study, and we calculated them for 2015 as moderating variables. First, we investigated the moderating effect of the built environment on the relationship between air pollution and mortality. Then, we further explored the heterogeneity in different areas classified by building height and density.

Addressing these issues will be beneficial to reveal the mechanism of the built environment on mortality, and the findings can provide support for urban planning to reduce the adverse effect from air pollution.

2. Materials and Methods

2.1. Study Area

China has suffered severe air pollution for the last decade because of fast urbanization and industrialization, particularly in the economically developed eastern regions such as the Yangtze River Delta (YRD) region [26]. This study was conducted in Nanjing (Figure 1), one of the cities in the YRD region, the capital of Jiangsu province in China. Jiangsu Province is one of the most developed provinces in China. Nanjing ($31°14'$N–$32°37'$N, $118°22'$E–$119°14'$E) had experienced a rapid urbanization process with a population of 8.4 million by the end of 2018. At the same time, because Nanjing is surrounded by mountains on three sides, which is not conducive to the diffusion of pollutants, the concentrations of air pollutants in Nanjing is higher than the national average. Nanjing has a subtropical monsoon climate.

2.2. Data Collection

The data used in this study include road data, building information, normalized difference vegetation index (NDVI), land use and land cover data, point of interest (POI), air pollution (NO_2, $PM_{2.5}$, and O_3), population density, and mortality data. Road data from OpenStreetMap were used to calculate the road length and road crossing at a grid scale.

The building dataset was acquired from BIGEMAP, which contained detailed information on building space coordinates, footprints, and the number of floors in a building. Point of interest (POI) data from BIGEMAP were adopted to measure the diversity of the POI and the number in the grid. NDVI was used to measure surface greenness in this study. The study calculated the proportion of construction land, forestland, and water area within the grid based on the land cover data from Resource and Environment Science and Date Center. This study considered three air pollutants i.e., O_3, $PM_{2.5}$, and NO_2. This study took into account population density as a covariate in the relationship between air pollution and mortality. Based on the address information, this study calculated the deaths in the grid scale. The sources of the data are shown in Table 1. The overall framework of this paper is shown in Figure 2.

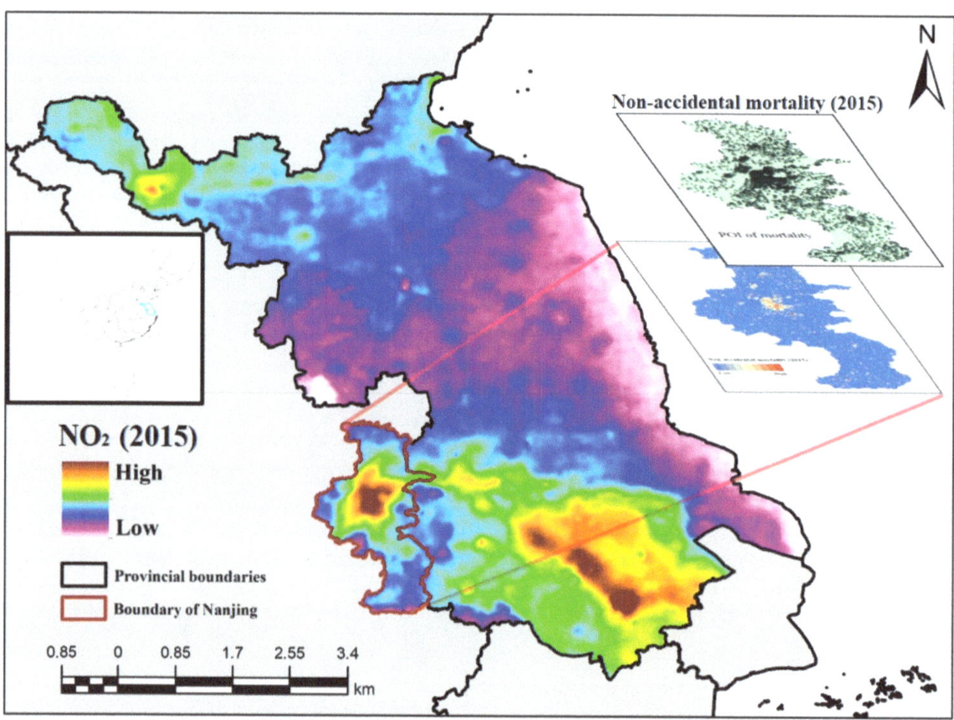

Figure 1. Location of the study area.

Table 1. Date, resolution, and sources.

Data	Resolution	Sources
Road data		OpenStreetMap (http://download.geofabrik.de/asia/china.html# (accessed on 10 February 2022))
Building dataset Point of interest	point	BIGEMAP (http://www.bigemap.com/ (accessed on 16 August 2020))
NDVI	1 km × 1 km	Resource and Environment Science and Date Center (https://www.resdc.cn/DOI/DOI.aspx?DOIID=49 (accessed on 4 June 2021))
Land cover	30 m	Resource and Environment Science and Date Center (https://www.resdc.cn/data.aspx?DATAID=184 (accessed on 27 April 2020))
Air pollution (O_3, $PM_{2.5}$, and NO_2)	1 km × 1 km	[27,28]
Population density Mortality	1 km × 1 km	WorldPop (www.worldpop.org (accessed on 7 February 2022)) Jiangsu Provincial Center for Disease Prevention and Control [29,30]

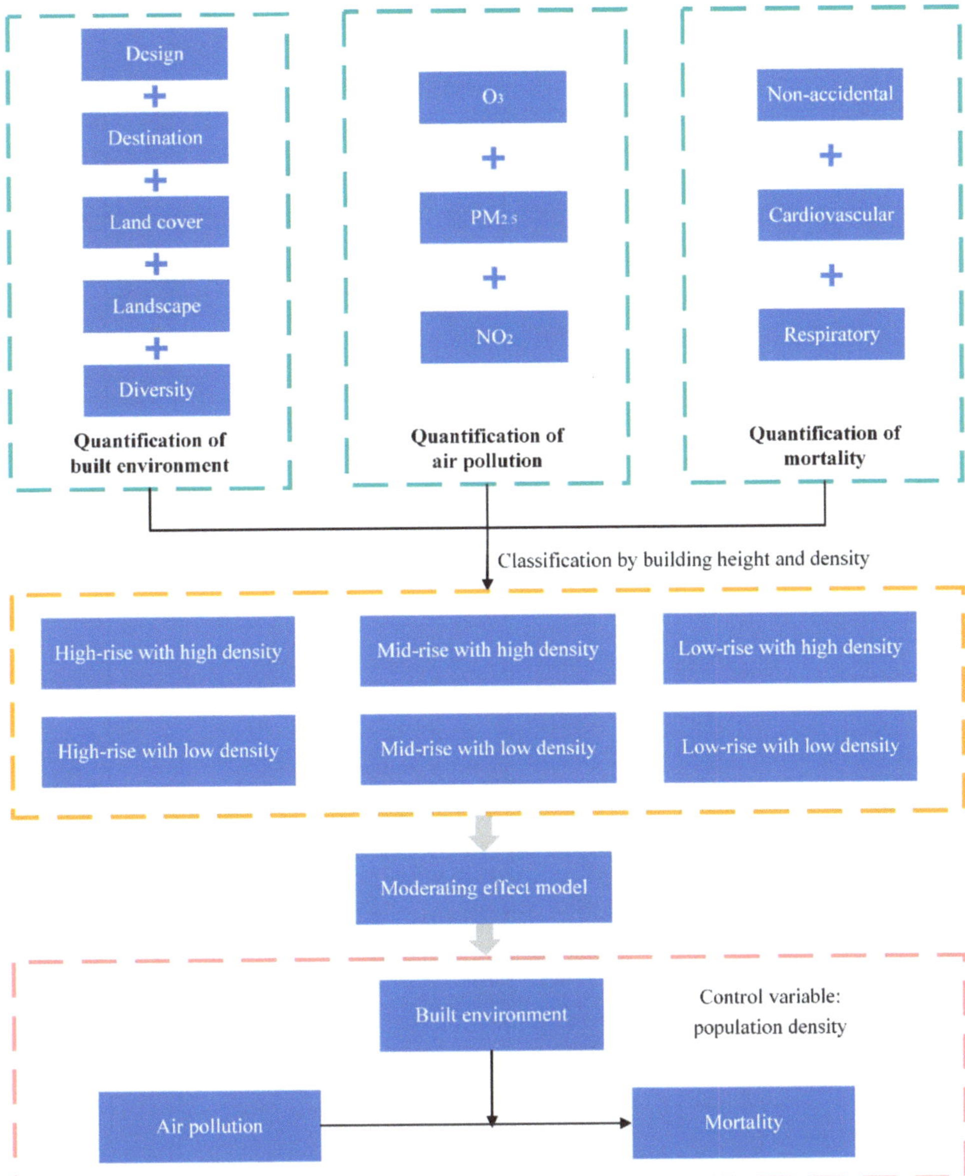

Figure 2. Overall framework of this study.

2.3. Measurement of Built Environment

In this study, 24 features were calculated to interpret the built environment. We operationalized the measurements of the built environment at the grid scale according to the framework in Table 2. All 24 features were divided into five categories including design, destination, land use, landscape, and diversity. The factors of design, destination, and land cover were calculated using the ArcGIS Desktop platform (version 10.8). The landscape indexes and SHDI of each grid were calculated by the Fragstats 4.2. The diversity of the built environment was calculated by the entropy method [31] based on the diversity of the design and diversity of POI and SHDI. This study used standardized indexes to calculate the diversity of design. The diversity of design and diversity of POI were calculated using the following Hill numbers equation [32]:

$$D = \left(\sum_{i=1}^{n} P_i^q \right)^{1/(1-q)} \tag{1}$$

where D is diversity, n indicates the number of the types, P_i is proportion of the category occupied by type i, and $q = 2$ (according to the Gini–Simpson concentration index).

Table 2. Description of 24 indexes of the built environment.

Category	Index	Description of Index
Design	Building density (BD)	The ratio of the building footprint area to grid area
	Building height (BH)	Average of all building heights within the grid
	Standard deviation of building height (SDBH)	The standard deviation of all building heights within the grid
	NDVI	The greenness within the grid
	Road length (RL)	The total length of roads within the grid
	Road crossing (RCS)	Number of road crossings within the grid
Destination	Hygiene facility (HF)	Number of hospital and health care facilities within the grid
	Government agency (GA)	Number of government agencies within the grid
	Residential community (RCM)	Number of residential communities within the grid
	Industrial park (IP)	Number of industrial parks within the grid
	Traffic facility (TF)	Number of traffic facilities within the grid
	Gas station (GS)	Number of gas stations within the grid
	Catering facility (CF)	Number of catering service facilities within the grid
Land cover	Construction land (CL)	Proportion of construction land within the grid
	Forestland (FL)	Proportion of forestland within the grid
	Water area (WA)	Proportion of water area within the grid
Landscape	AI	Aggregation index
	ED	Edge density
	PAFRAC	Perimeter-area fractal dimension
	PD	Patch density
Diversity	Diversity of built environment (DBE)	Diversity of built environment factors
	Diversity of design (DD)	Diversity of six indexes of design
	Diversity of POI (DPOI)	The POI mix
	Shannon's diversity index (SHDI)	Diversity of land cover

2.4. Air Pollution and Mortality from 2007 to 2015

High resolution exposure data are helpful to accurately assess the effect of air pollution on human health. Based on the high-resolution spatiotemporal model and data of previous studies [27,28], we calculated the annual O_3, $PM_{2.5}$, and NO_2 concentration from 2007 to 2015 at 1 km $\times$ 1 km grid scale in Nanjing.

Mortality data were classified into non-accidental causes (A00-R99), cardiovascular diseases (I00-I99), and respiratory diseases (J00-J99) based on the International Statistical Classification of Diseases and Related Health Problems 10th Revision (https://icd.who.int/browse10/2010/en#/IX (accessed on 14 June 2021)). Based on the addressed information, we calculated the gridded death counts of non-accidental, cardiovascular diseases, and respiratory diseases. The high death toll may be related to the high population base. Meanwhile, higher population density is related to higher mortality [33,34]. Hence, we selected population density as a control variable to eliminate the effect of population on mortality in this study.

2.5. Statistical Analysis

This study derived the air pollution and mortality in Nanjing for each year from 2007 to 2015. Then, correlation analysis was used to explore the relationships between the air pollution and mortality. In the correlation analysis, this study deleted the sample with zero death. Based on the correlation test, the study further used regression models to control the effect of population on deaths. Pearson's correlation coefficient can measure the correlation between two variables, x and y [14].

Because of the availability of data, the built environment indexes were calculated for 2015. This study assumed that the built environment remains unchanged. Then, we explored the spatial correlation between air pollution and the built environment and mortality and the built environment, respectively. Local indicators of spatial association (LISA) can describe the spatial correlation between the spatial distribution of different variables [35]. The study used GeoDa 1.14 to explore the bivariate spatial correlation. Local indicators of spatial association were calculated as follows [36]:

$$I_i = \frac{x_i - \overline{x}}{\delta^2} \sum_{j=1, j \neq 1}^{n} w_{ij} \left(x_j - \overline{x} \right) \tag{2}$$

$$\delta^2 = \frac{1}{n} \sum_{i=1}^{n} (x_i - \overline{x})^2 \tag{3}$$

$$\overline{x} = \frac{1}{n} \sum_{i=1}^{n} x_i \tag{4}$$

where I_i is the local Moran index of grid i; w_{ij} is the spatial weight between grid i and j; n is the number of spatial grids adjacent to grid i; x_i and x_j are the values of grid i and j, respectively; $\overline{x}$ is the variable average; and δ^2 indicates the variance.

Hierarchical regression was used to explore the moderating effect of the built environment in the relationship between air pollution and mortality. Within the model, air pollution was the independent variable (x) and mortality was the dependent variable (y), with population density as the covariate. The built environment was defined as the moderator (w). The study further conducted subgroup analyses by building density and height [37]. The moderating effect was confirmed by a significant interaction between the independent variable and the moderator. Before modeling, we normalized the variables and centered to zero. The moderation analysis was conducted in PROCESS, a plug-in for SPSS 25.0. All of the analyses were accepted with significance at $\alpha = 0.05$.

3. Results

3.1. Spatial Patterns of the Built Environment

The spatial patterns of the built environment are shown in Figure 3. It was observed that built environment indexes have a high level of agglomeration. The diversity of the built environment is higher in the middle part of Nanjing. Close to the city center, the urbanization level and development intensity were higher. High-intensity development usually means higher building density, building height, and road density, while greenness is relatively low.

3.2. Pairwise Correlation between Air Pollution, Mortality, and the Built Environment

The results of the correlation analysis of air pollution and mortality are shown in Figure 4, and the detailed results of the regression analysis are shown in the Table 3. Statistically positive associations were found between air pollution and mortality. NO_2 shows the strongest positive association with mortality, especially for respiratory death. The direct correlations between air pollution and mortality have an indicative character. However, there are some other factors that can also contribute to mortality, such as built environment factors.

The spatial correlation between air pollution and the built environment and mortality and the built environment is shown in Figure S1. It is clear that the built environment has agglomeration relationships with air pollution and mortality. High–high (H–H) areas indicate high air pollution or high mortality with a high built environment level. High–low

(H–L) areas indicate high air pollution or high mortality with a low built environment level. Low–high (L–H) areas indicate low air pollution or low mortality with high built environment level. Low–low (L–L) areas indicate low air pollution or low mortality with a low built environment level.

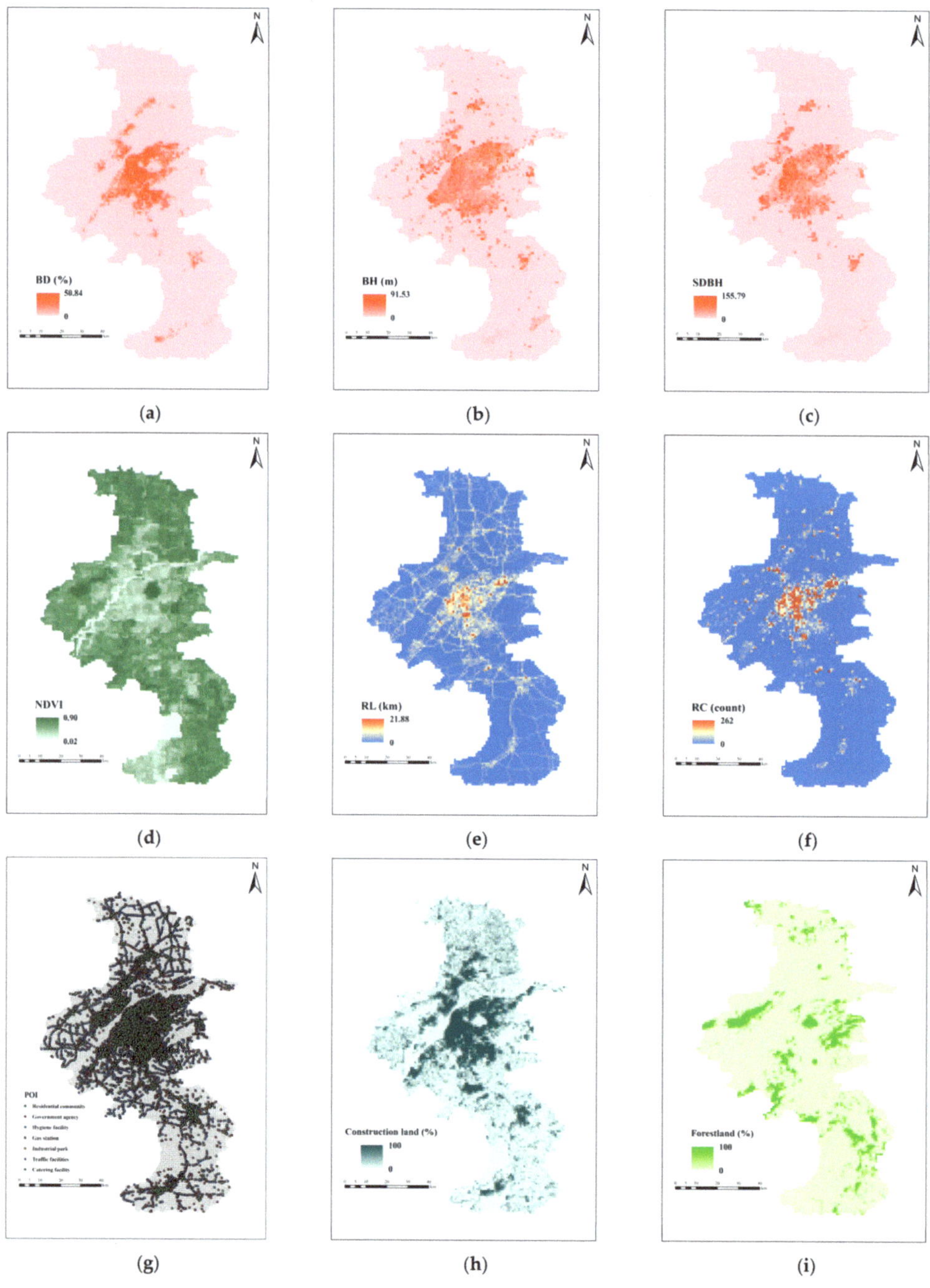

(a) (b) (c)

(d) (e) (f)

(g) (h) (i)

Figure 3. *Cont.*

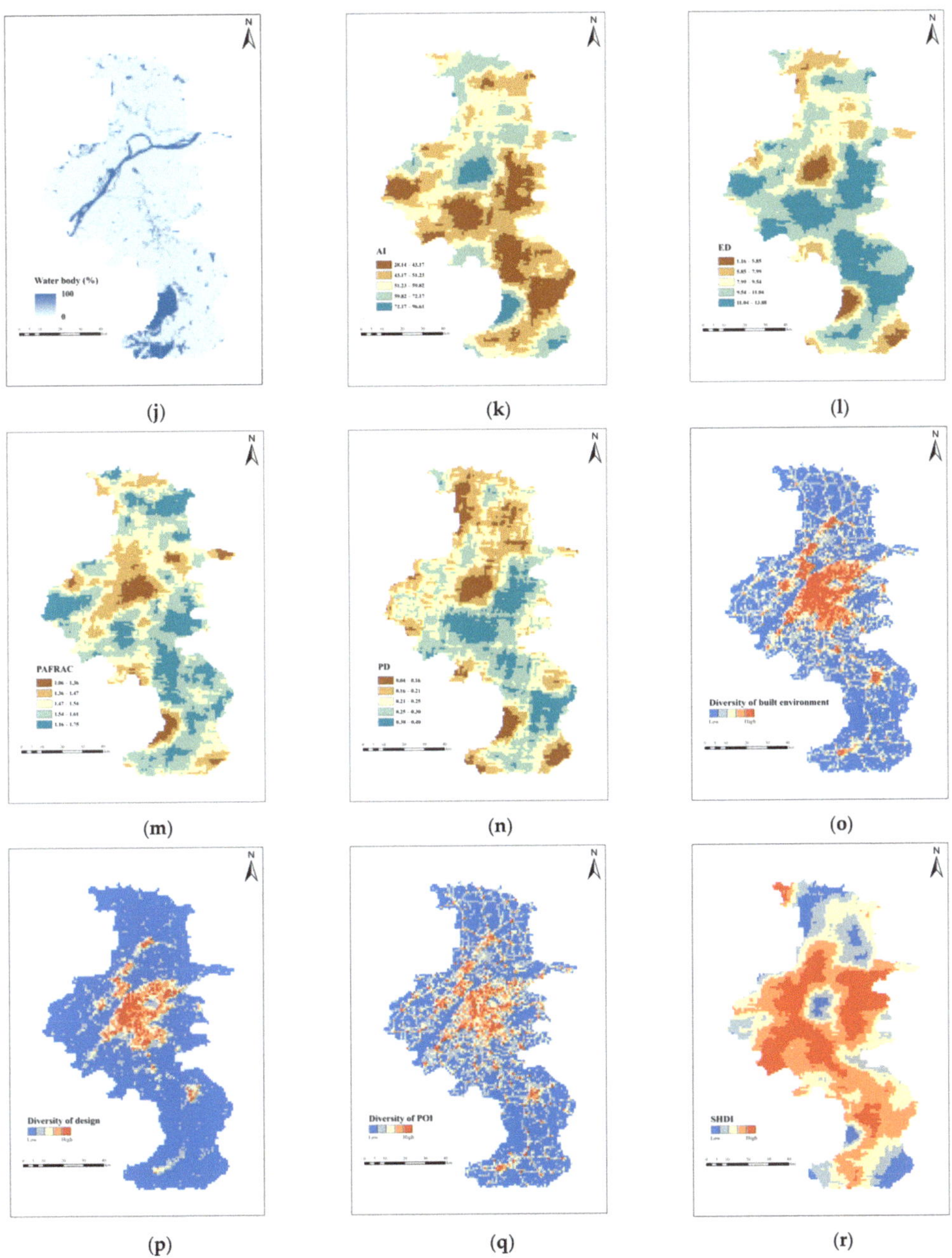

Figure 3. Spatial patterns of the built environment. (**a**) Building density; (**b**) building height; (**c**) standard deviation of building height; (**d**) NDVI; (**e**) road length; (**f**) road crossing; (**g**) POI of seven types; (**h**) proportion of construction land; (**i**) proportion of forestland; (**j**) proportion of water body; (**k**) aggregation index; (**l**) edge density; (**m**) perimeter-area fractal dimension; (**n**) patch density; (**o**) diversity of the built environment; (**p**) diversity of design; (**q**) diversity of design; (**q**) diversity of POI; (**r**) diversity of land cover.

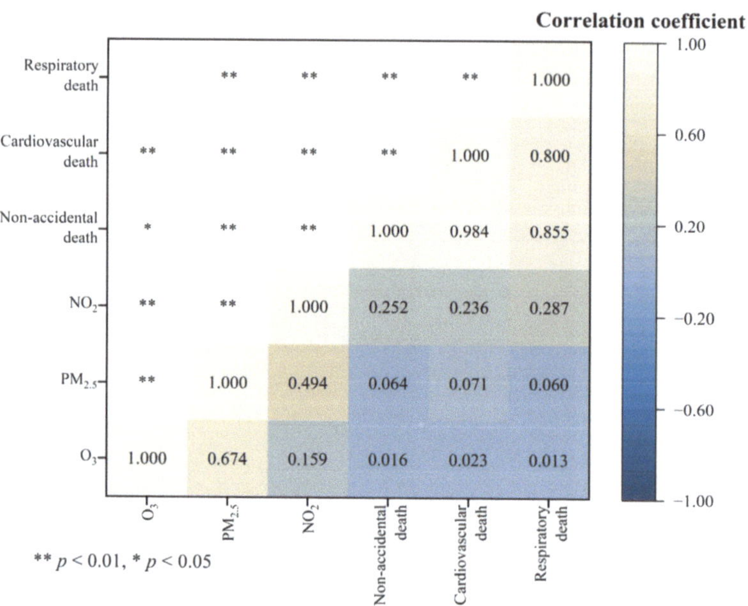

Figure 4. Correlation coefficient (n = 17,715).

Table 3. Results of regression models.

Independent variable	B	S.E.	β	p
Dependent variable: non-accidental mortality				
O_3	0.125	0.033	0.026	0.000
Population density	0.003	0.000	0.426	0.000
$PM_{2.5}$	0.133	0.028	0.032	0.000
Population density	0.003	0.000	0.423	0.000
NO_2	0.256	0.043	0.047	0.000
Population density	0.003	0.000	0.402	0.000
Dependent variable: cardiovascular mortality				
O_3	0.080	0.019	0.033	0.000
Population density	0.001	0.000	0.392	0.000
$PM_{2.5}$	0.084	0.016	0.040	0.000
Population density	0.001	0.000	0.388	0.000
NO_2	0.106	0.025	0.040	0.000
Population density	0.001	0.000	0.370	0.000
Dependent variable: respiratory mortality				
O_3	0.012	0.008	0.015	0.139
Population density	0.000	0.000	0.460	0.000
$PM_{2.5}$	0.007	0.007	0.010	0.304
Population density	0.000	0.000	0.459	0.000
NO_2	0.029	0.009	0.039	0.000
Population density	0.000	0.000	0.438	0.000

B is the unstandardized coefficient; S.E. is the standard error; β is the standardized coefficient.

3.3. The Moderation Analysis

The coefficient of the interaction item "built environment × air pollution" was used as an indicator to judge the moderating effect. The moderating effects of 24 built environment indexes on the relationship between air pollution (O_3, $PM_{2.5}$, and NO_2) and cause-specific mortality are shown in Figure 5. The results shown in Figure S2 reveal the moderating effect in six classes, which include high-rise with high-density, high-rise with low-density, mid-rise with high density, mid-rise with low density, low-rise with high density, and low-rise with high density. Respective moderating effects of the built environment on the relationship between air pollution (O_3, $PM_{2.5}$, and NO_2) and cause-specific mortality are illustrated below.

For the relationship between O_3 and cause-specific mortality, indexes like diversity of the built environment, diversity of design, diversity of POI, building density, building height, SDBH, and construction land have significant positive associations with three cause-specific mortalities, while NDVI has a negative moderating effect. Road length has a significant positive association with non-accidental and cardiovascular mortality. The moderating effect is significant in high-rise with high density areas. We found that a hygiene facility can inhibit the effect of O_3 on cause-specific mortality in mid-rise with high density areas. The moderating effects detected in respiratory and cardiovascular mortality are higher than in non-accidental mortality.

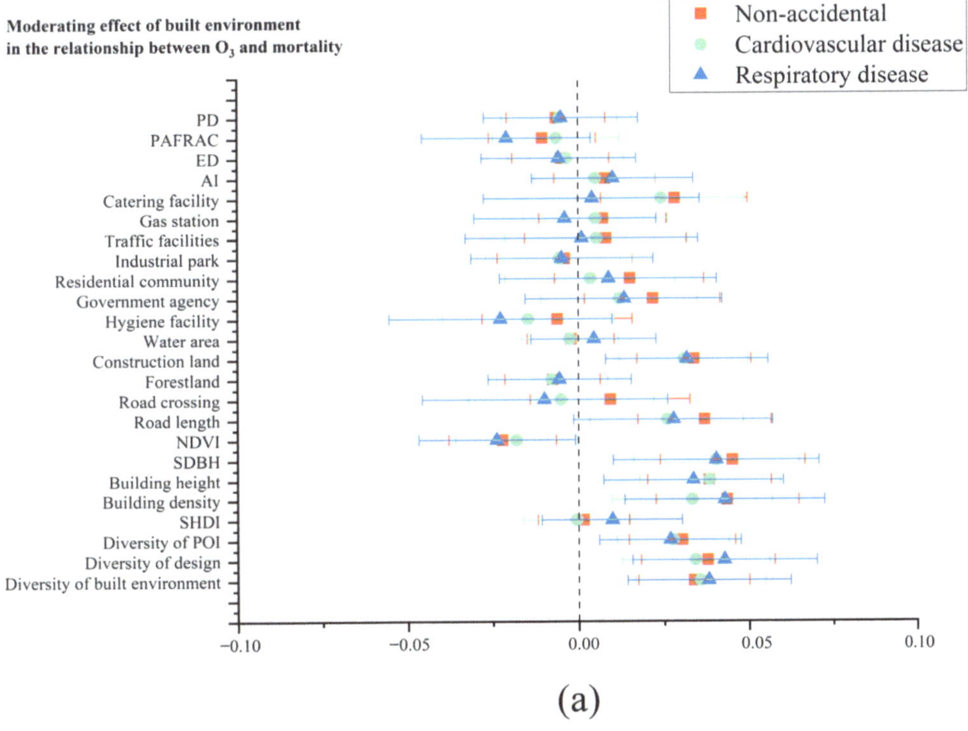

(a)

Figure 5. *Cont.*

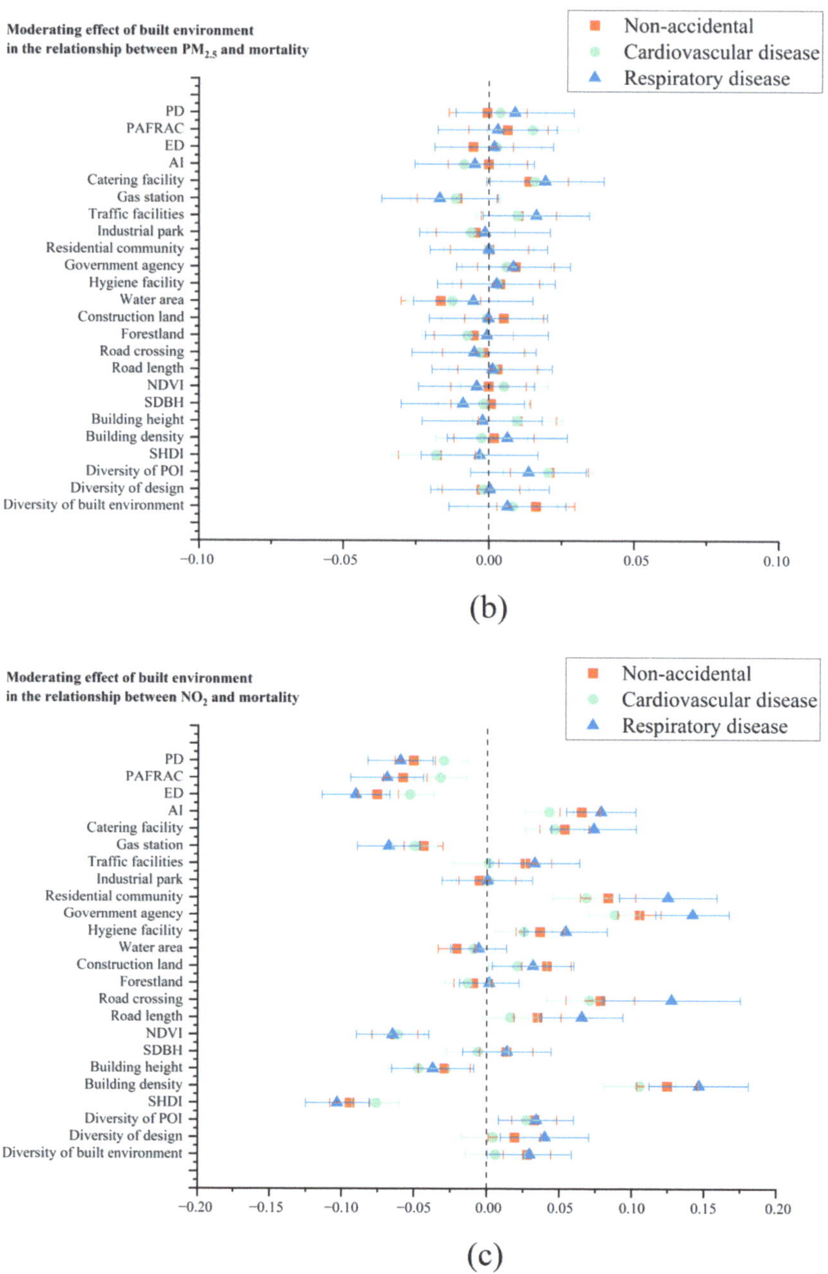

Figure 5. Moderating effect of the built environment in the relationship between air pollution and mortality in an unclassified area. (**a**) Moderating effect of the built environment in the relationship between O_3 and mortality; (**b**) moderating effect of the built environment in the relationship between $PM_{2.5}$ and mortality; (**c**) moderating effect of the built environment in the relationship between NO_2 and mortality.

The effects of $PM_{2.5}$ on non-accidental and cardiovascular mortality are moderated by the diversity of POI, SHDI, and catering facility, where SHDI shows a negative moderating effect. The moderating effects of SHDI and the catering facility are higher in cardiovascular than in non-accidental mortality. The water area has a significant negative moderating effect in the relationship between $PM_{2.5}$ and non-accidental mortality. In high-rise with high density and mid-rise with high density areas, the moderating effects of all built environment indexes are insignificant. The moderating effects of building density, road length, road crossing, hygiene facility, government agency, residential community, traffic facility, catering facility, AI, ED, PAFRAC, and PD are highest for respiratory mortality in low-rise with high density areas. We found that a higher building height can inhibit the effect of $PM_{2.5}$ on cause-specific mortality in high-rise with low density areas. Water area, government agency, and residential community have a negative effect on the relationship between $PM_{2.5}$ and mortality.

On the whole, the moderating effects in the relationship between NO_2 and mortality are higher than O_3 and $PM_{2.5}$. Diversity of POI, building density, road crossing, construction land, hygiene facility, government agency, residential community, catering facility, and AI have a positive moderating effect on the relationship between NO_2 and three cause-specific mortality, while SHDI, building height, NDVI, gas station, ED, PAFRAC, and PD have a negative effect. The effect of NO_2 on mortality is most significantly moderated by building density. The coefficient of the interaction of "building density $\times$ NO_2" is 0.125 (95%CI: 0.104–0.146) for non-accidental mortality, 0.106 (95%CI: 0.081–0.130) for cardiovascular mortality, and 0.146 (95%CI: 0.112–0.180) for respiratory mortality, with all *p*-values < 0.0001. Although the moderating effect of construction land is positive and not significant for forestland, SHDI indicates that the diversity of land cover has a negative moderating effect on the relationship between NO_2 and cause-specific mortality. Positive moderating effects for non-accidental and respiratory mortality were detected on the diversity of the built environment, diversity of design, road length, and traffic facility. The effects were higher for respiratory mortality. In mid-rise with high-density areas, building density, hygiene facility, government agency, residential community, and catering facility have a negative moderating effect on mortality. The building height and SDBH show a negative moderating effect for three cause-specific mortality in low-rise with high-density areas. The residential community can inhibit the effect of NO_2 on mortality on high-rise with low-density and mid-rise with low-density areas. The relationship between NO_2 and mortality can be inhibited by the diversity of the built environment, diversity of POI, and SHDI in the low-rise with low-density areas.

4. Discussion

In this study, we first assessed the associations among the built environment, air pollution, and mortality from different causes such as non-accident, cardiovascular disease and respiratory disease. Then, we analyzed how the built environment affected the relationship between air pollution and mortality. The results show that the built environment has a significant moderating effect on the relationship between air pollution and mortality. The moderating effect is stronger in NO_2 than $PM_{2.5}$ and O_3. We found some heterogeneity across different areas classified by building height and density.

4.1. Which Built Environment Can Moderate the Effect of Air Pollution on Mortality?

As the first step, the study analyzed the relationship between air pollution and mortality by correlation analysis and linear regression. This study confirmed that air pollutants have an adverse effect on human health. The results were consistent with most existing air pollution and health studies [22,38,39]. We have found a statistically significant association between air pollution and mortality, especially in NO_2. Then, we explored the spatial agglomeration between air pollution and the built environment and mortality and the built environment. According to the spatial correlation (Figure S1), a high built environment may lead to high or low air pollution and mortality; the pairwise relationship between built environment, air pollution, and mortality is not absolute. Then, we took the built environment as the moderating variable to explore how the built environment affects the impact of air pollution on mortality.

The results showed that the built environment played a moderating role in the relationship between air pollution and non-accident mortality (Figure 5). NDVI can reduce the effect of O_3 on all-cause mortality, which means the effect of O_3 on all-cause mortality is lower for people living in areas with higher greenness. Many studies have indicated the benefit of greenness. For example, a cohort study of Hong Kong found that elders living in low greenness areas had a higher health risk than those living in higher greenness areas [3,17]. SHDI and water area can reduce the effect of $PM_{2.5}$ on all-cause mortality. SHDI, NDVI, water area, building height, gas station, ED, PAFRAC, and PD can reduce the effect of NO_2 on all-cause mortality. Landscape patterns can change the travel mode, industrial layout and climate conditions, further influencing human health [40]. Higher greenness and a larger water area can mitigate the adverse effects of the environment on human health [41,42]. Building height might mitigate the effect of NO_2 on all-cause mortality by providing shade to influence human comfort [43].

4.2. Heterogeneity of the Moderating Effect in Different Areas and Cause-Specific Mortality

We further discussed the moderating effect of the built environment for cause-specific mortality and different areas. The moderating effects of all built environment indexes on the relationship between three air pollutants and cause-specific mortality are shown in Figures 5 and S2. The moderating effect of the diversity of the built environment, design, and NDVI in the relationship between O_3 and mortality is the strongest for respiratory disease, while building height is the strongest for cardiovascular disease. The moderating effect of the built environment was not significant in the relationship of $PM_{2.5}$ and respiratory disease mortality. Generally, the moderating effect of the built environment in the relationship between NO_2 and mortality is the strongest for respiratory disease.

In high-rise with high-density areas, the moderating effect of the built environment is the strongest. SHDI, building height, water area, ED, PAFRAC, and PD can mitigate the effect of O_3 and NO_2 on non-accident, cardiovascular disease and respiratory disease mortality. We found that the moderating effect of NDVI and forestland in high-rise with high-density areas was not significant. The construction land can intensify the effect of O_3 and NO_2 on mortality. However, SHDI, which indicates the diversity of land cover, has a significant negative effect on the relationship between O_3 and NO_2 on mortality. The results show that the adverse effect of air pollution can be mitigated with the reasonable allocation of landscape and improvement in the diversity of land cover in high-rise with high-density areas. In mid-rise with high-density areas, the hygiene facility plays a vital role in mitigating the effect of O_3 and NO_2 on mortality. Government agency, residential community, and catering facility can also mitigate the effect of NO_2 on mortality in mid-rise with high-density areas. High accessibility to amenities and greater residential density mean more transport via walking, which can reduce the risk of all-cause mortality [11,44]. In low-rise with high-density areas, this study found that building height and SDBH can mitigate the effect of NO_2 on mortality. Building height and difference is associated with the wind environment and a high building height will form a strong angular flow area [2]. It is known that wind speed and air flow affect human comfort [45]. They can provide comfortability and health for pedestrians, as well as maintain a fresh air flow [46].

In high-rise with low-density areas, the diversity of design can mitigate the effect of $PM_{2.5}$ on non-accident mortality. Building density can mitigate the effect of $PM_{2.5}$ on three cause-specific mortality, and it is strongest for cardiovascular disease mortality. Hygiene facility, residential community, and traffic facilities can mitigate the effect of $PM_{2.5}$ on respiratory disease mortality. For NO_2, the moderating effect of the built environment is the strongest for respiratory disease mortality. In mid-rise with low-density areas, SHDI, NDVI, gas station, ED, PAFRAC, and PD have a negative effect on the relationship between NO_2 and mortality in mid-rise with low-density areas. The moderating effect of the built environment exited heterogeneity across different air pollutants. SHDI can mitigate the effect of NO_2 on mortality while intensifying the effect of O_3 on mortality. This heterogeneity is due to the correlation between pollutants. For instance, NO_2 and O_3 may be negatively correlated in urban areas [47]. In low-rise with low-density areas, we found that the diversity of POI can mitigate the effect of NO_2 on mortality, while it had a positive moderating effect in other areas. A higher diversity of facilities represents more activities and higher vitality [48]. High accessibility to facilities and mixed land use were associated with more transport via walking, and walking is an activity that can reduce the non-accident mortality risk [11,44].

We surprisingly found that greenness appeared to aggravate the effect of $PM_{2.5}$ and NO_2 on mortality in mid-rise with high-density and high-rise with low-density areas, respectively. A study in China found a nonlinear, inverted U-shaped association between greenness and health [18]. One potential explanation for the heterogeneity is that larger NDVI values are probably areas with a large proportion of wild environment and agricultural farmland, which means supporting service facilities and lower physical activity [49]. Hygiene facility, government agency, residential community, and catering facility can mitigate the effect of NO_2 on mortality in high-rise with low-density areas, while they appear to intensify the effect on the whole.

4.3. Implications for Urban Planning

Air pollution has been recognized as a global concern to human health for decades [47]. In recent years, the built environment, being the human-made physical environment surroundings and conditions, was confirmed to be linked with human health [9,50]. Based on the moderating effect analysis in Nanjing city, we put forward the following suggestions for mortality risk mitigation measures from the perspective of the built environment:

In general, greenness can help effectively mitigate the effect of O_3 and NO_2 on cause-specific mortality, including non-accident, cardiovascular disease, and respiratory disease. Controlling the building density may help effectively reduce the adverse effect of air pollution. However, it is difficult to reduce the building density on an original basis. In contrast, optimizing landscape patterns and facilities can mitigate the effect of air pollution on mortality. Landscapes such as ED, PAFRAC, and PD have a negative effect in the relationship between NO_2 and mortality. The diversity of land cover can mitigate the effect of NO_2 on mortality, as well as the effect of $PM_{2.5}$ on non-accident and cardiovascular disease mortality. We can increase the proportion of grassland, forestland, and water area appropriately to improve the landscape patterns.

Moreover, the built environment as a mitigation measure to moderate the effect of air pollution on mortality should consider heterogeneity between areas. In this study, we also provided suggestions for different building height and density areas. We can improve the diversity of land cover, especially water area, to mitigate the mortality risk from air pollution in high-rise with high-density areas. The adverse effect of air pollution can be mitigated by hygiene facility, government agency, residential community, and catering facility in mid-rise with high-density areas. We can focus on the SDBH and PAFRAC in low-rise with high-density areas, controlling the effect of air pollution on mortality by increasing the complexity of the landscape. In low density areas, landscape configuration plays a vital role in mitigating the effect of air pollution on mortality. In high-rise with low-density areas, respiratory disease mortality was more severely affected by the built environment. Traffic facility has a negative effect on the relationship between $PM_{2.5}$ and respiratory disease mortality. The adverse effect from $PM_{2.5}$ and O_3 in mid-rise with low-density areas can be mitigated by placing more government agencies. In low-rise with low-density areas, the effect of air pollution on mortality can be moderated by increasing the diversity of land cover, greenness, and gas stations. Our results highlight that the effects of the built environment vary across different areas classified by building density and height.

4.4. Limitations and Future Studies

In this paper, a moderating effect model was used to explored the relationship among built environment, air pollution, and mortality. Nevertheless, it is important to note that there are several limitations. Firstly, only the spatial built environment was quantified, and the change in the built environment over time was not taken into consideration in the moderating effect model. Secondly, we ignored the effects of meteorology and seasonal changes, which can modify the impact on health. Climate and weather changes may have an impact on aero-allergens through variability in meteorological factors (temperature, wind, rain, storm, blizzard, and so on), thus resulting in health risks [51]. Increased temperatures in autumn significantly contributed to the increase in ozone-related health effects [52]. There is a seasonal variation in the effect of particulate air pollution on mortality, with the greatest effect occurring in winter and summer in China [53]. Lastly, our study did not analyze the moderating effect of all built environment factors systematically, partly because too many variables can cause instability in the model.

Therefore, in future research, it is necessary to pay attention to the seasonal differences in the moderating effects of the built environment, and analyze the effects of the built environment combination on air pollution and mortality from a systematic perspective. At the same time, we are also searching for a suitable algorithm model to explore the best built environment combination to reduce air pollution concentration and mortality from the perspective of optimal design.

5. Conclusions

Based on an empirical study of Nanjing in China from 2007 to 2015, this paper revealed the moderating effect of the built environment on the relationship between air pollution and mortality. The result was obtained by the moderating effect model based on hierarchical regression. In particular, the greenness, water body, and landscape elements can effectively mitigate the adverse effect of air pollution, while the intensity of the development may aggravate it. The sensitivity of different populations to air pollution varies, and the moderating effect of the built environment is also heterogeneous. We should pay more attention to the diversity of land cover. A single land cover type may not significantly reduce the adverse impact of air pollutants, or even aggravate it. For example, forestland, construction land, and water body have no significant impact on the association between $PM_{2.5}$ and cardiovascular mortality, while their diversity can mitigate the effect of $PM_{2.5}$ on cardiovascular mortality.

Identifying the heterogeneity of the moderating effect of the built environment can help to formulate more accurate mitigation planning. The moderating effect of the built environment is the strongest in areas with high building density and height. Water body and landscape patterns can effectively mitigate the adverse effect of air pollution. A hygiene facility can mitigate the association between air pollution and mortality in mid-rise with high-density areas. At the same time, the height of the building also has a mitigating effect. It is worth noting that government agency can improve the air-pollution-related mortality in mid-rise areas, while it may aggravate the adverse effects of air pollution in other areas. Our findings suggest that the built environment could be integrated into urban planning to mitigate the effect of air pollution on human health. In addition, the development of mitigating measures needs to take into account the heterogeneity of areas and consider the mix of built environment factors.

Supplementary Materials: The following supporting information can be downloaded at: https://www.mdpi.com/article/10.3390/rs14153703/s1, Figure S1: Spatial correlation of air pollution (O3, PM2.5 and NO2) and mortality with built environment respectively; Figure S2: Moderating effect of built environment in the relationship between air pollution and mortality in (1) unclassified area and (2) areas classified by building density and height. Table S1: The percentage of H-H type grids corresponding to Figure S1; Table S2: The percentage of L-L type grids corresponding to Figure S1; Table S3: The percentage of L-H type grids corresponding to Figure S1; Table S4: The percentage of H-L type grids corresponding to Figure S1; Table S5: The percentage of not significant type grids corresponding to Figure S1.

Author Contributions: Conceptualization, P.W. and C.W.; methodology, C.W.; software, Y.S.; validation, C.W. and Y.S.; formal analysis, C.W.; investigation, J.W. and Y.W.; resources, L.H.; data curation, C.W.; writing—original draft preparation, C.W., Y.S., J.W. and Y.W.; writing—review and editing, Y.S. and J.W.; visualization, J.W.; supervision, C.W. and Y.S.; project administration, P.W.; funding acquisition, P.W. All authors have read and agreed to the published version of the manuscript.

Funding: This work was supported by the National Natural Science Foundation of China [Grant No. 51908249], the Natural Science Foundation of the Jiangsu Higher Education Institutions of China [Grant No. 19KJB560012], the High-level Scientific Research Foundation for the introduction of talent for Jiangsu University [Grant No. 18JDG038], and the Innovative Approaches Special Project of the Ministry of Science and Technology of China [Grant No. 2020IM020300].

Data Availability Statement: The data used in this study are presented in Table 1.

Conflicts of Interest: The authors declare no conflict of interest.

References

1. Xu, H.; Wang, Q.; Zhu, H.; Zhang, Y.; Ma, R.; Ban, J.; Li, T. Air Pollution Control of China Brought Significant Health and Economic Benefit but the Health Burden Is Still Heavy and Keeps Rising. *SSRN Electron. J.* **2022**. [CrossRef]
2. Yang, J.; Shi, B.; Zheng, Y.; Shi, Y.; Xia, G. Urban Form and Air Pollution Disperse: Key Indexes and Mitigation Strategies. *Sustain. Cities Soc.* **2020**, *57*, 101955. [CrossRef]
3. Sun, S.; Sarkar, C.; Kumari, S.; James, P.; Cao, W.; Lee, R.S.Y.; Tian, L.; Webster, C. Air Pollution Associated Respiratory Mortality Risk Alleviated by Residential Greenness in the Chinese Elderly Health Service Cohort. *Environ. Res.* **2020**, *183*, 109139. [CrossRef]
4. Michetti, M.; Gualtieri, M.; Anav, A.; Adani, M.; Benassi, B.; Dalmastri, C.; D'Elia, I.; Piersanti, A.; Sannino, G.; Zanini, G.; et al. Climate Change and Air Pollution: Translating Their Interplay into Present and Future Mortality Risk for Rome and Milan Municipalities. *Sci. Total Environ.* **2022**, *830*, 154680. [CrossRef] [PubMed]
5. WHO. *WHO Global Air Quality Guidelines: Particulate Matter (PM2.5 and PM10), Ozone, Nitrogen Dioxide, Sulfur Dioxide and Carbon Monoxide*; World Health Organization: Geneva, Switzerland, 2021; ISBN 9789240034228.
6. Atkinson, R.W.; Kang, S.; Anderson, H.R.; Mills, I.C.; Walton, H.A. Epidemiological Time Series Studies of PM2.5 and Daily Mortality and Hospital Admissions: A Systematic Review and Meta-Analysis. *Thorax* **2014**, *69*, 660–665. [CrossRef] [PubMed]
7. Stafoggia, M.; Bellander, T. Short-Term Effects of Air Pollutants on Daily Mortality in the Stockholm County—A Spatiotemporal Analysis. *Environ. Res.* **2020**, *188*, 109854. [CrossRef] [PubMed]
8. Travert, A.S.; Annerstedt, K.; Daivadanam, M. Built Environment and Health Behaviors: Deconstructing the Black Box of Interactions—A Review of Reviews. *Int. J. Environ. Res. Public Health* **2019**, *16*, 1454. [CrossRef]
9. Lin, J.; Leung, J.; Yu, B.; Woo, J.; Kwok, T.; Ka-Lun Lau, K. Socioeconomic Status as an Effect Modifier of the Association between Built Environment and Mortality in Elderly Hong Kong Chinese: A Latent Profile Analysis. *Environ. Res.* **2021**, *195*, 110830. [CrossRef]
10. Epstein, S.A.; Lee, S.M.; Katzenstein, A.S.; Carreras-Sospedra, M.; Zhang, X.; Farina, S.C.; Vahmani, P.; Fine, P.M.; Ban-Weiss, G. Air-Quality Implications of Widespread Adoption of Cool Roofs on Ozone and Particulate Matter in Southern California. *Proc. Natl. Acad. Sci. USA* **2017**, *114*, 8991–8996. [CrossRef]
11. Frank, L.D.; Iroz-Elardo, N.; MacLeod, K.E.; Hong, A. Pathways from Built Environment to Health: A Conceptual Framework Linking Behavior and Exposure-Based Impacts. *J. Transp. Health* **2019**, *12*, 319–335. [CrossRef]
12. Sinharay, R.; Gong, J.; Barratt, B.; Ohman-Strickland, P.; Ernst, S.; Kelly, F.J.; Zhang, J.; Collins, P.; Cullinan, P.; Chung, K.F. Respiratory and Cardiovascular Responses to Walking down a Traffic-Polluted Road Compared with Walking in a Traffic-Free Area in Participants Aged 60 Years and Older with Chronic Lung or Heart Disease and Age-Matched Healthy Controls: A Randomised, Crosso. *Lancet* **2018**, *391*, 339–349. [CrossRef]
13. Wu, C.; Kim, I.; Chung, H. The Effects of Built Environment Spatial Variation on Bike-Sharing Usage: A Case Study of Suzhou, China. *Cities* **2021**, *110*, 103063. [CrossRef]
14. Yuan, M.; Yin, C.; Sun, Y.; Chen, W. Examining the Associations between Urban Built Environment and Noise Pollution in High-Density High-Rise Urban Areas: A Case Study in Wuhan, China. *Sustain. Cities Soc.* **2019**, *50*, 101678. [CrossRef]
15. Chen, E.; Ye, Z.; Wu, H. Nonlinear Effects of Built Environment on Intermodal Transit Trips Considering Spatial Heterogeneity. *Transp. Res. Part D Transp. Environ.* **2021**, *90*, 102677. [CrossRef]
16. Opdam, P. Implementing Human Health as a Landscape Service in Collaborative Landscape Approaches. *Landsc. Urban Plan.* **2020**, *199*, 103819. [CrossRef]
17. Brochu, P.; Jimenez, M.P.; James, P.; Kinney, P.L.; Lane, K. Benefits of Increasing Greenness on All-Cause Mortality in the Largest Metropolitan Areas of the United States Within the Past Two Decades. *Front. Public Health* **2022**, *10*. [CrossRef]
18. Huang, B.; Yao, Z.; Pearce, J.R.; Feng, Z.; James Browne, A.; Pan, Z.; Liu, Y. Non-Linear Association between Residential Greenness and General Health among Old Adults in China. *Landsc. Urban Plan.* **2022**, *223*, 104406. [CrossRef]

19. Sarkar, C.; Zhang, B.; Ni, M.; Kumari, S.; Bauermeister, S.; Gallacher, J.; Webster, C. Environmental Correlates of Chronic Obstructive Pulmonary Disease in 96,779 Participants from the UK Biobank: A Cross-Sectional, Observational Study. *Lancet Planet. Health* **2019**, *3*, e478–e490. [CrossRef]

20. Gronlund, C.J.; Zanobetti, A.; Wellenius, G.A.; Schwartz, J.D.; O'Neill, M.S. Vulnerability to Renal, Heat and Respiratory Hospitalizations During Extreme Heat Among U.S. Elderly. *Clim. Chang.* **2016**, *136*, 631–645. [CrossRef]

21. Xu, Y.; Dadvand, P.; Barrera-Gómez, J.; Sartini, C.; Marí-Dell'Olmo, M.; Borrell, C.; Medina-Ramón, M.; Sunyer, J.; Basagaña, X. Differences on the Effect of Heat Waves on Mortality by Sociodemographic and Urban Landscape Characteristics. *J. Epidemiol. Community Health* **2013**, *67*, 519–525. [CrossRef]

22. Zhang, Z.; Wang, J.; Kwong, J.C.; Burnett, R.T.; van Donkelaar, A.; Hystad, P.; Martin, R.V.; Bai, L.; McLaughlin, J.; Chen, H. Long-Term Exposure to Air Pollution and Mortality in a Prospective Cohort: The Ontario Health Study. *Environ. Int.* **2021**, *154*, 106570. [CrossRef]

23. Giles-Corti, B.; Vernez-Moudon, A.; Reis, R.; Turrell, G.; Dannenberg, A.L.; Badland, H.; Foster, S.; Lowe, M.; Sallis, J.F.; Stevenson, M.; et al. City Planning and Population Health: A Global Challenge. *Lancet* **2016**, *388*, 2912–2924. [CrossRef]

24. Patino, J.E.; Hong, A.; Duque, J.C.; Rahimi, K.; Zapata, S.; Lopera, V.M. Built Environment and Mortality Risk from Cardiovascular Disease and Diabetes in Medellín, Colombia: An Ecological Study. *Landsc. Urban Plan.* **2021**, *213*, 104126. [CrossRef]

25. Engineer, A.; Gualano, R.J.; Crocker, R.L.; Smith, J.L.; Maizes, V.; Weil, A.; Sternberg, E.M. An Integrative Health Framework for Wellbeing in the Built Environment. *Build. Environ.* **2021**, *205*, 108253. [CrossRef]

26. Dai, L.; Zhang, L.; Chen, D.; Zhao, Y. Assessment of Carbonaceous Aerosols in Suburban Nanjing under Air Pollution Control Measures: Insights from Long-Term Measurements. *Environ. Res.* **2022**, *212*, 113302. [CrossRef] [PubMed]

27. Huang, C.; Hu, J.; Xue, T.; Xu, H.; Wang, M. High-Resolution Spatiotemporal Modeling for Ambient PM2.5 Exposure Assessment in China from 2013 to 2019. *Environ. Sci. Technol.* **2021**, *55*, 2152–2162. [CrossRef] [PubMed]

28. Wang, Y.; Huang, C.; Hu, J.; Wang, M. Development of High-Resolution Spatio-Temporal Models for Ambient Air Pollution in a Metropolitan Area of China from 2013 to 2019. *Chemosphere* **2022**, *291*, 132918. [CrossRef] [PubMed]

29. Ma, Y.; Zhou, L.; Chen, K. Burden of Cause-Specific Mortality Attributable to Heat and Cold: A Multicity Time-Series Study in Jiangsu Province, China. *Environ. Int.* **2020**, *144*, 105994. [CrossRef]

30. Chen, K.; Bi, J.; Chen, J.; Chen, X.; Huang, L.; Zhou, L. Influence of Heat Wave Definitions to the Added Effect of Heat Waves on Daily Mortality in Nanjing, China. *Sci. Total Environ.* **2015**, *506–507*, 18–25. [CrossRef]

31. Wang, P.; Qiao, W.; Wang, Y.; Cao, S.; Zhang, Y. Urban Drought Vulnerability Assessment–A Framework to Integrate Socio-Economic, Physical, and Policy Index in a Vulnerability Contribution Analysis. *Sustain. Cities Soc.* **2020**, *54*, 102004. [CrossRef]

32. Frank, L.D.; Pivo, G. Impacts of Mixed Use and Density on Utilization of Three Modes of Travel: Single-Occupant Vehicle, Transit, and Walking. *Transp. Res. Rec.* **1994**, *1466*, 44–52.

33. Beenackers, M.A.; Oude Groeniger, J.; Kamphuis, C.B.M.; Van Lenthe, F.J. Urban Population Density and Mortality in a Compact Dutch City: 23-Year Follow-up of the Dutch GLOBE Study. *Health Place* **2018**, *53*, 79–85. [CrossRef]

34. Meijer, M.; Mette Kejs, A.; Stock, C.; Bloomfield, K.; Ejstrud, B.; Schlattmann, P. Population Density, Socioeconomic Environment and All-Cause Mortality: A Multilevel Survival Analysis of 2.7 Million Individuals in Denmark. *Health Place* **2012**, *18*, 391–399. [CrossRef]

35. Wu, S.; Zhou, S.; Bao, H.; Chen, D.; Wang, C.; Li, B.; Tong, G.; Yuan, Y.; Xu, B. Improving Risk Management by Using the Spatial Interaction Relationship of Heavy Metals and PAHs in Urban Soil. *J. Hazard. Mater.* **2019**, *364*, 108–116. [CrossRef]

36. Hou, X.; Wu, S.; Chen, D.; Cheng, M.; Yu, X.; Yan, D.; Dang, Y.; Peng, M. Can Urban Public Services and Ecosystem Services Achieve Positive Synergies? *Ecol. Indic.* **2021**, *124*, 107433. [CrossRef]

37. Stewart, I.D.; Oke, T.R. Local Climate Zones for Urban Temperature Studies. *Bull. Am. Meteorol. Soc.* **2012**, *93*, 1879–1900. [CrossRef]

38. Liu, M.; Huang, Y.; Ma, Z.; Jin, Z.; Liu, X.; Wang, H.; Liu, Y.; Wang, J.; Jantunen, M.; Bi, J.; et al. Spatial and Temporal Trends in the Mortality Burden of Air Pollution in China: 2004–2012. *Environ. Int.* **2017**, *98*, 75–81. [CrossRef]

39. So, R.; Andersen, Z.J.; Chen, J.; Stafoggia, M.; de Hoogh, K.; Katsouyanni, K.; Vienneau, D.; Rodopoulou, S.; Samoli, E.; Lim, Y.-H.; et al. Long-Term Exposure to Air Pollution and Mortality in a Danish Nationwide Administrative Cohort Study: Beyond Mortality from Cardiopulmonary Disease and Lung Cancer. *Environ. Int.* **2022**, *164*, 107241. [CrossRef]

40. Zhang, P.; Yang, L.; Ma, W.; Wang, N.; Wen, F.; Liu, Q. Spatiotemporal Estimation of the PM2.5 Concentration and Human Health Risks Combining the Three-Dimensional Landscape Pattern Index and Machine Learning Methods to Optimize Land Use Regression Modeling in Shaanxi, China. *Environ. Res.* **2022**, *208*, 112759. [CrossRef]

41. Aram, F.; Higueras García, E.; Solgi, E.; Mansournia, S. Urban Green Space Cooling Effect in Cities. *Heliyon* **2019**, *5*, e01339. [CrossRef]

42. Tan, X.; Sun, X.; Huang, C.; Yuan, Y.; Hou, D. Comparison of Cooling Effect between Green Space and Water Body. *Sustain. Cities Soc.* **2021**, *67*, 102711. [CrossRef]

43. Shareef, S.; Abu-Hijleh, B. The Effect of Building Height Diversity on Outdoor Microclimate Conditions in Hot Climate. A Case Study of Dubai-UAE. *Urban Clim.* **2020**, *32*, 100611. [CrossRef]

44. Li, J.; Auchincloss, A.H.; Hirsch, J.A.; Melly, S.J.; Moore, K.A.; Peterson, A.; Sánchez, B.N. Exploring the Spatial Scale Effects of Built Environments on Transport Walking: Multi-Ethnic Study of Atherosclerosis. *Health Place* **2021**, *73*, 102722. [CrossRef]

5. Ghasemi, Z.; Esfahani, M.A.; Bisadi, M. Promotion of Urban Environment by Consideration of Human Thermal & Wind Comfort: A Literature Review. *Procedia Soc. Behav. Sci.* **2015**, *201*, 397–408. [CrossRef]

6. Stathopoulos, T.; Wu, H.; Zacharias, J. Outdoor Human Comfort in an Urban Climate. *Build. Environ.* **2004**, *39*, 297–305. [CrossRef]

7. Huangfu, P.; Atkinson, R. Long-Term Exposure to NO_2 and O_3 and All-Cause and Respiratory Mortality: A Systematic Review and Meta-Analysis. *Environ. Int.* **2020**, *144*, 105998. [CrossRef]

8. Li, C.; Wang, Z.; Li, B.; Peng, Z.R.; Fu, Q. Investigating the Relationship between Air Pollution Variation and Urban Form. *Build. Environ.* **2019**, *147*, 559–568. [CrossRef]

9. Liu, Y.; Wang, R.; Xiao, Y.; Huang, B.; Chen, H.; Li, Z. Exploring the Linkage between Greenness Exposure and Depression among Chinese People: Mediating Roles of Physical Activity, Stress and Social Cohesion and Moderating Role of Urbanicity. *Health Place* **2019**, *58*, 102168. [CrossRef]

10. Rojas-Rueda, D.; Nieuwenhuijsen, M.J.; Gascon, M.; Perez-Leon, D.; Mudu, P. Green Spaces and Mortality: A Systematic Review and Meta-Analysis of Cohort Studies. *Lancet Planet. Health* **2019**, *3*, e469–e477. [CrossRef]

11. Dunea, D.; Liu, H.Y.; Iordache, S.; Buruleanu, L.; Pohoata, A. Liaison between Exposure to Sub-Micrometric Particulate Matter and Allergic Response in Children from a Petrochemical Industry City. *Sci. Total Environ.* **2020**, *745*, 141170. [CrossRef]

12. Guan, Y.; Xiao, Y.; Wang, F.; Qiu, X.; Zhang, N. Health Impacts Attributable to Ambient PM2.5 and Ozone Pollution in Major Chinese Cities at Seasonal-Level. *J. Clean. Prod.* **2021**, *311*, 127510. [CrossRef]

13. Chen, R.; Peng, R.D.; Meng, X.; Zhou, Z.; Chen, B.; Kan, H. Seasonal Variation in the Acute Effect of Particulate Air Pollution on Mortality in the China Air Pollution and Health Effects Study (CAPES). *Sci. Total Environ.* **2013**, *450–451*, 259–265. [CrossRef] [PubMed]

Article

Evaluating the Spatial Risk of Bacterial Foodborne Diseases Using Vulnerability Assessment and Geographically Weighted Logistic Regression

Wanchao Bian [1], Hao Hou [1], Jiang Chen [2], Bin Zhou [1], Jianhong Xia [3], Shanjuan Xie [1] and Ting Liu [1,*]

[1] Zhejiang Provincial Key Laboratory of Urban Wetlands and Regional Change, Hangzhou Normal University, Hangzhou 311121, China; 2019210214004@stu.hznu.edu.cn (W.B.); houhao@hznu.edu.cn (H.H.); zhoubin@hznu.edu.cn (B.Z.); shanj_x@hznu.edu.cn (S.X.)

[2] Zhejiang Provincial Center for Disease Control and Prevention, Hangzhou 310051, China; jchen@cdc.zj.cn

[3] School of Earth and Planetary Sciences, Curtin University, Perth 6845, Australia; c.xia@curtin.edu.au

* Correspondence: tingliu_hz@hznu.edu.cn

Abstract: Foodborne diseases are an increasing concern to public health; climate and socioeconomic factors influence bacterial foodborne disease outbreaks. We developed an "exposure–sensitivity–adaptability" vulnerability assessment framework to explore the spatial characteristics of multiple climatic and socioeconomic environments, and analyzed the risk of foodborne disease outbreaks in different vulnerable environments of Zhejiang Province, China. Global logistic regression (GLR) and geographically weighted logistic regression (GWLR) models were combined to quantify the influence of selected variables on regional bacterial foodborne diseases and evaluate the potential risk. GLR results suggested that temperature, total precipitation, road density, construction area proportions, and gross domestic product (GDP) were positively correlated with foodborne diseases. GWLR results indicated that the strength and significance of these relationships varied locally, and the predicted risk map revealed that the risk of foodborne diseases caused by *Vibrio parahaemolyticus* was higher in urban areas (60.6%) than rural areas (20.1%). Finally, distance from the coastline was negatively correlated with predicted regional risks. This study provides a spatial perspective for the relevant departments to prevent and control foodborne diseases.

Keywords: bacterial foodborne disease; global logistic regression; geographically weighted logistic regression; urban and rural areas; vulnerability

Citation: Bian, W.; Hou, H.; Chen, J.; Zhou, B.; Xia, J.; Xie, S.; Liu, T. Evaluating the Spatial Risk of Bacterial Foodborne Diseases Using Vulnerability Assessment and Geographically Weighted Logistic Regression. *Remote Sens.* **2022**, *14*, 3613. https://doi.org/10.3390/rs14153613

Academic Editors: Xinhu Li, Jinchao Song, Franz W. Gatzweiler, Peng Jia and Shihong Du

Received: 5 June 2022
Accepted: 26 July 2022
Published: 28 July 2022

Publisher's Note: MDPI stays neutral with regard to jurisdictional claims in published maps and institutional affiliations.

1. Introduction

Foodborne diseases are infectious or toxic diseases transmitted by the consumption of food [1] and are one of the most significant public health problems worldwide. According to a World Health Organization (WHO) Foodborne Disease Burden Epidemiology Reference Group (FERG) report, 600 million foodborne illnesses and 420,000 deaths were caused by global foodborne hazards in 2010 [2]. In China, studies have shown that 748 million cases of acute gastrointestinal illness and 420 million medical consultations occur annually throughout the country [3]. As a result, foodborne diseases bring significant socioeconomic burdens and hidden dangers to residential health. According to the national foodborne disease molecular tracing network established in 2013, Salmonella species, *Vibrio parahaemolyticus, Staphylococcus aureus*, and diarrheagenic *Escherichia coli* are the most common foodborne pathogens that cause outbreaks in China [4]. Among these, *V. parahaemolyticus* is a halophilic, gram-negative bacterium that has been the leading cause of foodborne disease outbreaks and cases of infectious diarrhea in China, especially in coastal regions [5]. Therefore, it is essential to analyze the influencing factors and risk of foodborne diseases caused by *V. parahaemolyticus*.

Based on surveillance data, many studies on foodborne diseases have explored their epidemiological characteristics [6–8] and many scholars have conducted research on foodborne diseases, such as exploring influencing factors and predicting infection risks [9–12]. Chen et al. conducted a multivariable logistic regression analysis to analyze the association between food-handling behaviors and foodborne acute gastroenteritis in Anhui, China [9]. Zhang et al. used several machine learning models (e.g., support vector machine, random forest, and XGBoost) to study foodborne disease outbreaks across China and identify their confounding factors [10]. Wang et al. applied a Bayesian nowcasting model to forecast the total daily number of foodborne disease cases [11]. Li et al. used the autoregressive integrated moving average (ARIMA) model to predict foodborne disease incidence in Shenzhen City [12]. However, these studies assumed each factor affected the diseases uniformly, ignoring geographical variations in the influencing factors. As a spatial regression method, geographically weighted logical regression (GWLR) allows the intensity of these factors and their relative importance to vary geographically [13] and has been widely used in epidemiological studies of infectious diseases, such as thrombocytopenia syndrome, dengue, and malaria [14–16]. For instance, Zhou et al. found that, compared to the non-spatial logistical regression, the GWLR model offers better understanding of the geographical variations of the risk factors associated with infection of hepatitis C virus [17]. Using GWLR to explore influencing factors can provide a unique spatial perspective.

Vulnerability, which comprises exposure, sensitivity, and adaptability [18], plays a vital role in global environmental change and sustainability research (e.g., flood, heat waves, dengue, and SARS-CoV-2 infections) [19–23]. The concept of vulnerability first appeared in the study of natural hazards [24]. Then it gradually developed into an interdisciplinary and multiscale direction. Adger considered vulnerability as " ... the state of susceptibility to harm from exposure to stresses associated with environmental and social change and from the absence of capacity to adapt" [24]. The above definition is widely recognized [25–27]. Exposure is defined as the proximity of people or systems to external disturbances [28]. Climate variables, such as temperature and precipitation patterns, extreme weather events, and ocean warming, have complex effects on the food chain, thus affecting the occurrence of foodborne diseases, especially those caused by bacteria. Studies have shown that rising temperatures and heavy rainfall may increase the number of foodborne disease cases [29–31]. Sensitivity can be understood as the degree of a system being easily disturbed [26]. The effect of foodborne diseases on crowds varies and the degree of this effect depends on age, sex, food preferences, and food-handling behaviors [32,33]. Osei-Tutu et al. found that the most affected group were those between the ages of 15 and 34 in Accra, Ghana [34]. Moreover, the degree of impact also differs between urban and rural areas; for example, Czerwinski et al. found that the incidence of foodborne botulism among rural residents was more than twice as high as that in urban areas [35]. Adaptability reflects the ability of a system to adapt and adjust to external disturbances [28]. The higher the level of economic medical development, the stronger the ability to deal with health threats. Xiao et al. found that per capita gross domestic product (GDP) was negatively associated with disease incidence [36]. The selection of vulnerability indicators varies depending on the physical attributes of the event; however, few studies have assessed foodborne diseases and vulnerability together. Therefore, we propose a comprehensive foodborne disease vulnerability assessment framework to identify the dominant influencing factors.

Zhejiang Province is an important part of the Yangtze River Delta urban agglomeration, which belongs to the typical subtropical monsoon climate and has a wide variety of aquatic products. Geographical and climatic conditions are suitable for the growth of microorganisms. Among the identified causes of foodborne disease outbreaks, the number of foodborne illnesses caused by bacterial pathogen infections were the largest [6]. Previous evidence has shown that *V. parahaemolyticus* was responsible for the largest number of outbreaks in Zhejiang Province from 2010 to 2014 [37]. Taking Zhejiang as a case study, this study aimed to screen the influencing factors of foodborne diseases, based on the vulnerability assessment framework, and investigate the specific relationship between these factors and the positive foodborne disease cases caused by *V. parahaemolyticus*. Furthermore, the GWLR model was combined with foodborne diseases to identify the relative geographical importance of environmental and sociodemographic variables. Finally, we produced a map of the predicted probability of foodborne diseases to determine the spatial epidemiological risk.

2. Materials and Methods

2.1. Study Area

Zhejiang Province is on the southeast coast of China (Figure 1) and covers 101,800 km^2 with a long 1805 km zigzag-shaped coastline. It has a subtropical monsoon climate with hydrothermal conditions that are conducive to microorganism growth. The province includes 11 prefecture-level cities and has experienced significant economic development With rapid socioeconomic development, regional relationships grow closer, and personal dietary structures become richer. Zhejiang Province has a high incidence of foodborne diseases, especially bacterial foodborne diseases caused by *V. parahaemolyticus*. According to the Zhejiang Province Foodborne Disease Monitoring and Reporting System, the detection rate of foodborne diseases caused by *V. parahaemolyticus* has recently increased. Zhejiang Province first set up sentinel hospitals to conduct foodborne disease surveillance and reporting in 2010. To ensure that each district and county can be effectively monitored, 101 sentinel hospitals are located in 89 districts and counties in Zhejiang Province. It should be noted that sentinel hospitals were added in areas with a large resident population. However, some problems remain unsolved, such as imperfect monitoring mechanisms, disunity of information construction standards, and inadequate data utilization. Overall, it is essential to more effectively mine information based on existing monitoring data. To carry out more detailed research, we used the ArcGIS fishnet tool to generate the $0.1° \times 0.1°$ grid data.

This study uses the township administrative region as the primary division unit, dividing the study area into urban and rural areas (Figure 1). The division principles were defined based on the "Provisions on Statistical Division of Urban and Rural Areas" designated by the National Bureau of Statistics [38]. Urban areas are those that house municipal district governments and other subdistrict offices under the jurisdiction of the district, along with town governments and other neighborhood committee areas under the jurisdiction of the town. Rural areas refer to the regions outside of these urban areas. Notably, since regular grids and irregular administrative boundaries do not always fit well, some grids required manual judgement when dividing the urban and rural areas according to local knowledge. For example, street administrative divisions are incomplete in a grid. When the proportion of urban administrative areas in a grid was greater than 70%, we defined the grid as an urban area. In short, there was an initial division of urban and rural areas based on the administrative division data of the town, and then a more detailed judgment was made with the help of remote sensing images and local knowledge.

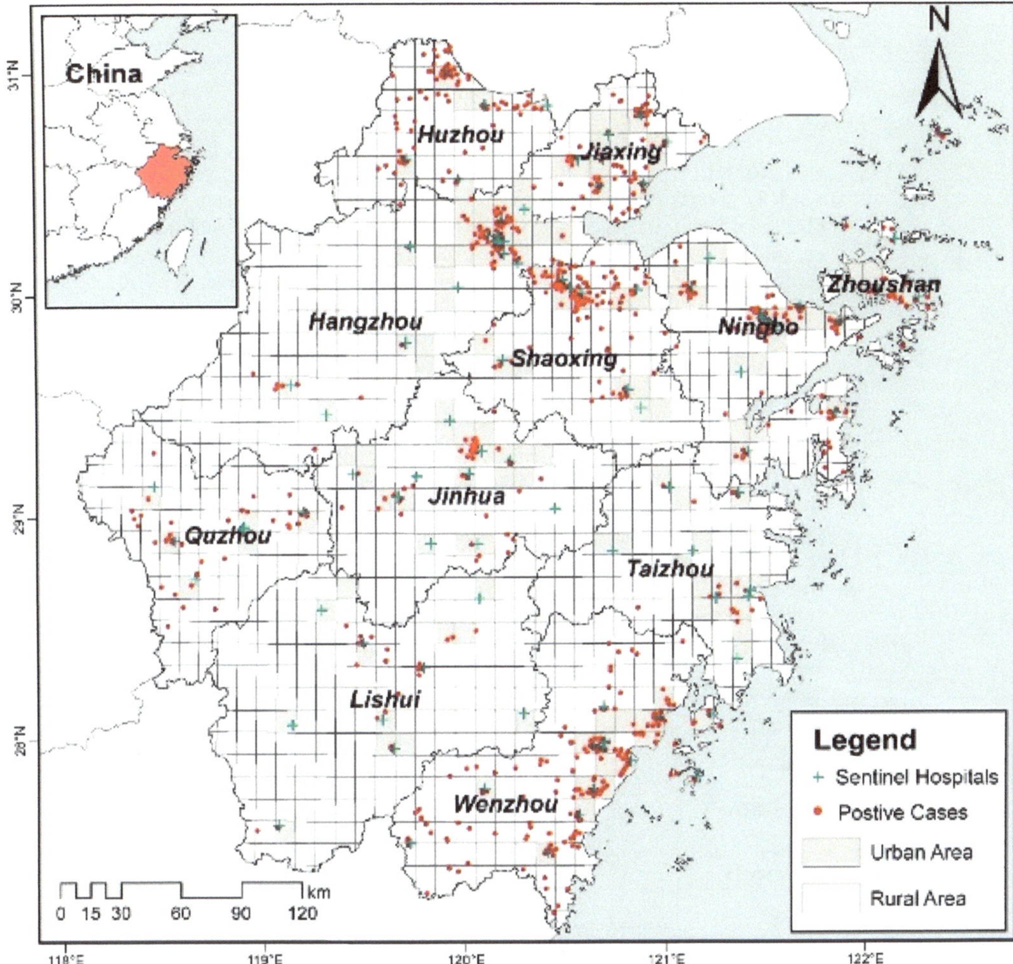

Figure 1. Locations of the study area (0.1° × 0.1° grid size) and positive cases caused by *V. parahaemolyticus*.

2.2. Data Source

Meteorological data, including dew point temperature, temperature, surface net solar radiation, total precipitation, daily maximum temperature, and daily minimum temperature, were obtained from the European Center for Medium-Range Weather Forecasts (ECMWF).

Road data were derived from OpenStreetMap (https://www.openstreetmap.org (accessed on 23 March 2021)), and the hospital-related points of interest (POI) data were extracted from Amap application programming interface (API). The population density data came from the Gridded Population of the World (https://sedac.ciesin.columbia.edu/data/collection/gpw-v4 (accessed on 23 July 2022)). Based on the Seventh National Census of China in 2020, the population raster data were adjusted [39]. The GDP spatial distribution kilometer grid data, annual normalized difference vegetation index (NDVI), spatial distribution data, and the China land use data were downloaded from the Resource and Environmental Science and Data Center of the Chinese Academy of Sciences (https://www.resdc.cn (accessed on 28 April 2021)).

Data on foodborne diseases caused by *V. parahaemolyticus* were collected from the "Zhejiang Province Foodborne Disease Surveillance and Reporting System", which contained data on 31,932 tested cases came from 101 sentinel hospitals in 2018, including attributes of date, gender, age, address, occupation, and detection results.

2.3. Foodborne Diseases Vulnerability Assessment Framework

According to the definition of vulnerability, this study proposes an assessment framework for foodborne disease vulnerability based on exposure, sensitivity, and adaptability (Table 1). Previous studies have shown that climate change will have a complex impact on the persistence and dispersal of foodborne pathogens [40]; therefore, our exposure indices here were focused on various meteorological indicators. Urbanization affects consumption patterns and food production processes, which can increase the risk of foodborne diseases [41]; therefore, our sensitivity indices were focused on road density and construction area proportions. As indispensable social resources for combating diseases, hospitals and health institutes are crucial to maintaining personal health [42]. Additionally, medical resources are closely related to regional economic development; therefore, our adaptability indices are focused on regional medical time costs and GDP. The vulnerability of foodborne diseases mentioned in this paper refers to the relationship between the "human–environment" system and foodborne diseases; that is, the susceptibility of the state of the system to harm from exposure to stresses associated with foodborne diseases and from the absence of the capacity to adapt. All data were processed using the same grid size ($0.1° \times 0.1°$) in ArcGIS as shown in Figure 2. The numerical differences among each variable are shown in Table 2.

Table 1. Evaluation index system of foodborne diseases vulnerability.

Criterion	Index	Source	Resolution	Year
Exposure	Wind Speed (m/s) Dewpoint Temperature (K) Temperature (K) Surface Net Solar Radiation (KJ/m²) Total Precipitation (m) Daily Maximum Temperature (K) Daily Minimum Temperature (K)	ERA5-Land (https://www.ecmwf.int/ (accessed on 20 October 2021))	$0.1° \times 0.1°$	2018
Sensitivity	Road Density (km/km²)	Road Data (https://www.openstreetmap.org/ (accessed on 23 March 2021))	Vector	2021
	Proportion of Construction Area (%)	Land Use Data (https://www.resdc.cn/ (accessed on 28 April 2021))	1 km	2015
	Rural Areas	Administrative Division Data (http://www.ngcc.cn/ngcc/ (accessed on 6 November 2021))	Vector	2018
	Population Density (people/km²)	Grid Population Density (https://sedac.ciesin.columbia.edu/data/collection/gpw-v4 (accessed on 23 July 2022))	1 km	2020
	NDVI	Grid NDVI (https://www.resdc.cn/ (accessed on 28 April 2021))	1 km	2018
Adaptability	Medical Cost (h)	POI from Amap (https://restapi.amap.com/v3/place/text (accessed on 15 May 2013))	Vector	2012
	GDP (million yuan/km²)	Grid GDP (https://www.resdc.cn/ (accessed on 28 April 2021))	1 km	2015

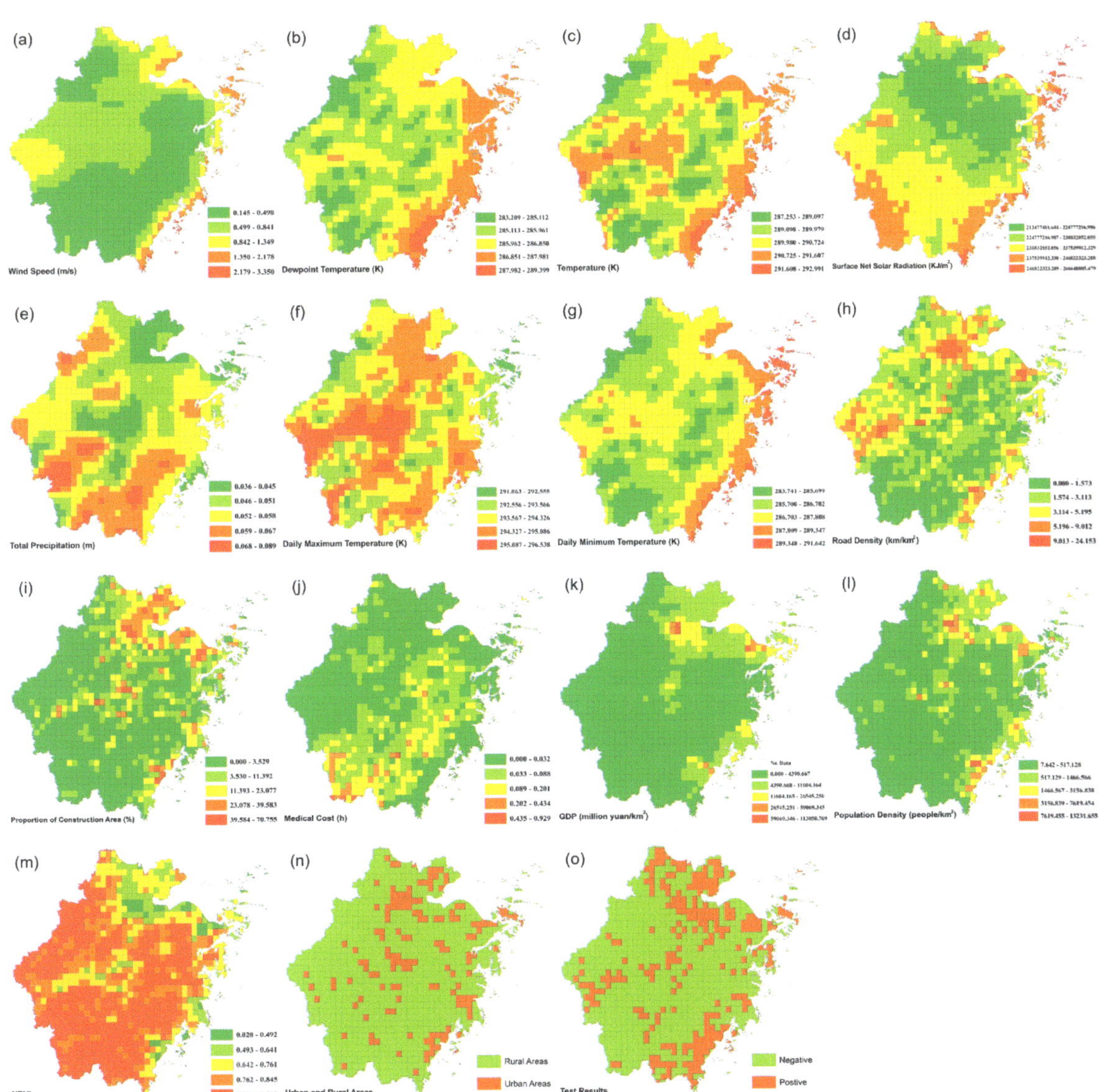

Figure 2. Spatial distribution of variables and detection results of *V. parahaemolyticus*, including (**a**) wind speed, (**b**) dewpoint temperature, (**c**) temperature, (**d**) surface net solar radiation, (**e**) total precipitation, (**f**) daily maximum temperature, (**g**) daily minimum temperature, (**h**) road density, (**i**) proportion of construction area, (**j**) medical cost, (**k**) GDP, (**l**) population density, (**m**) NDVI, (**n**) distribution of urban and rural areas, and (**o**) test results.

Table 2. Mean and standard deviations of urban and rural areas.

Index	Total	Urban Area	Rural Area
	964 Grids	155 Grids	809 Grids
Wind Speed (m/s)	0.655 (0.405)	0.767 (0.456)	0.633 (0.391)
Dewpoint Temperature (K)	286.204 (0.980)	286.673 (0.792)	286.114 (0.987)
Temperature (K)	290.307 (0.892)	290.788 (0.592)	290.215 (0.911)
Surface Net Solar Radiation (KJ/m^2)	231,698.574 (8818.364)	229,737.638 (9464.728)	232,074.279 (8644.563)
Total Precipitation (m)	0.052 (0.008)	0.049 (0.006)	0.053 (0.008)
Daily Maximum Temperature (K)	294.143 (0.875)	294.423 (0.830)	294.089 (0.874)
Daily Minimum Temperature (K)	287.010 (1.316)	287.614 (1.150)	286.894 (1.315)
Road Density (km/km^2)	3.105 (2.288)	5.486 (3.544)	2.649 (1.597)
Proportion of Construction Area (%)	7.946 (11.614)	22.643 (15.630)	5.131 (8.052)
Population Density (people/km^2)	638.452 (1098.019)	2048.545 (2105.047)	407.526 (546.304)
NDVI	0.783 (0.121)	0.655 (0.130)	0.807 (0.102)
Medical Cost (h)	0.041 (0.063)	0.017 (0.038)	0.046 (0.065)
GDP (million yuan/km^2)	4762.486 (9651.622)	10,687.459 (15866.161)	3627.293 (7417.539)

Note: Standard deviations are in parentheses.

2.4. Global Logistic Regression

We applied classic logistic regression to explore the relationship between vulnerability to environmental factors and foodborne diseases caused by *V. parahaemolyticus*. The logistic regression model is a generalized linear model with a binomial distribution for the dependent variable [43]. The dependent variable of the logistic regression in this study was the presence or absence of foodborne disease cases caused by *V. parahaemolyticus*. When $Y = 1$, there were positive cases in the grid; otherwise, $Y = 0$. The independent variables were temperature, total precipitation, road density, proportion of the construction area, distribution of urban and rural areas, and GDP. The classic logistic regression is called global logistic regression (GLR), expressed as follows:

$$logit(Y) \ = \ \beta_0 + \sum_{n\,=\,1}^{k} \beta_n X_n + \varepsilon \tag{1}$$

where β_n is the regression coefficient of independent variable X_n, β_0 is the intercept, logit (Y) is a linear combination function of the covariates, and ε is the error term. To detect and reduce the multicollinearity of these independent variables before regression modeling, we calculated the variance inflation factor (VIF). When VIF > 10, collinearity in the explanatory variables was considered problematic [44,45]. The probability $(Y = 1)$ can be calculated as follows:

$$P(Y \ = \ 1) \ = \ \frac{\exp\left(\beta_0 + \sum_{n\,=\,1}^{k} \beta_n X_n\right)}{1 + \exp\left(\beta_0 + \sum_{n\,=\,1}^{k} \beta_n X_n\right)} \tag{2}$$

where $P\,(Y = 1)$ represents the probability of detecting foodborne disease cases caused by *V. parahaemolyticus*. Areas with larger probability values represent a higher risk of foodborne diseases. Thus, we can identify the risk of foodborne diseases in the analysis area.

2.5. Geographically Weighted Logistic Regression Model

Figure 2 shows the geospatial heterogeneity of independent and dependent variables. Therefore, using the GWLR model is instrumental for considering the spatial dependence and capturing spatial variations. The GWLR model is a local regression method for investigating spatial non-stationarity [46], and it can explore the variation of the coefficient of each covariate geographically. For GWLR, the variables involved in the operation of the model were the same as those previously described. GWLR model is expressed as follows:

$$y = \beta_0(u_j, v_j) + \sum_{n=1}^{k} \beta_n(u_j, v_j) X_{nj} + \varepsilon_j \tag{3}$$

where u_j and v_j are the spatial coordinates of grid j, $\beta_n(u_j, v_j)$ is the regression coefficient of the independent variable X_n at location j, and ε_j is the error term specific to location j. Similar to the GLR model, the probability that ($Y = 1$) is expressed as follows:

$$P(Y = 1) = \frac{\exp\left[\beta_0(u_j, v_j) + \sum_{n=1}^{k} \beta_n(u_j, v_j) X_{nj}\right]}{1 + \exp\left[\beta_0(u_j, v_j) + \sum_{n=1}^{k} \beta_n(u_j, v_j) X_{nj}\right]} \tag{4}$$

To compare the performances of the GLR and GWLR models, we used the deviance, corrected Akaike information criterion (AICc), and area under the receiver operating characteristic curve (AUC) to evaluate the model fitness and prediction accuracy. For example, the lower the deviance and AICc, the better the model fits the data [47,48]; the higher the AUC, the better the prediction accuracy of the model [47].

3. Results

3.1. Global Logistic Regression

Based on correlation analysis, we eliminated some variables because of the multi-collinearity problem. For instance, proportion of construction area, population density, and NDVI were significantly collinear; dewpoint temperature, temperature, and daily minimum temperature were highly correlated. After evaluating the performance of the models from the perspective of collinearity, temperature, total precipitation, road density, proportion of construction area, dummy variable for rural areas, and GDP were included as the dependent variables in the regression model. Table 3 shows that the VIF values of these covariates (VIF = 1.443, 1.412, 1.820, 2.207, 1.503, and 1.456, respectively) are all smaller than the preselected threshold. The results of the GLR model showed that except for dummy variable for rural areas, most selected variables had a significant positive association with foodborne disease cases caused by *V. parahaemolyticus* ($p < 0.05$), which indicated that as temperature, total precipitation, road density, construction area proportions, and GDP increased, the probability of a grid converting from negative to positive also increased (Table 3). Furthermore, the positive effects of these independent variables on the presence or absence of foodborne disease cases, from strong to weak, were construction area proportions, GDP, road density, temperature, and total precipitation.

Table 3. Parameter estimates for the global logistic regression model.

Variable	β	S.E	z-Value	p	Exp(β)	VIF
Temperature	0.390	0.104	3.747	<0.001	1.476	1.443
Total Precipitation	0.262	0.099	2.652	0.008	1.300	1.412
Road Density	0.272	0.142	1.914	0.056	1.312	1.820
Proportion of Construction Area	0.373	0.122	3.053	0.002	1.452	2.207
Is Rural Areas	−0.924	0.231	−4.007	<0.001	0.397	1.503
GDP	0.559	0.212	2.644	0.008	1.750	1.456
Intercept	−1.222	0.094	−13.000	<0.001	0.295	-
AICc	952.390		Deviance	938.390		
AUC	0.772					

3.2. Geographically Weighted Logistic Regression

To capture geographical spatial variations, we applied GWLR to the same dataset of 964 grids, which showed a clear improvement over the GLR model, as shown in Tables 3 and 4. While the AICc and deviance values for the GWLR model (AICc = 874.659; deviance = 760.530) were much lower than those of the GLR model (AICc = 952.390; deviance = 938.390), which meant that the GWLR model had a much better model fit, the higher AUC value of the GLR model (AUC = 0.871) compared to that of the GWLR model (AUC = 0.772) suggested that it had a higher prediction accuracy for foodborne diseases.

Table 4. Summary statistics for geographically weighted logistic regression parameter estimates.

Variable	Mean	STD	Min	Max	% −	% +
Temperature	0.458	0.469	−0.491	1.693	16.5%	83.5%
Total Precipitation	0.297	0.637	−0.830	1.982	37.4%	62.6%
Road Density	0.461	1.076	−1.790	2.072	32.4%	67.6%
Proportion of Construction Area	0.273	0.506	−0.712	1.777	29.0%	71.0%
Is Rural Areas	−1.324	0.745	−3.187	0.389	97.9%	2.1%
GDP	1.218	1.768	−6.442	7.535	12.5%	87.5%
Intercept	−0.001	1.131	−3.697	4.111	51.1%	48.9%
AICc	874.659		Deviance	760.530		
AUC	0.871					

Parameter estimates and pseudo-t-statistics for each grid were generated using the software package GWR4. The summary descriptive statistics of the local parameter coefficients are shown in Table 4, suggesting that temperature, total precipitation, road density, construction area proportions, and GDP each have negative and positive parameter values. The majority of the local parameter coefficients for all the variables were positive except for the dummy variable for rural areas.

The spatial distributions of the generated coefficients and t-statistics surfaces with a grid size of $0.1° \times 0.1°$ are shown in Figures 3 and 4, respectively. Figure 4 shows that all the selected variables had certain areas where they were not statistically significant. For example, temperature had a significant positive effect on foodborne diseases in the northwestern and southeastern portions of the study area, while total precipitation had a larger significant positive impact area. In significantly affected areas, only the coefficient of road density and rural areas were negative. The construction area proportions had a positive relationship, mainly in the northern regions of the study area. The significantly positive influence areas of GDP extend in a strip from the northwest to the center of Zhejiang Province.

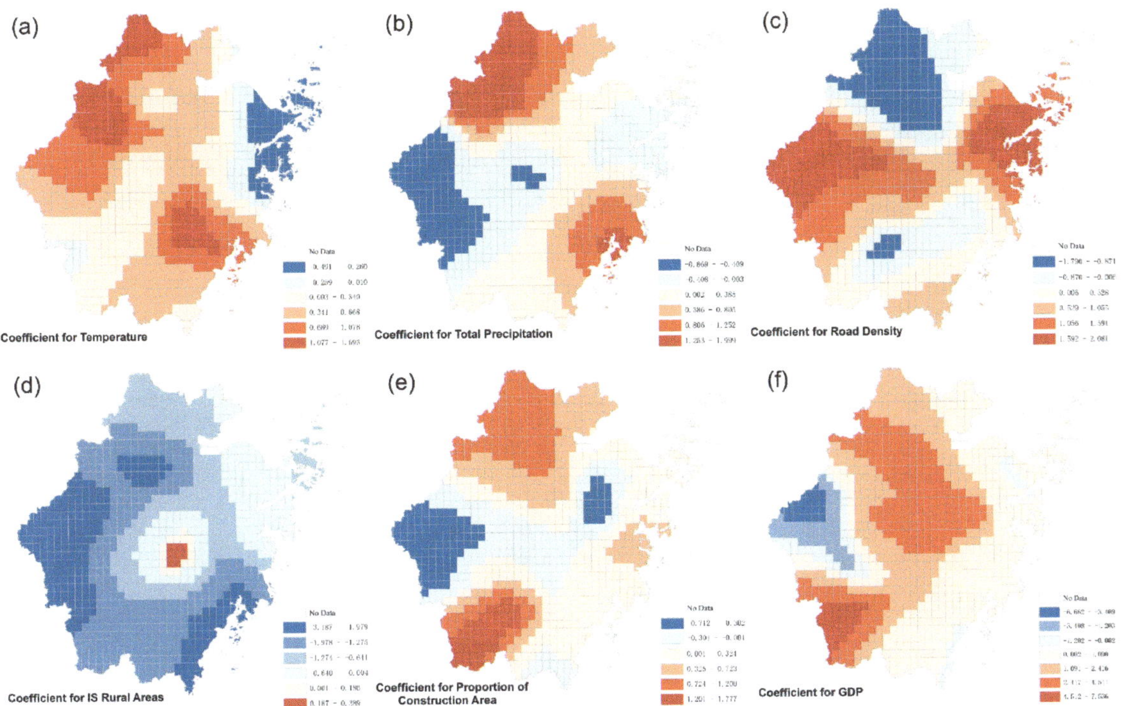

Figure 3. Spatial variation of regression coefficients in geographically weighted logistic regression model, including (**a**) temperature, (**b**) total precipitation, (**c**) road density, (**d**) dummy variable for rural areas, (**e**) proportion of construction area, and (**f**) GDP.

3.3. Mapping the Risk of Foodborne Diseases

Figure 5 illustrates a map of the predicted probability of foodborne disease infection based on the GLR and GWLR models. We divided the risk of foodborne diseases into five grades, from low to high, based on GWLR. The areas of each risk region were 55.7%, 20.1%, 12.3%, 6.7%, and 5.2%, respectively (Table 5). Moreover, this map indicates that the risks were higher in urban areas (60.6%) than in rural areas (20.1%). Compared with the GLR, the GWLR risk map predicted a higher probability of cases in some regions (e.g., Wenzhou in the southern part of Zhejiang Province) and a lower probability in some areas (e.g., the western part of Hangzhou).

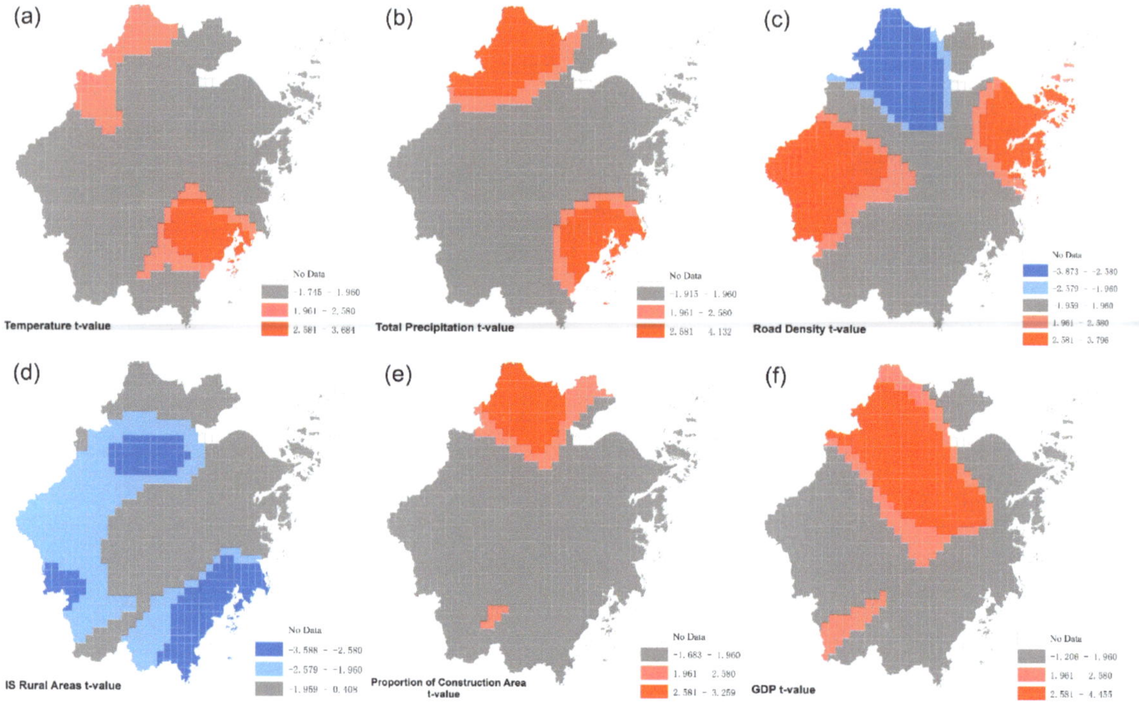

Figure 4. Spatial variation of t-value in geographically weighted logistic regression model, including (**a**) temperature, (**b**) total precipitation, (**c**) road density, (**d**) dummy variable for rural areas, (**e**) proportion of construction area, and (**f**) GDP.

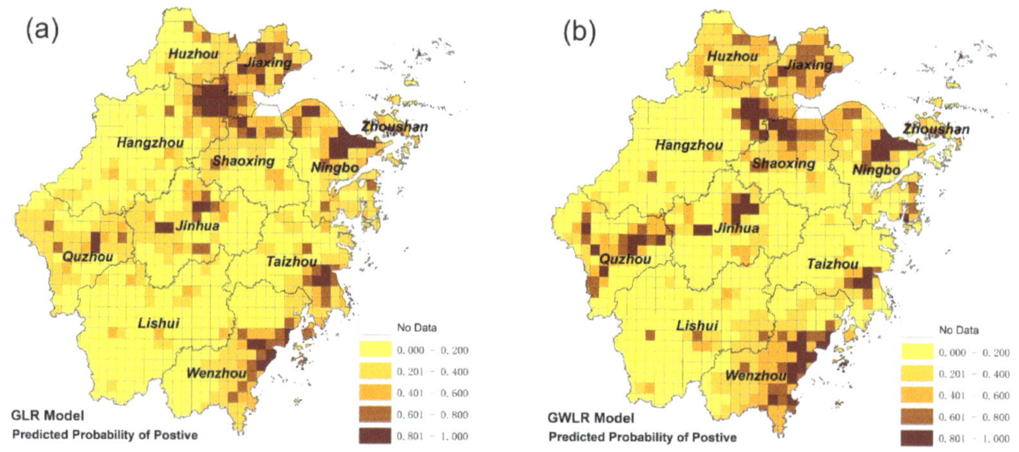

Figure 5. Predicted risk map of foodborne disease cases caused by *V. parahaemolyticus*. (**a**) Global logistic regression model; (**b**) geographically weighted logistic regression model.

Table 5. Percentage of foodborne disease risk areas and average prediction probability based on geographically weighted logistic regression model.

Grade	Total Area	Urban Area	Rural Area
Very low (0–0.2)	55.7%	11.4%	63.2%
Low (0.2–0.4)	20.1%	15.8%	20.8%
Middle (0.4–0.6)	12.3%	17.1%	11.5%
High (0.6–0.8)	6.7%	25.3%	3.6%
Very High (0.8–1.0)	5.2%	30.4%	0.9%
Average Prediction probability	26.0%	60.6%	20.1%

Furthermore, we divided Zhejiang Province into different areas according to their distance from the coastline and calculated the average prediction probability for each area. The calculation results are shown in Figure 6. The changing trend of the prediction probability is similar to that of the actual average detection rate. The distance from the coastline had a negative association with the average prediction probability around Zhejiang Province, except for an area 90–120 km and 180–240 km from the coastline.

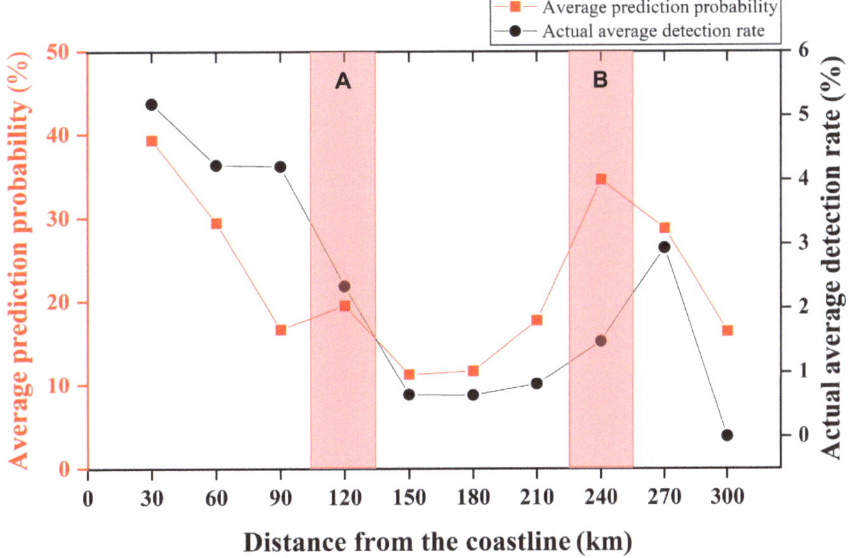

Figure 6. Average of prediction probability within a certain distance and actual average detection rate. Note: The corresponding areas near the peak (A and B) are located in Jinhua and Quzhou, respectively.

4. Discussion

Based on the vulnerability assessment framework and variable screening method, Tables 3 and 5 show that regional temperature, total precipitation, road density, construction area proportions, rural areas, and GDP have a significant effect on the positive detection of *V. parahaemolyticus* in the GLR or GWLR models. Positive cases of foodborne diseases were positively correlated with air temperature, which is consistent with the findings of Hsiao et al. [49], who reported similar findings in Taiwan [49]. However, these positive relationships differ from the findings of Shih et al. [50], in that the detection rate of *V. parahaemolyticus* was negatively correlated with average daily rainfall [50]. One possible reason for this disparity is that the very humid plum rain season occurs in June and July in Zhejiang Province, which brings abundant precipitation, making it easier for bacteria to breed. Road density and construction area proportions also had significant positive effects on foodborne diseases, while the occurrence of rural areas had a negative relationship with positive cases occurrence. It is consistent with the view proposed by Prinsen [41] that urbanization can affect the risk of foodborne diseases. Based on these correlations, urban areas in Zhejiang Province should be more heavily studied than rural areas. Additionally, there was a positive correlation between GDP and foodborne diseases, suggesting that the prevention and control of foodborne diseases should not be neglected when pursuing economic development. However, this differed from the findings of Yang et al. [51], who reported that the incidence of foodborne diseases had a negative correlation with GDP in Jinan. This discrepancy could be because the attitudes and behavior of foodborne disease patients in choosing healthcare vary due to different socioeconomic and cultural backgrounds; for example, rural residents are less likely to visit hospitals when they suffer from foodborne diseases.

The GLR model ignores geographical variations in the relationships between the dependent variable and covariates, whereas the GWLR model can detect this spatial variability [52]. Temperature and total precipitation only showed significant positive effects in some areas of northwest and southeast Zhejiang. The mostly non-significant relationship and differences from others may have been a consequence of the spatiotemporal scales of the meteorological data. A significant and positive relationship between road density and the presence or absence of positive cases was found in the western and eastern regions while an outlier occurred in northern Zhejiang, and there was a negative correlation between road density and *V. parahaemolyticus* detection rate. Although the developed road network can promote food transportation, the supervision of food hygiene quality in these areas near the main urban area of the provincial capital city is more stringent, and the daily dietary hygiene habits of residents are healthier. The proportion of construction area had a positive relationship, predominantly in the northern and southwestern parts of the study area. Furthermore, the rural areas had a significant negative relationship in the west and southeast of Zhejiang Province. The significantly positive influence area of GDP extends in a strip from the northwest to the center of Zhejiang Province. The coefficients and significance of the proportion of construction areas and GDP varied geographically. The influence of these factors on foodborne diseases caused by *V. parahaemolyticus* varied geographically. Furthermore, our findings verify that GWLR can provide improvements and additional perspectives over classic non-spatial regression models for eco epidemiological studies on bacterial foodborne diseases [53]. As the relative importance of each independent variable differed geographically, public health practitioners can identify the most important influencing factors and develop public health interventions for various regions more precisely.

GWLR had more advantages in model performance compared to GLR. Owing to the spatial heterogeneity of the driving mechanism, non-spatial regression models may perform poorly [54]. A comparison of the model performances between the GLR and GWLR is shown in Tables 3 and 4. The AICc and deviance values of GWLR were 874.659 and 760.530, respectively, which were lower than those of the GLR model, indicating that GWLR performed better than the GLR model in quantifying the impact of selected variables on foodborne diseases. Additionally, the AUC of GWLR was 0.871, which was much larger than that of the GLR model, suggesting that GWLR had higher prediction accuracy for the probability of positive foodborne disease cases. The evaluation indices of the GWLR model were all better than those of the GLR model, similar to the findings of previous studies [53,55], suggesting that the influence of spatial geographical location on the results of the dependent variables should be considered when fitting data with spatial structure. However, this GWLR approach used the annual total value or average value of covariables to model the relationship between *V. parahaemolyticus* detection information and vulnerability environmental factors in different regions, ignoring the seasonal variation characteristics of foodborne diseases. It is difficult to predict the epidemic trends of foodborne diseases according to the climate change and other risk factors. Similar to the study on the spatial trends in Salmon infection in Spain [56], we paid attention to the geospatial variation of bacterial pathogen detection, and confirmed the existence of the spatial difference in the risk of bacterial pathogen infection at the province level or grid level. Under the constraints of limited resource input and environmental improvement, it is important to evaluate the spatial risk of *V. parahaemolyticus* infection in Zhejiang province to help develop local public health strategies, which is the main contribution of this study.

According to the predicted risk maps of foodborne diseases caused by *V. parahaemolyticus* (Figure 5), the prediction probability in urban areas (60.6%) was higher than that in the rural areas (20.1%). One reason for this phenomenon is that the percentage of seafood intake is lower in rural residents than in urban residents [57]. Outbreaks of foodborne diseases caused by *V. parahaemolyticus* are associated with dietary habits of seafood consumption. *V. parahaemolyticus* foodborne disease risks differed between urban and rural areas, which have also been reported by other researchers; for example, using simple linear regression and locally weighted regression, Ford et al. found significantly higher rates of *V. parahaemolyticus* serotype typhimurium in urban areas than rural areas [58]. Therefore, relevant departments should pay more attention to urban areas than to rural areas. Stricter food regulations, such as the step-by-step seafood safety regulations in Japan, from the production to the consumption stages, are recommended [59]. Considering factors such as medical distance and cost, there are also patients in rural areas who do not visit the hospital after falling ill. Therefore, the monitoring, prevention, and control of foodborne diseases requires efforts from both urban and rural departments.

However, not all urban areas have high prediction probabilities and not all rural areas have low prediction probabilities; for example, the prediction probability of the coastal areas that belonged to rural areas in the northeast and southeast of Zhejiang Province was also high, whereas the prediction probability of some grids belonging to urban areas in the middle of Zhejiang Province was low. Additionally, Figure 6 shows that, as the distance from the coast increased, the prediction probability decreased overall, except for some regions, which was consistent with a study conducted in the littoral domain, where *V. parahaemolyticus* caused outbreaks of most foodborne diseases [6]. One possible reason for the high prediction probability in inland areas is that *V. parahaemolyticus* is occasionally detected in other foods. Remarkably, *V. parahaemolyticus* contamination has been found at a high rate in aquatic products as well as in ready-to-eat (RTE) foods, such as cooked meat, roasted poultry, and cold vegetable dishes in sauce, which are popular in China [60]. Jinhua, 90–180 km from the coastline, is famous for its traditional pickled food, Jinhua ham; Quzhou, 210–270 km from the coastline, is famous for its special stewed meat. Therefore, relevant departments should not ignore other foods while addressing seafood contamination.

5. Conclusions

In this study, we proposed a foodborne disease vulnerability assessment framework, and foodborne disease vulnerability environments were comprehensively described using various types of geographic data. This study combined the GWLR model with foodborne diseases for the first time to analyze the spatial epidemiological risk of foodborne diseases caused by *V. parahaemolyticus*.

We found that temperature, total precipitation, road density, the proportion of construction area, and GDP are important environmental indicators that affect foodborne diseases. Additionally, the GWLR model had better model fitness and higher prediction accuracy than the GLR model. Compared with the GLR model, the GWLR model can consider the spatial heterogeneity of selected independent variables and their relative geographical importance. The significant relationship between foodborne diseases and these covariates was mostly positive throughout Zhejiang Province, except that road density also had a negative relationship in the northern part of the study area. Furthermore, our model can effectively predict the foodborne disease risks, and our predicted risk map showed that urban areas had a higher overall probability of positive cases. Although foodborne diseases caused by *V. parahaemolyticus* are related to the distance from the coastline, the supervision of food and residential dietary hygiene habits also contributed to an increased risk of foodborne diseases. Generally, this study provides guidance and geographical support for relevant government departments to prevent and control foodborne diseases in Zhejiang Province.

The limitations and future perspectives should be addressed to better understand these findings. First, urban and rural divisions in this study were based on administrative divisions, and a mixed area may still exist. Determining how to divide urban and rural areas more effectively would be instrumental for future work. Second, the spatiotemporal resolutions of meteorological and socio-economic data limit the spatial scale and neglects seasonal changes in the results. Except for meteorological data, other vulnerability environmental assessment data show the annual average distribution of factors in each region, which are cross-sectional data. Future studies should include higher spatial resolution and dynamically changing time-series data, and then carry out spatiotemporal risk prediction research. Third, the proposed foodborne disease vulnerability assessment framework must be further improved. Based on the perspective of exposure, sensitivity, and adaptability, related information such as age structure, food consumption structure, and fiscal medical hygienic expenditure to the index framework should be included in future studies, aiming to describe the vulnerability of the Zhejiang Province environments from a more comprehensive perspective.

Author Contributions: Conceptualization, W.B. and T.L.; methodology, W.B., H.H., B.Z. and T.L.; software, W.B.; validation, J.C. and S.X.; formal analysis, W.B., H.H. and T.L.; investigation, J.C., W.B. and T.L.; resources, J.C.; data curation, W.B. and T.L.; writing—original draft preparation, W.B and B.Z.; writing—review and editing, H.H. and J.X.; visualization, W.B.; supervision, H.H. and T.L.; project administration, T.L.; funding acquisition, T.L. All authors have read and agreed to the published version of the manuscript.

Funding: This research was funded by Hangzhou Science and Technology Development Plan (Grant No. 20201203B141), the National Natural Science Foundation of China (Grant No. 41101371).

Data Availability Statement: The data used to support the findings of this study are available from the corresponding author upon request.

Acknowledgments: The authors would like to thank of Zhejiang Provincial Center for Disease Control and Prevention for providing data for the study.

Conflicts of Interest: The authors declare no conflict of interest. The funders had no role in the design of the study; in the collection, analyses, or interpretation of data; in the writing of the manuscript, or in the decision to publish the results.

References

1. World Health Organization. *Foodborne Disease Outbreaks: Guidelines for Investigation and Control*; World Health Organization: Geneva, Switzerland, 2008.
2. World Health Organization. *WHO Estimates of the Global Burden of Foodborne Diseases: Foodborne Disease Burden Epidemiology Reference Group 2007–2015*; World Health Organization: Geneva, Switzerland, 2015.
3. Chen, Y.; Yan, W.; Zhou, Y.; Zhen, S.; Zhang, R.; Chen, J.; Liu, Z.; Cheng, H.; Liu, H.; Duan, S.; et al. Burden of Self-reported Acute Gastrointestinal Illness in China: A Population-based Survey. *BMC Public Health* **2013**, *13*, 456. [CrossRef]
4. Li, W.; Wu, S.; Fu, P.; Liu, J.; Han, H.; Bai, L.; Pei, X.; Li, N.; Liu, X.; Guo, Y. National Molecular Tracing Network for Foodborne Disease Surveillance in China. *Food Control* **2018**, *88*, 28–32. [CrossRef]
5. Pang, R.; Li, Y.; Chen, M.; Zeng, H.; Lei, T.; Zhang, J.; Ding, Y.; Wang, J.; Wu, S.; Ye, Q.; et al. A Database for Risk Assessment and Comparative Genomic Analysis of Foodborne Vibrio Parahaemolyticus in China. *Sci. Data* **2020**, *7*, 321. [CrossRef] [PubMed]
6. Chen, L.; Sun, L.; Zhang, R.; Liao, N.; Qi, X.; Chen, J. Surveillance for Foodborne Disease Outbreaks in Zhejiang Province, China, 2015–2020. *BMC Public Health* **2022**, *22*, 135. [CrossRef]
7. Swoveland, J.L.; Stewart, L.K.; Eckmann, M.K.; Gee, R.; Allen, K.J.; Vandegrift, C.M.; Olson, G.; Kang, M.G.; Tran, M.L.; Melius, E.; et al. Laboratory Review of Foodborne Disease Investigations in Washington State 2007–2017. *Foodborne Pathog. Dis.* **2019**, *16*, 513–523. [CrossRef] [PubMed]
8. Lee, J.K.; Kwak, N.S.; Kim, H.J. Systemic Analysis of Foodborne Disease Outbreak in Korea. *Foodborne Pathog. Dis.* **2016**, *13*, 101–107. [CrossRef] [PubMed]

9. Chen, Y.J.; Wen, Y.F.; Song, J.G.; Chen, B.F.; Ding, S.S.; Ding, L.; Dai, J.J. The Correlation Between Family Food Handling Behaviors and Foodborne Acute Gastroenteritis: A Community-oriented, Population-based Survey in Anhui, China. *BMC Public Health* **2018**, *18*, 1290. [CrossRef] [PubMed]
10. Zhang, P.; Cui, W.; Wang, H.; Du, Y.; Zhou, Y. High-Efficiency Machine Learning Method for Identifying Foodborne Disease Outbreaks and Confounding Factors. *Foodborne Pathog. Dis.* **2021**, *18*, 590–598. [CrossRef] [PubMed]
11. Wang, X.L.; Zhou, M.Q.; Jia, J.Z.; Geng, Z.; Xiao, G.X. A Bayesian Approach to Real-Time Monitoring and Forecasting of Chinese Foodborne Diseases. *Int. J. Environ. Res. Public Health* **2018**, *15*, 1740. [CrossRef] [PubMed]
12. Li, S.; Peng, Z.; Zhou, Y.; Zhang, J. Time Series Analysis of Foodborne Diseases During 2012–2018 in Shenzhen, China. *J. Consum. Prot. Food Saf.* **2021**, *17*, 83–91. [CrossRef]
13. Lesiv, M.; Moltchanova, E.; Schepaschenko, D.; See, L.; Shvidenko, A.; Comber, A.; Fritz, S. Comparison of Data Fusion Methods Using Crowdsourced Data in Creating a Hybrid Forest Cover Map. *Remote Sens.* **2016**, *8*, 261. [CrossRef]
14. Yasuo, K.; Nishiura, H. Spatial Epidemiological Determinants of Severe Fever with Thrombocytopenia Syndrome in Miyazaki, Japan: A GWLR Modeling Study. *BMC Infect. Dis.* **2019**, *19*, 498. [CrossRef] [PubMed]
15. Imran, M.; Hamid, Y.; Mazher, A.; Ahmad, S.R. Geo-spatially Modelling Dengue Epidemics in Urban Cities: A Case Study of Lahore, Pakistan. *Geocarto. Int.* **2021**, *36*, 197–211. [CrossRef]
16. Manyangadze, T.; Mavhura, E.; Mudavanhu, C.; Pedzisai, E. An Exploratory Analysis of the Spatial Variation of Malaria Cases and Associated Household Socio-economic Factors in Flood-prone Areas of Mbire district, Zimbabwe. *GeoJournal* **2021**, 1–16. [CrossRef]
17. Zhou, Y.B.; Wang, Q.X.; Yang, M.X.; Gong, Y.H.; Yang, Y.; Nie, S.J.; Liang, S.; Nan, L.; Coatsworth, A.; Yang, A.H.; et al. Geographical variations of risk factors associated with HCV infection in drug users in southwestern China. *Epidemiol. Infect.* **2016**, *144*, 1291–1300. [CrossRef] [PubMed]
18. Birch, E.L. Climate Change 2014: Impacts, Adaptation, and Vulnerability. *J. Am. Plann. Assoc.* **2014**, *80*, 184–185. [CrossRef]
19. Rathi, S.K.; Chakraborty, S.; Mishra, S.K.; Dutta, A.; Nanda, L. A Heat Vulnerability Index: Spatial Patterns of Exposure, Sensitivity and Adaptive Capacity for Urbanites of Four Cities of India. *Int. J. Environ. Res. Public Health* **2022**, *19*, 283. [CrossRef]
20. Udayanga, L.; Gunathilaka, N.; Iqbal, M.C.M.; Abeyewickreme, W. Climate Change Induced Vulnerability and Adaption for Dengue Incidence in Colombo and Kandy Districts: The Detailed Investigation in Sri Lanka. *Infect. Dis. Poverty* **2020**, *9*, 102. [CrossRef] [PubMed]
21. De Andrade, M.M.N.; Szlafsztein, C.F. Vulnerability Assessment Including Tangible and Intangible Components in the Index Composition: An Amazon Case Study of Flooding and Flash Flooding. *Sci. Total Environ.* **2018**, *630*, 903–912. [CrossRef] [PubMed]
22. Mitrica, B.; Mocanu, I.; Grigorescu, I.; Dumitrascu, M.; Pistol, A.; Damian, N.; Serban, P.R. Population Vulnerability to the SARS-CoV-2 Virus Infection. A County-Level Geographical-Methodological Approach in Romania. *GeoHealth* **2021**, *5*, e2021GH000461. [CrossRef] [PubMed]
23. Ahmad, I.; Wang, X.; Waseem, M.; Zaman, M.; Aziz, F.; Khan, R.Z.N.; Ashraf, M. Flood Management, Characterization and Vulnerability Analysis Using an Integrated RS-GIS and 2D Hydrodynamic Modelling Approach: The Case of Deg Nullah, Pakistan. *Remote Sens.* **2022**, *14*, 2138. [CrossRef]
24. Adger, W.N. Vulnerability. *Glob. Environ. Change* **2006**, *16*, 268–281. [CrossRef]
25. Zhang, Y.; Shen, J.; Li, Y. Atmospheric Environment Vulnerability Cause Analysis for the Beijing-Tianjin-Hebei Metropolitan Region. *Int. J. Environ. Res. Public Health* **2018**, *15*, 128. [CrossRef] [PubMed]
26. Chen, Y.; Liu, T.; Ge, Y.; Xia, S.; Yuan, Y.; Li, W.; Xu, H. Examining Social Vulnerability to Flood of Affordable Housing Communities in Nanjing, China: Building Long-term Disaster Resilience of Low-income Communities. *Sustain. Cities Soc.* **2021**, *71*, 102939. [CrossRef]
27. Swami, D.; Parthasarathy, D. Dynamics of Exposure, Sensitivity, Adaptive Capacity and Agricultural Vulnerability at District Scale for Maharashtra, India. *Ecol. Indic.* **2021**, *121*, 107206. [CrossRef]
28. He, C.; Ma, L.; Zhou, L.; Kan, H.; Zhang, Y.; Ma, W.; Chen, B. Exploring the Mechanisms of Heat Wave Vulnerability at the Urban Scale Based on the Application of Big Data and Artificial Societies. *Environ. Int.* **2019**, *127*, 573–583. [CrossRef]
29. Morral-Puigmal, C.; Martinez-Solanas, E.; Villanueva, C.M.; Basagana, X. Weather and Gastrointestinal Disease in Spain: A Retrospective Time Series Regression Study. *Environ. Int.* **2018**, *121*, 649–657. [CrossRef] [PubMed]
30. Kim, Y.S.; Park, K.H.; Chun, H.S.; Choi, C.; Bahk, G.J. Correlations Between Climatic Conditions and Foodborne Disease. *Food Res. Int.* **2015**, *68*, 24–30. [CrossRef]
31. Bari, L.; Yeasmin, S.; Kawamoto, S. Impact of Climate Change on Foodborne Pathogens and Diseases. *J. Jpn. Soc. Food. Sci.* **2008**, *55*, 264–269. [CrossRef]
32. Strassle, P.D.; Gu, W.; Bruce, B.B.; Gould, L.H. Sex and Age Distributions of Persons in Foodborne Disease Outbreaks and Associations with Food Categories. *Epidemiol. Infect.* **2019**, *147*, e200. [CrossRef]
33. Chen, Y.J.; Wen, Y.F.; Song, J.G.; Chen, B.F.; Wang, L.; Ding, S.S.; Ding, L.; Dai, J.J. Food Handling Behaviors Associated with Reported Acute Gastrointestinal Disease That May Have Been Caused by Food. *J. Food Prot.* **2019**, *82*, 494–500. [CrossRef] [PubMed]
34. Osei-Tutu, B.; Anto, F. Trends of Reported Foodborne Diseases at the Ridge Hospital, Accra, Ghana: A Retrospective Review of Routine Data from 2009–2013. *BMC Infect. Dis.* **2016**, *16*, 139. [CrossRef]
35. Czerwinski, M.; Czarkowski, M.P.; Kondej, B. Foodborne Botulism in Poland in 2017. *Prz. Epidemiol.* **2019**, *73*, 445–450. [CrossRef]

36. Xiao, G.X.; Xu, C.D.; Wang, J.F.; Yang, D.Y.; Wang, L. Spatial-temporal Pattern and Risk Factor Analysis of Bacillary Dysentery in the Beijing-Tianjin-Tangshan Urban Region of China. *BMC Public Health* **2014**, *14*, 998. [CrossRef] [PubMed]
37. Chen, J.; Zhang, R.; Qi, X.; Zhou, B.; Wang, J.; Chen, Y.; Zhang, H. Epidemiology of Foodborne Disease Outbreaks Caused by Vibrio Parahaemolyticus During 2010–2014 in Zhejiang Province, China. *Food Control* **2017**, *77*, 110–115. [CrossRef]
38. National Bureau of Statistics of China. Provisions on the Statistical Division of Urban and Rural Areas (for Trial Implementation). Available online: http://www.stats.gov.cn/tjsj/pcsj/rkpc/5rp/html/append7.htm (accessed on 22 December 2021).
39. Bai, H.M.; Shi, Y.L.; Seong, M.S.; Gao, W.K.; Li, Y.H. Influence of Spatial Resolution on Satellite-Based PM2.5 Estimation Implications for Health Assessment. *Remote Sens.* **2022**, *14*, 2933. [CrossRef]
40. Hellberg, R.S.; Chu, E. Effects of Climate Change on the Persistence and Dispersal of Foodborne Bacterial Pathogens in the Outdoor Environment: A review. *Crit. Rev. Microbiol.* **2016**, *42*, 548–572. [CrossRef] [PubMed]
41. Prinsen, G.; Benschop, J.; Cleaveland, S.; Crump, J.A.; French, N.P.; Hrynick, T.A.; Mariki, B.; Mmbaga, B.T.; Sharp, J.P.; Swai, E.S. et al. Meat Safety in Tanzania's Value Chain: Experiences, Explanations and Expectations in Butcheries and Eateries. *Int. J. Environ. Res. Public Health* **2020**, *17*, 82833. [CrossRef] [PubMed]
42. Cutter, S.L.; Finch, C. Temporal and Spatial Changes in Social Vulnerability to Natural Hazards. *Proc. Natl. Acad. Sci. USA* **2008**, *105*, 2301–2306. [CrossRef]
43. Wu, W.; Zhang, L. Comparison of Spatial and Non-spatial Logistic Regression Models for Modeling the Occurrence of Cloud Cover in North-eastern Puerto Rico. *Appl. Geogr.* **2013**, *37*, 52–62. [CrossRef]
44. Zuur, A.F.; Ieno, E.N.; Elphick, C.S. A Protocol for Data Exploration to Avoid Common Statistical Problems. *Methods Ecol. Evol* **2010**, *1*, 3–14. [CrossRef]
45. Ye, X.P.; Yu, X.P.; Wang, T.J. Investigating Spatial Non-stationary Environmental Effects on the Distribution of Giant Pandas in the Qinling Mountains, China. *Glob. Ecol. Conserv.* **2020**, *21*, e00894. [CrossRef]
46. Lu, B.; Charlton, M.; Harris, P.; Fotheringham, A.S. Geographically Weighted Regression with a Non-Euclidean Distance Metric A Case Study Using Hedonic House Price Data. *Int. J. Geogr. Inf. Sci.* **2014**, *28*, 660–681. [CrossRef]
47. Han, H.; Jang, K.-M.; Chung, J.-S. Selecting Suitable Sites for Mountain Ginseng (Panax ginseng) Cultivation by Using Geographically Weighted Logistic Regression. *J. Mt. Sci.* **2017**, *14*, 492–500. [CrossRef]
48. Yang, L.; Yu, K.; Ai, J.; Liu, Y.; Yang, W.; Liu, J. Dominant Factors and Spatial Heterogeneity of Land Surface Temperatures in Urban Areas: A Case Study in Fuzhou, China. *Remote Sens.* **2022**, *14*, 1266. [CrossRef]
49. Hsiao, H.I.; Jan, M.S.; Chi, H.J. Impacts of Climatic Variability on Vibrio Parahaemolyticus Outbreaks in Taiwan. *Int. J. Environ. Res. Public Health* **2016**, *13*, 20188. [CrossRef]
50. Shih, Y.J.; Chen, J.S.; Chen, Y.J.; Yang, P.Y.; Kuo, Y.J.; Chen, T.H.; Hsu, B.M. Impact of Heavy precipitation Events on Pathogen Occurrence in Estuarine Areas of the Puzi River in Taiwan. *PLoS ONE* **2021**, *16*, e0256266. [CrossRef]
51. Yang, L.; Sun, Y.B.; Zhong, Q.; Duan, D.S.; Liu, S.Q.; Zhang, Y. Epidemiological Characteristics and Spatio-temporal Patterns of Foodborne Diseases in Jinan, Northern China. *Biomed. Environ. Sci.* **2019**, *32*, 309–313. [CrossRef]
52. Zhang, L.; Wei, Y.; Meng, R. Spatiotemporal Dynamics and Spatial Determinants of Urban Growth in Suzhou, China. *Sustainability* **2017**, *9*, 30393. [CrossRef]
53. Mayfield, H.J.; Lowry, J.H.; Watson, C.H.; Kama, M.; Nilles, E.J.; Lau, C.L. Use of Geographically Weighted Logistic Regression to Quantify Spatial Variation in the Environmental and Sociodemographic Drivers of Leptospirosis in Fiji: A Modelling Study. *Lancet Planet. Health* **2018**, *2*, 223–232. [CrossRef]
54. Li, H.; Peng, J.; Yanxu, L.; Yi'na, H. Urbanization Impact on Landscape Patterns in Beijing City, China: A Spatial Heterogeneity Perspective. *Ecol. Indic.* **2017**, *82*, 50–60. [CrossRef]
55. Clary, C.; Lewis, D.J.; Flint, E.; Smith, N.R.; Kestens, Y.; Cummins, S. The Local Food Environment and Fruit and Vegetable Intake: A Geographically Weighted Regression Approach in the ORiEL Study. *Am. J. Epidemiol.* **2016**, *184*, 837–846. [CrossRef] [PubMed]
56. Teng, K.T.Y.; Aviles, M.M.; Ugarte-Ruiz, M.; Barcena, C.; de la Torre, A.; Lopez, G.; Moreno, M.A.; Dominguez, L.; Alvarez, J. Spatial Trends in Salmonella Infection in Pigs in Spain. *Front. Vet. Sci.* **2020**, *7*, 345. [CrossRef] [PubMed]
57. Yan, Y.; You, L.; Wang, X.; Zhang, Z.; Li, F.; Wu, H.; Wu, M.; Zhang, J.; Wu, J.; Chen, C.; et al. Iodine Nutritional Status, the Prevalence of Thyroid Goiter and Nodules in Rural and Urban Residents: A Cross-sectional Study from Guangzhou, China. *Endocr. Connect.* **2021**, *10*, 1550–1559. [CrossRef] [PubMed]
58. Ford, M.W.; Odoi, A.; Majowicz, S.E.; Michel, P.; Middleton, D.; Ciebin, B.; Dore, K.; McEwen, S.A.; Aramini, J.A.; Deeks, S.; et al. A Descriptive Study of Human Salmonella Serotype Typhimurium Infections Reported in Ontario from 1990 to 1998. *Can. J. Infect. Dis. Med. Microbiol.* **2003**, *14*, 267–273. [CrossRef]
59. Hara-Kudo, Y.; Kumagai, S. Impact of Seafood Regulations for Vibrio Parahaemolyticus Infection and Verification by Analyses of Seafood Contamination and Infection. *Epidemiol. Infect.* **2014**, *142*, 2237–2247. [CrossRef] [PubMed]
60. Xie, T.; Xu, X.; Wu, Q.; Zhang, J.; Cheng, J. Prevalence, Molecular Characterization, and Antibiotic Susceptibility of Vibrio Parahaemolyticus from Ready-to-Eat Foods in China. *Front. Microbiol.* **2016**, *7*, 17–23. [CrossRef] [PubMed]

MDPI AG
Grosspeteranlage 5
4052 Basel
Switzerland
Tel.: +41 61 683 77 34

Remote Sensing Editorial Office
E-mail: remotesensing@mdpi.com
www.mdpi.com/journal/remotesensing